Grid-based Nonlinear Estimation and Its Applications

Bin Jia
Intelligent Fusion Technology, Inc., Germantown, Maryland, USA

Ming Xin
Department of Mechanical and Aerospace Engineering, University of Missouri, Columbia, Missouri, USA

CRC Press
Taylor & Francis Group
Boca Raton London New York

CRC Press is an imprint of the
Taylor & Francis Group, an **informa** business
A SCIENCE PUBLISHERS BOOK

CRC Press
Taylor & Francis Group
6000 Broken Sound Parkway NW, Suite 300
Boca Raton, FL 33487-2742

First issued in paperback 2021

Version Date: 20181226

ISBN-13: 978-0-367-77995-5 (pbk)
ISBN-13: 978-1-138-72309-2 (hbk)

Library of Congress Cataloging-in-Publication Data

Names: Jia, Bin, 1982- author. | Xin, Ming, 1972- author.
Title: Grid-based nonlinear estimation and its applications / Bin Jia (Intelligent Fusion Technology, Inc., Germantown, Maryland, USA), Ming Xin (Department of Mechanical and Aerospace Engineering [University of Missouri]).
Description: Boca Raton, FL : CRC Press, Taylor & Francis Group, 2019. | "A science publishers book." | Includes bibliographical references and index.
Identifiers: LCCN 2018051023 | ISBN 9781138723092 (hardback)
Subjects: LCSH: Estimation theory. | Nonlinear theories.
Classification: LCC QA276.8 .J53 2019 | DDC 519.5/44--dc23
LC record available at https://lccn.loc.gov/2018051023

Visit the Taylor & Francis Web site at
http://www.taylorandfrancis.com

and the CRC Press Web site at
http://www.crcpress.com

Preface

Estimation (prediction, filtering, and smoothing) is of great importance to virtually all engineering and science disciplines that require inference, learning, information fusion, identification, and retrieval of unknown dynamic system states and parameters. The theory and practice of linear estimation of Gaussian dynamic systems have been well established and successful. Although optimal estimation of nonlinear dynamic systems has been studied for decades, it is still a very challenging and unresolved problem. In general, approximations are inevitable in order to design any applicable nonlinear estimation algorithm.

This monograph aims to provide a unified nonlinear estimation framework from the Bayesian perspective and to develop systematic estimation algorithms with grid-based numerical rules. It presents a common approach to the nonlinear Gaussian estimation, by emphasizing that all Gaussian-assumed estimation algorithms in the literature distinguish themselves merely by the way of approximating Gaussian weighted integrals. A variety of numerical rules to generate deterministic grid points and weights can be utilized to approximate such Gaussian integrals and yield a family of nonlinear Gaussian estimation algorithms, which can be unified in the same Bayesian framework. A unique feature of this book is to reveal the close relationships among these nonlinear estimators such as the unscented Kalman filter, cubature Kalman filter, Gauss-Hermite quadrature filter, and sparse-grid quadrature filter. It is shown that by certain grid selection rules, one filter can be constructed from the other, or vice versa. The readers can learn the advantages and disadvantages of these estimation techniques and have a guideline to select the suitable one to solve their problems, because estimation accuracy and computation complexity can be analytically given and easily controlled. Several important applications are presented to demonstrate the capability of the grid-based nonlinear estimation including multiple sensor estimation, uncertainty propagation, target tracking, and navigation. The integration of the grid-based estimation techniques and the parallel computing models, such as MapReduce and graphics processing units (GPU), is briefly introduced. Pseudo-code is provided for the key algorithms, allowing interested readers to develop their own programs for their estimation problems.

Contents

1 Introduction

In a data inundated world, extracting useful information from the data becomes indispensable in all science and engineering disciplines and even in people's daily life. Systems that extract signals from noisy measurements have to face the challenges of sensor constraints, random disturbances, and lack of precise knowledge of the underlying physical process. Estimation theories and techniques are at the heart of such data processing systems. The data processing methods can be traced back to Gauss (Gelb 1974), who originated the deterministic least-squares method and employed it in a relatively simple orbit determination problem. After more than 100 years, Fisher (Fisher 1912) developed the maximum likelihood estimation method based on the probability density functions. As a remarkable milestone, Wiener (Wiener 1949) proposed statistically optimal filters in the frequency domain for the continuous-time problem. Nevertheless, it provided optimal estimates only in the steady-state constrained by statistically stationary processes. The discrete-time estimation problem was investigated by Kolmogorov (Kolmogorov 1941) during the same time period. The most significant breakthrough of the estimation theory took place when Kalman (Kalman 1960) introduced the state-space model and optimal recursive filtering techniques based on the time domain formulations. It is this celebrated Kalman filter that makes possible digital computer implementation of the estimation algorithm, and the tremendous success in a wide range of applications.

An estimator is a procedure that processes measurement to deduce an estimate of the state or parameters of a system from the prior (deterministic or statistic) knowledge of the system, measurement, and initial conditions. The estimation is usually obtained by minimizing the estimation error in a well-defined statistical sense. Three types of estimation problems have been extensively investigated based on the times at which the measurement is obtained and estimation is processed. When an estimate is obtained at the same time as the last measurement point, it is a *filtering* problem; when the estimation takes place within the time interval of available measurement data, it is a *smoothing* problem; and when the estimation is processed at the time after the last available measurement, it is a *prediction* problem. This book will cover all these three estimation problems with relatively more attention to the first one since a large spectrum of estimation applications is the filtering problem.

In most estimation applications, measurements and observations are expressed as numerical quantities and they typically exhibit uncertain variability every time they are repeated. These numerical-valued random quantities are usually named as random variables. The concept of the random variable is central to all the estimation concepts. For example, based on a probability space on which the random variable is defined, probability distributions and probability density functions can be defined.

1.1 Random Variables and Random Process

Definition 1.1. **Random Vector** $\mathbf{x} = \left[\mathrm{x}_1, \cdots, \mathrm{x}_n\right]^T$

Given a probability space $(\Omega, \mathcal{A}, \mathcal{P})$ where Ω is the sample space, $\mathcal{A}$ is the σ-algebra of Ω, and $\mathcal{P}$ is a probability measure on $\mathcal{A}$, a random vector $\mathbf{x}(\bullet): \Omega \to \mathbb{R}^n$ is a real-valued function that maps a sample point, $\omega \in \Omega$ into a point in $\mathbb{R}^n$ such that $A = \{\omega : \mathbf{x}(\omega) \leq \boldsymbol{x}\} \in \mathcal{A}$ for any $\boldsymbol{x} = \left[x_1, \cdots, x_n\right]^T \in \mathbb{R}^n$.

Note that the lowercase and boldface letter denotes the random vector while the lowercase, boldface, and italic letter denotes the realization of the random vector. A random variable is a univariate random vector and is denoted by a non-boldface letter.

Definition 1.2. **Probability Distribution Function**

A real scalar-valued function:

$$F_{\mathbf{x}}(\boldsymbol{x}) \triangleq P\left(\{\omega : \mathbf{x}(\omega) \leq \boldsymbol{x}\}\right) = P\left(\mathrm{x}_1 < x_1, \mathrm{x}_2 < x_2, \cdots, \mathrm{x}_n < x_n\right)$$

is called a probability distribution function where P denotes the probability. It is also referred to as the joint probability distribution function of $\mathrm{x}_1, \cdots, \mathrm{x}_n$.

Definition 1.3. **Probability Density Function (PDF)**

The derivative of the probability distribution function $F_{\mathbf{x}}(\boldsymbol{x})$ is called the probability density function $p(\boldsymbol{x})$ of the random vector $\boldsymbol{x}$, i.e.,

$$p(\boldsymbol{x}) \triangleq \frac{dF_{\mathbf{x}}(\boldsymbol{x})}{d\boldsymbol{x}} \tag{1.1}$$

or
$$F_{\mathbf{x}}(\boldsymbol{x}) = \int_{-\infty}^{x_1}\int_{-\infty}^{x_2}\cdots\int_{-\infty}^{x_n} p(u_1,\cdots,u_n)du_1\cdots du_n \tag{1.2}$$

In the subsequent text, we will use a simpler notation

$$\int_{-\infty}^{x_1}\int_{-\infty}^{x_2}\cdots\int_{-\infty}^{x_n} p(u_1,\cdots,u_n)du_1\cdots du_n = \int_{-\infty}^{\boldsymbol{x}} p(\mathbf{u})d\mathbf{u} \tag{1.3}$$

to represent an integral with respect to a random vector. $p(\boldsymbol{x})$ satisfies $\int_{-\infty}^{\infty} p(\boldsymbol{x})d\boldsymbol{x} = 1$.

Definition 1.4. **Expectation**

The expectation or mean of a random vector $\mathbf{x} = [\mathrm{x}_1,\cdots,\mathrm{x}_n]^T$ is defined to be the vector of expectations of its elements; i.e.,

$$E[\mathbf{x}] = \begin{bmatrix} E[\mathrm{x}_1] \\ \vdots \\ E[\mathrm{x}_n] \end{bmatrix} \triangleq \begin{bmatrix} \int_{-\infty}^{\infty}\cdots\int_{-\infty}^{\infty} x_1 p(\boldsymbol{x})dx_1\cdots dx_n \\ \vdots \\ \int_{-\infty}^{\infty}\cdots\int_{-\infty}^{\infty} x_n p(\boldsymbol{x})dx_1\cdots dx_n \end{bmatrix} \tag{1.4}$$

In a vector representation,

$$E[\mathbf{x}] \triangleq \int_{-\infty}^{\infty} \boldsymbol{x} p(\boldsymbol{x})d\boldsymbol{x} \triangleq \boldsymbol{m}_{\mathbf{x}} \tag{1.5}$$

The expectation satisfies the linearity property:

$$E[c\mathbf{x}] = cE[\mathbf{x}] \tag{1.6}$$

$$E[\mathbf{x}+\mathbf{y}] = E[\mathbf{x}] + E[\mathbf{y}] \tag{1.7}$$

where c is a constant scalar. A more general linearity can be derived from (1.6) and (1.7)

$$E[\mathbf{AXB} + \mathbf{C}] = \mathbf{A}E[\mathbf{X}]\mathbf{B}+\mathbf{C} \tag{1.8}$$

where **A**, **B**, and **C** are constant matrices and **X** is a random matrix. Assume that these matrices have compatible dimensions.

The expectation of a function **y(x)** of the random vector **x** is

$$E[\mathbf{y}] \triangleq \int_{-\infty}^{\infty} \mathbf{y}(\boldsymbol{x})p(\boldsymbol{x})d\boldsymbol{x} \triangleq \boldsymbol{m}_{\mathbf{y}} \tag{1.9}$$

$E[\cdot]$ can be used to derive other statistics of the random variable/vector. For a random variable x,

$$E[\mathrm{x}^2] = \int_{-\infty}^{\infty} x^2 p(x)dx \tag{1.10}$$

is called the mean square or second moment,

$$E\left[x^n\right] = \int_{-\infty}^{\infty} x^n p(x)dx \tag{1.11}$$

is the nth order moment, and

$$E\left[\left(x-m_x\right)^n\right] = \int_{-\infty}^{\infty} \left(x-m_x\right) p(x)dx \tag{1.12}$$

is the nth order central moment.

Definition 1.5. **Conditional Probability Density** can be represented by

$$p(x \mid y) \triangleq \frac{p(x,y)}{p(y)} \tag{1.13}$$

Another expression of the conditional probability density is in the form of Bayes' rule:

$$p(x \mid y) = \frac{p(x,y)}{p(y)} = \frac{p(y \mid x)p(x)}{p(y)} = \frac{p(y \mid x)p(x)}{\int_{-\infty}^{\infty} p(y \mid x)p(x)dx} \tag{1.14}$$

where the denominator can be regarded as a term to normalize the expression, so that the total integral of the conditional probability density is unity. Note that the integrand in the denominator is of the same form as the numerator. Through the conditional probability density, inter-relationship among random variables can be revealed.

Definition 1.6. **Independence**

Given a probability space $(\Omega, \mathcal{A}, \mathcal{P})$, two random vectors **x** and **y** are independent if

$$\mathcal{P}\left(\left\{\omega : \mathbf{x}(\omega) \in A \text{ and } \mathbf{y}(\omega) \in B\right\}\right) = \mathcal{P}\left(\left\{\omega : \mathbf{x}(\omega) \in A\right\}\right)\mathcal{P}\left(\left\{\omega : \mathbf{y}(\omega) \in B\right\}\right) \tag{1.15}$$

for all $A, B \in \mathcal{A}$. In terms of probability distribution functions,

$$F_{\mathbf{x},\mathbf{y}}(x,y) = F_{\mathbf{x}}(x)F_{\mathbf{y}}(y) \tag{1.16}$$

or the probability density function

$$p(x,y) = p(x)p(y) \tag{1.17}$$

Definition 1.7. **Correlation**

The correlation matrix or the second moment of **x** is defined by

$$\mathcal{C}(\mathbf{x}) \triangleq E\left[\mathbf{x}\mathbf{x}^T\right] \tag{1.18}$$

with the (i, j) entry $E\left[x_i x_j\right]$ representing the correlation between x_i and x_j. Note that $\mathcal{C}$ is a symmetric matrix. $E\left[x_i x_j\right] = E[x_i]E[x_j]$ if x_i and x_j are uncorrelated.

If two random variables x_i and x_j are independent, they are uncorrelated. The property of being uncorrelated is weaker than the property of independence. For independent random variables, $E\left[f(x_i)g(x_j)\right] = E[f(x_i)]E[g(x_j)]$ for any functions $f(x_i)$ and $g(x_j)$, whereas for uncorrelated random variables, it is only required that this holds for $f(x_i) = x_i$ and $g(x_j) = x_j$.

Definition 1.8. **Covariance**

Define the covariance matrix or the second central moment of **x** by

$$\mathbf{P}(\mathbf{x}) \triangleq E\left[\left(\mathbf{x} - \boldsymbol{m}_\mathbf{x}\right)\left(\mathbf{x} - \boldsymbol{m}_\mathbf{x}\right)^T\right] \tag{1.19}$$

$\mathbf{P}_{ij}(\mathbf{x}) = E\left[\left(x_i - m_{x_i}\right)\left(x_j - m_{x_j}\right)\right]$ is called the cross-covariance of the *i*th and *j*th components of **x**.

It can be shown that

$$\mathbf{P}(\mathbf{x}) = \mathcal{C}(\mathbf{x}) - \boldsymbol{m}_\mathbf{x}\boldsymbol{m}_\mathbf{x}^T \tag{1.20}$$

We see that the covariance and correlation matrices are equal if and only if the mean vector is zero. Note that the diagonal entries of **P**(**x**) are the **variance** of each random variable x_i, i.e.,

$Var(x_i) \triangleq E\left[\left(x_i - E\left[x_i\right]\right)^2\right] = \mathbf{P}_{ii}(\mathbf{x})$. For $i \neq j$, $\mathbf{P}_{ij}(\mathbf{x}) = 0$ if and only if x_i and x_j are uncorrelated. Thus, **P**(**x**) is a diagonal matrix if x_i and x_j are uncorrelated. The square root of a variance $\mathbf{P}_{ii}(\mathbf{x})$ is termed the **standard deviation** of x_i.

If a random vector **x** has a covariance matrix **P**(**x**), **y** = **Ax** has a covariance matrix $\mathbf{AP}(\mathbf{x})\mathbf{A}^T$.

Cross-covariance

If $\mathbf{x} = \left[x_1, \cdots, x_n\right]^T$ and $\mathbf{z} = \left[z_1, \cdots, z_p\right]^T$ are both random vectors with respective means $\boldsymbol{m}_\mathbf{x}$ and $\boldsymbol{m}_\mathbf{z}$, then their cross-covariance matrix is the $n \times p$ matrix

$$\mathbf{P}(\mathbf{x},\mathbf{z}) \triangleq E\left[\left(\mathbf{x} - \boldsymbol{m}_\mathbf{x}\right)\left(\mathbf{z} - \boldsymbol{m}_\mathbf{z}\right)^T\right] \tag{1.21}$$

The two random vectors $\mathbf{x} = \left[x_1, \cdots, x_n\right]^T$ and $\mathbf{z} = \left[z_1, \cdots, z_p\right]^T$ are uncorrelated if $\mathbf{P}(x_i, z_j) = 0$ for all $i = 1, \cdots, n$ and all $j = 1, \cdots, p$. This is equivalent to the condition that $\mathbf{P}(\mathbf{x},\mathbf{z}) = 0$ be a $n \times p$ zero matrix.

Definition 1.9. **Stochastic Process**
A stochastic process is a generalization of a random variable and can be considered as an infinite collection of random variables. It is a set of random variables, $\mathbf{x}(\omega, t)$, indexed by the time t, and defined on a probability space $(\Omega, \mathcal{A}, \mathcal{P})$.

The random process can be thought of as a function of the sample point ω and time t (Maybeck 1979). A random process is a set of time functions, each of which is one possible outcome, ω, out of Ω. A random variable is a real number, while a random process is a continuously indexed collection of real numbers or a real-valued function. If t is fixed, $\mathbf{x}(\omega, t)$ results in a random variable. If ω is fixed, $\mathbf{x}(\omega, t)$ becomes a realization of the random process. In estimation or filtering, the statistical properties of interest for random processes include how their statistics and joint statistics are related across time or between components of vectors. To simplify the notation without causing confusion, we will denote a random process by $\mathbf{x}(t)$.

The statistical properties of a random process can be similarly defined as those for random variables/vectors by including the time t.
The expected value or mean can be given by

$$E[\mathbf{x}(t)] \triangleq \int_{-\infty}^{\infty} x(t)p(x(t))dx(t) \triangleq \boldsymbol{m}_{\mathbf{x}}(t) \tag{1.22}$$

A process $\mathbf{x}(t)$ is considered *time-independent* if for any distinct times $t_1, t_2, t_3, \cdots, t_i$,

$$p(x(t_1), x(t_2), \cdots, x(t_i)) = p(x(t_1))p(x(t_2))\cdots p(x(t_i)) \tag{1.23}$$

The *autocorrelation* of the random process $\mathbf{x}(t)$ between any two times t_1 and t_2 can be defined as

$$\mathcal{C}(\mathbf{x}) \triangleq E\left[\mathbf{x}(t_1)\mathbf{x}^T(t_2)\right] \tag{1.24}$$

The diagonal elements of $\mathcal{C}(\mathbf{x})$ when $t_1 = t_2 = t$, is the *average power* of $\mathbf{x}(t)$, i.e., $E[\mathbf{x}^2(t)]$.
The *autocovariance* of $\mathbf{x}(t)$ can be defined by

$$P(\mathbf{x}) \triangleq E\left[\left(\mathbf{x}(t_1) - \boldsymbol{m}_{\mathbf{x}}(t_1)\right)\left(\mathbf{x}(t_2) - \boldsymbol{m}_{\mathbf{x}}(t_2)\right)^T\right] \tag{1.25}$$

The cross-correlation between two random processes $\mathbf{x}(t) \in \mathbb{R}^n$ and $\mathbf{z}(t) \in \mathbb{R}^p$ is defined by an $n \times p$ matrix

$$\mathcal{C}(\mathbf{x}, \mathbf{z}) = E\left[\mathbf{x}(t_1)\mathbf{z}^T(t_2)\right] \tag{1.26}$$

The cross-covariance between $\mathbf{x}(t)$ and $\mathbf{z}(t)$ can be similarly defined by an $n \times p$ matrix

$$P(\mathbf{x},\mathbf{z}) \triangleq E\left[\left(\mathbf{x}(t_1) - \boldsymbol{m}_{\mathbf{x}}(t_1)\right)\left(\mathbf{z}(t_2) - \boldsymbol{m}_{\mathbf{z}}(t_2)\right)^T\right] \tag{1.27}$$

Two random processes $\mathbf{x}(t)$ and $\mathbf{z}(t)$ are uncorrelated if their cross-covariance matrix is identically zero for all t_1 and t_2. They are called orthogonal if their cross-correlation matrix is identically zero. A white noise is an example of an uncorrelated random process.

A particularly important characteristic of a random process is whether or not it is stationary. Stationarity is the property of a random process that guarantees that its statistical properties, such as its mean, variance, and moments will not change over time. The random process $\mathbf{x}(t)$ is *strict-sense stationary* if its PDF is invariant with respect to the time shift (Papoulis 2002), i.e.,

$$p(x(t_1), x(t_2), \cdots, x(t_i)) = p(x(t_1 + \tau), x(t_2 + \tau), \cdots, x(t_i + \tau)) \tag{1.28}$$

The random process $\mathbf{x}(t)$ is *wide-sense stationary* if

$$E\left[\mathbf{x}(t)\right] = constant \tag{1.29a}$$

$$E[\mathbf{x}(t_1)\mathbf{x}^T(t_2)] = \boldsymbol{C}(t_2 - t_1) = \boldsymbol{C}(\tau) \tag{1.29b}$$

where $\boldsymbol{C}$ is a matrix with its value depending only on the time difference $t_2 - t_1 = \tau$.

A wide-sense stationary process implies that its first-order and second-order statistics of $\mathbf{x}(t)$ are independent of the time origin, while a strict-sense stationary process implies that all statistics do not depend on the time origin.

A rather strictly stationary process of particular interest is the *ergodic processes* (Gelb 1974). A process is ergodic if any statistic calculated by averaging over all members of the ensemble of samples at a fixed time can be calculated equivalently by time-averaging over any single representative member of the ensemble. In other words, a sampled function $\mathbf{x}(t)$ is ergodic if its time-averaged statistics equal its ensemble averages. In practice, almost all empirical results for stationary processes are derived from tests of a single function with the ergodic assumption.

One of the most important random processes in estimation is the Markov process. A random process $\mathbf{x}(t)$ is called a Markov process if its future state distribution, conditioned on knowledge of its present state, does not rely on its previous states. Specifically, for the times $t_1 < t_2 < t_3 < \cdots < t_i$,

$$P\left(\mathbf{x}(t_i) \le x_i \mid \mathbf{x}(t_{i-1}), \cdots, \mathbf{x}(t_1)\right) = P\left(\mathbf{x}(t_i) \le x_i \mid \mathbf{x}(t_{i-1})\right) \tag{1.30}$$

Similarly, for discrete-time systems, a random process $\mathbf{x}_k$ is called a Markov sequence if

$$P(\mathbf{x}_i \le x_i \mid \mathbf{x}_k; k \le i-1) = P(\mathbf{x}_i \le x_i \mid \mathbf{x}_{i-1}) \tag{1.31}$$

The solution to a general first-order differential equation

$$\frac{d\mathbf{x}(t)}{dt} = \mathbf{g}(\mathbf{x},t) + \mathbf{w}(t) \tag{1.32}$$

with an independent random process $\mathbf{w}(t)$ as a forcing function is a Markov process. If we impose the restriction that the PDF of $\mathbf{w}(t)$ and consequently $\mathbf{x}(t)$ are Gaussian, the process $\mathbf{x}(t)$ is called a Gauss-Markov process.

1.2 Gaussian Distribution

The most important probability density function of the random variable is the Gaussian or normal distribution. As a consequence of the central limit theorem, the Gaussian density is a good approximation of probability density function involving a sum of many independent random variables, which is true whether the random variables are continuous or discrete. The Gaussian random variable/vector is very useful and of practical importance to the estimation problem because (1) it provides an adequate model for many random behaviors exhibited in nature, (2) it is a tractable mathematical model for estimation algorithms. A normal distribution of a random variable can be represented by

$$p(x) = \frac{1}{\sqrt{2\pi}\sigma} exp\left[-\frac{1}{2}\left(\frac{x-m_x}{\sigma}\right)^2\right] \tag{1.33}$$

where m_x is the mean of the random variable x, σ is its standard deviation, and σ^2 is its variance. *exp* denotes the exponential function. We usually use $N(m_x, \sigma^2)$ to represent the normal distribution. If $m_x = 0$ and $\sigma^2 = 1$, it is called a standard normal distribution, $x \sim N(0,1)$. As can be seen, the Gaussian density function is completely defined by the two parameters m_x and σ^2. Unlike other density functions, higher order moments are not required to generate a complete description of the Gaussian density function.
The moments of a standard normal distribution can be easily calculated

$$E\left[x^n\right] = \begin{cases} 1\cdot 3\cdots(n-3)(n-1), & n \text{ even} \\ 0, & n \text{ odd} \end{cases} \tag{1.34}$$

Univariate Gaussian or normal random variables can be generalized to random vectors. The probability density function of a Gaussian (normal) random vector can be represented by

$$p(x) = \frac{1}{(2\pi)^{n/2}|\mathbf{P}|^{1/2}} exp\left\{-\frac{1}{2}[x - m_x]^T \mathbf{P}^{-1}[x - m_x]\right\} \quad (1.35)$$

where $|\cdot|$ denotes the determinant of a matrix. The covariance matrix **P** is assumed to be positive definite.

Definition 1.10. **Jointly Gaussian Distribution**

A random vector $\mathbf{x} = [\mathrm{x}_1, \cdots, \mathrm{x}_n]^T$ is Gaussian or normal if every linear combination of its components of **x**, $\sum_{i=1}^{n} c_i \mathrm{x}_i$, is a scalar Gaussian random variable. $\mathrm{x}_1, \cdots, \mathrm{x}_n$ is also called jointly Gaussian or jointly normal. Here, any constant random variable is considered to be Gaussian in order for this definition to make sense when all $c_i = 0$ or when **x** has a singular covariance matrix.

If $\mathbf{x} = [\mathrm{x}_1, \cdots, \mathrm{x}_n]^T$ is a Gaussian random vector with mean $\boldsymbol{m}_\mathbf{x}$ and covariance matrix $\mathbf{P}_\mathbf{x}$, we denote it as $\mathbf{x} \sim N(\boldsymbol{m}_\mathbf{x}, \mathbf{P}_\mathbf{x})$.

If x_i are independent, every linear combination of x_i is Gaussian. Then, $\mathbf{x} = [\mathrm{x}_1, \cdots, \mathrm{x}_n]^T$ is a Gaussian vector.

Any subvector of a Gaussian vector is a Gaussian vector.

If $\mathbf{x} = [\mathrm{x}_1, \cdots, \mathrm{x}_n]^T$ is a Gaussian random vector with mean $\boldsymbol{m}_\mathbf{x}$ and covariance matrix $\mathbf{P}_\mathbf{x}$, $\mathbf{Cx} + \boldsymbol{b}$ is a Gaussian random vector for any $p \times n$ matrix **C** and any p-vector $\boldsymbol{b}$.

Note that $\mathbf{y} = \mathbf{Cx}$ being Gaussian does not necessarily imply that **x** is Gaussian. If **C** is invertible, $\mathbf{x} = \mathbf{C}^{-1}\mathbf{y}$ is Gaussian.

If the components of a random vector are uncorrelated, then its covariance matrix is diagonal. In general, this does not mean that the components of the random vector are independent. However, if **x** is a Gaussian random vector, then its components are independent. In other words, two jointly Gaussian random vectors that are uncorrelated are also independent.

If **x** is a Gaussian random vector, $\mathbf{x} \sim N(\boldsymbol{m}_\mathbf{x}, \mathbf{P}_\mathbf{x})$, the linear transformation y $\mathbf{y} = \left[\sqrt{\mathbf{P}_\mathbf{x}}\right]^{-1}(\mathbf{x} - \boldsymbol{m}_\mathbf{x})$ results in a random vector **y** that follows a standard normal distribution, i.e., $\mathbf{y} \sim N(\mathbf{0}, I)$. This transformation is called the decorrelating transformation (Gubner 2006, Kailath 2000), or stochastic decomposition (Arasaratnam and Haykin 2007). **y** will be a Gaussian random vector with uncorrelated and therefore independent components. $\sqrt{\mathbf{P}_\mathbf{x}}$ is the square root of the covariance matrix $\mathbf{P}_\mathbf{x}$, which can be obtained by many matrix decomposition techniques that factorizes a covariance matrix $\mathbf{P}_\mathbf{x}$ in the form of $\mathbf{P}_\mathbf{x} = \mathbf{SS}^T$ in which $\mathbf{S} = \sqrt{\mathbf{P}_\mathbf{x}}$, e.g., the Cholesky decomposition, the singular value decomposition (SVD) and the eigenvector decomposition.

1.3 Bayesian Estimation

There are two commonly used models for the estimation procedure. The first model is called non-Bayesian or Fisher estimation (Bar-Shalom et al. 2001) assuming that the parameters to be estimated are nonrandom and constant in the observation interval but the observations are noisy. The second model is called Bayesian estimation assuming that the parameters to be estimated are random variables with a prior probability distribution, and the observations are noisy as well.

Bayesian estimation is more powerful in state estimation for dynamic systems. It is based on the Bayes' rule

$$p(x \mid y) = \frac{p(y \mid x)p(x)}{p(y)} \tag{1.36}$$

to estimate the random parameter **x**, from the noisy observation $\mathbf{y} = y$. The estimation procedure starts with a prior statistics or belief statement of the random parameter **x**, i.e., $p(x)$. As measurements are made, the prior PDF $p(x)$ is transformed to the posterior distribution function $p(x|y)$ utilizing the likelihood $p(y|x)$. $p(y)$ is called the *evidence* that scales the posterior to assure its integral is unity. As more measurements are used, a posterior distribution $p(x|y)$ is improved such that it results in a sharper peak closer to the true parameter/state. Once the posterior distribution is determined, all statistical inferences or estimates can be made. This procedure can be illustrated in the following discrete-time dynamic state estimation problem.

Consider a class of nonlinear discrete-time dynamical systems:

$$\mathbf{x}_k = \boldsymbol{f}(\mathbf{x}_{k-1}) + \boldsymbol{v}_{k-1} \tag{1.37}$$

with an observation/measurement function that connects the state vector with the observation:

$$\mathbf{y}_k = \boldsymbol{h}(\mathbf{x}_k) + \boldsymbol{n}_k \tag{1.38}$$

where $\mathbf{x}_k \in \mathbb{R}^n; \mathbf{y}_k \in \mathbb{R}^m$; $\boldsymbol{v}_{k-1}$ and $\boldsymbol{n}_k$ are independent white Gaussian process noise and measurement noise with covariance $\mathbf{Q}_{k-1}$ and $\mathbf{R}_k$, respectively.

It is assumed that the discrete-time dynamic process (1.37) is a first-order Markov process in which the current state is dependent only on the previous state. The specific estimation problem is estimating the state vector $\mathbf{x}_k$ based on the sequence of observation vectors $\mathbf{y}_{1:k} \triangleq \{\mathbf{y}_1, \mathbf{y}_1, \cdots \mathbf{y}_k\}$.

The estimation can be performed based on the Bayesian estimation framework by turning the problem into an estimation of the conditional posterior density $p(\mathbf{x}_k \mid \mathbf{y}_{1:k})$. Recursive prediction and update procedures can be developed for estimation of $p(\mathbf{x}_k \mid \mathbf{y}_{1:k})$.

Applying Bayes' rule (1.36) to the posterior distribution $p(\mathbf{x}_k \mid \mathbf{y}_{1:k})$ yields

$$p(\mathbf{x}_k \mid \mathbf{y}_{1:k}) = \frac{p(\mathbf{y}_{1:k} \mid \mathbf{x}_k)p(\mathbf{x}_k)}{p(\mathbf{y}_{1:k})} \tag{1.39}$$

$p(\mathbf{x}_k \mid \mathbf{y}_{1:k})$ is the PDF of $\mathbf{x}_k$ conditioned on all observations up to and including the current observation. Equation (1.39) can be rewritten as

$$p(\mathbf{x}_k \mid \mathbf{y}_{1:k}) = \frac{p(\mathbf{y}_k, \mathbf{y}_{1:k-1} \mid \mathbf{x}_k)p(\mathbf{x}_k)}{p(\mathbf{y}_k, \mathbf{y}_{1:k-1})} = \frac{p(\mathbf{y}_k \mid \mathbf{y}_{1:k-1}, \mathbf{x}_k)p(\mathbf{y}_{1:k-1} \mid \mathbf{x}_k)p(\mathbf{x}_k)}{p(\mathbf{y}_k \mid \mathbf{y}_{1:k-1})p(\mathbf{y}_{1:k-1})} \tag{1.40}$$

Applying Bayes' rule to $p(\mathbf{y}_{1:k-1} \mid \mathbf{x}_k)$ yields

$$\begin{aligned} p(\mathbf{x}_k \mid \mathbf{y}_{1:k}) &= \frac{p(\mathbf{y}_k \mid \mathbf{y}_{1:k-1}, \mathbf{x}_k)p(\mathbf{x}_k \mid \mathbf{y}_{1:k-1})p(\mathbf{y}_{1:k-1})p(\mathbf{x}_k)}{p(\mathbf{y}_k \mid \mathbf{y}_{1:k-1})p(\mathbf{y}_{1:k-1})p(\mathbf{x}_k)} \\ &= \frac{p(\mathbf{y}_k \mid \mathbf{y}_{1:k-1}, \mathbf{x}_k)p(\mathbf{x}_k \mid \mathbf{y}_{1:k-1})}{p(\mathbf{y}_k \mid \mathbf{y}_{1:k-1})} \\ &= \frac{p(\mathbf{y}_k \mid \mathbf{x}_k)p(\mathbf{x}_k \mid \mathbf{y}_{1:k-1})}{p(\mathbf{y}_k \mid \mathbf{y}_{1:k-1})} \end{aligned} \tag{1.41}$$

Note that the last equality is obtained because, from the observation Eq. (1.38), the observation at time k does not depend on the observations at times $1:k-1$.
Since the denominator $p(\mathbf{y}_k \mid \mathbf{y}_{1:k-1})$ satisfies

$$p(\mathbf{y}_k \mid \mathbf{y}_{1:k-1}) = \int p(\mathbf{y}_k \mid \mathbf{x}_k)p(\mathbf{x}_k \mid \mathbf{y}_{1:k-1})d\mathbf{x}_k \tag{1.42}$$

Equation (1.41) becomes

$$p(\mathbf{x}_k \mid \mathbf{y}_{1:k}) = \frac{p(\mathbf{y}_k \mid \mathbf{x}_k)p(\mathbf{x}_k \mid \mathbf{y}_{1:k-1})}{\int p(\mathbf{y}_k \mid \mathbf{x}_k)p(\mathbf{x}_k \mid \mathbf{y}_{1:k-1})d\mathbf{x}_k} \tag{1.43}$$

where $p(\mathbf{y}_k \mid \mathbf{x}_k)$ is the likelihood function, $p(\mathbf{x}_k \mid \mathbf{y}_{1:k-1})$ can be obtained from the Chapman–Kolmogorov equation (Papoulis 2002)

$$p(\mathbf{x}_k \mid \mathbf{y}_{1:k-1}) = \int p(\mathbf{x}_k \mid \mathbf{x}_{k-1})p(\mathbf{x}_{k-1} \mid \mathbf{y}_{1:k-1})d\mathbf{x}_{k-1} \tag{1.44}$$

where $p(\mathbf{x}_k \mid \mathbf{x}_{k-1})$ is the predictive PDF.

Equations (1.43) and (1.44) establish a recursive Bayesian filtering procedure. To initialize the filter, a prior posterior density $p(\mathbf{x}_{k-1} \mid \mathbf{y}_{1:k-1})$ can be assumed. The conditional PDF $p(\mathbf{x}_k \mid \mathbf{y}_{1:k-1})$ satisfies the Chapman-Kolmogorov Eq. (1.44). When the observation at time k is available, the posterior conditional PDF $p(\mathbf{x}_k \mid \mathbf{y}_{1:k})$ can be calculated from the Eq. (1.43).

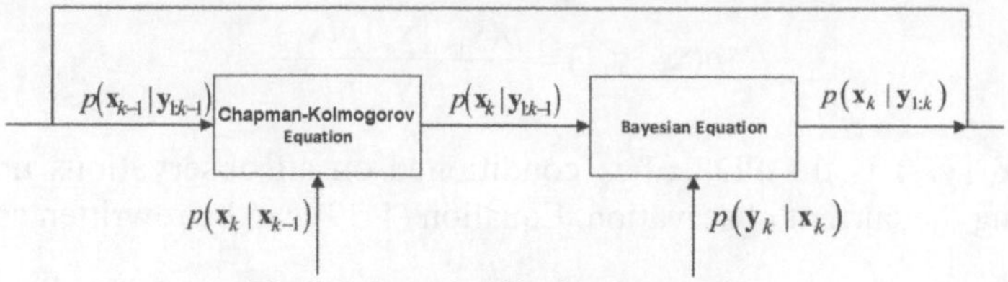

Fig. 1.1 Bayesian filtering framework.

Equation (1.44) is called the prediction formula while Eq. (1.43) is called the update formula. The moments, such as the mean and covariance, can be calculated from the PDF $p(\mathbf{x}_k \mid \mathbf{y}_{1:k})$. The filtering algorithm described previously can be presented in a block diagram, as shown in Fig. 1.1. Initially, $p(\mathbf{x}_{k-1} \mid \mathbf{y}_{1:k-1})$ is assumed to be known at the time $k-1$, and then $p(\mathbf{x}_k \mid \mathbf{y}_{1:k-1})$ can be computed using Eq. (1.44). When the measurement at time k arrives, the posterior density $p(\mathbf{x}_k \mid \mathbf{y}_{1:k})$ at time k is calculated by the Bayesian update formula in Eq. (1.43). Then, the mean and covariance of the state at time k can be calculated from $p(\mathbf{x}_k \mid \mathbf{y}_{1:k})$. State estimation can be recursively carried out by this procedure.

The recursive propagation of the posterior density given by Eqs. (1.43) and (1.44) is only a conceptual solution, in the sense that (in general) it cannot be determined analytically. Thus, one has to use approximations or suboptimal Bayesian algorithms. In this book, one of the focuses is to use various grid methods to approximate such a Bayesian estimation algorithm.

References

Arasaratnam, I. and S. Haykin. 2007. Discrete nonlinear filtering algorithms using Gauss-Hermite quadrature. Proceedings of the IEEE 95: 953–977.

Bar-Shalom, Y., X. Li and T. Kirubarajan. 2001. Estimation with Application to Tracking and Navigation: Theory, Algorithms and Software. John Wiley & Sons, Inc., New York.

Fisher, R.A. 1912. On an absolute criterion for fitting frequency curves. Messenger of Math 41: 155.

Gelb, A. 1974. Applied Optimal Estimation. MIT Press, Cambridge.

Gubner, J.A. 2006. Probability and Random Processes for Electrical and Computer Engineers, Cambridge University Press, New York.

Kailath, T., A.H. Sayed and B. Hassibi. 2000. Linear Estimation. Prentice Hall, Upper Saddle River.

Kalman, R.E. 1960. A new approach to linear filtering and prediction problem. Journal of Basic Engineering, pp. 35–46.

Kolmogorov, A.N. 1941. Interpolation and Extrapolation von Stationaren Zufalligen Folgen. Bull. Acad. Sci. USSR, Ser. Math. 5: 3–14.

Maybeck, P.S. 1979. Stochastic Models, Estimation, and Control, Vol. 1, Mathematics in Science and Engineering Series. Academic Press, New York.

Papoulis, A. 2002. Probability, Random Variables, and Stochastic Processes. 4th Edition, McGraw-Hill, New York.

Wiener, N. 1949. The Extrapolation, Interpolation and Smoothing of Stationary Time Series. John Wiley & Sons, Inc., New York.

Linear Estimation of Dynamic Systems

2

In the previous chapter, we described the estimation problem from the Bayesian perspective. An estimation problem is usually hard to solve analytically for general stochastic dynamic systems, so only conceptual algorithms are given therein. The difficulties stem from the nonlinear dynamics and computation of statistics that involves integrals of large dimensionality with respect to arbitrary PDF. However, under the Gaussian and linearity assumptions, the estimation problems become tractable. The celebrated Kalman filter gained its great success because the recursive Kalman filtering algorithm formulated in the time-domain state-space representation for linear dynamic systems with Gaussian statistics can be readily implemented in digital computers. Numerous books on the Kalman filter have been published. Hence, this chapter provides a concise review of the Kalman filter without delving into its theoretical development and properties. We first present the most widely used discrete-time linear Kalman filter in Section 2.1, and then derive one of its variants, the information Kalman filter in Section 2.2. Although the Kalman filter was not originally derived from the Bayesian estimation, we will show in Section 2.3 that the Kalman filter is equivalent to the Bayesian estimation for the Gaussian and linear system. The continuous-time Kalman-Bucy filter is also summarized at the end of this chapter for completeness despite that it is not much used in practice.

2.1 Linear Discrete-Time Kalman Filter

Consider a linear dynamic system described by the discrete-time state-space representation

$$\mathbf{x}_k = \mathbf{F}_{k-1}\mathbf{x}_{k-1} + \mathbf{v}_{k-1} \tag{2.1}$$

$$\mathbf{y}_k = \mathbf{H}_k \mathbf{x}_k + \mathbf{n}_k \tag{2.2}$$

where $\mathbf{x}_k \in \mathbb{R}^{n_x}$ is the system state vector and $\mathbf{y}_k \in \mathbb{R}^{n_p}$ is the measurement, n_x and n_p are the state and measurement dimensions, respectively. The additive process noise $\mathbf{v}_{k-1} \in \mathbb{R}^{n_x}$ and measurement noise $\mathbf{n}_k \in \mathbb{R}^{n_p}$ are assumed to be zero-mean, Gaussian, white noise.
The covariance matrices for $\mathbf{v}_k$ and $\mathbf{n}_k$ are given by

$$E\left[\mathbf{v}_i \mathbf{v}_k^T\right] = \begin{cases} \mathbf{Q}_k, & i = k \\ 0, & i \neq k \end{cases} \qquad E\left[\mathbf{n}_i \mathbf{n}_k^T\right] = \begin{cases} \mathbf{R}_k, & i = k \\ 0, & i \neq k \end{cases} \tag{2.3}$$

It is assumed that $\mathbf{v}_k$, $\mathbf{n}_k$, and $\mathbf{x}_0$ are uncorrelated with one another, i.e.,

$$E\left[\mathbf{v}_i \mathbf{n}_k^T\right] = 0, \text{ for all } k \text{ and } i, \quad E\left[\mathbf{v}_k \mathbf{x}_0^T\right] = 0, \text{ for all } k, \quad E\left[\mathbf{n}_k \mathbf{x}_0^T\right] = 0, \text{ for all } k \tag{2.4}$$

It can be shown that the discrete-time state-space system (2.1), (2.2) is a Gauss-Markov model. The filtering problem is to recover the system state $\mathbf{x}_k$ from the measurement $\mathbf{y}_k$. There are several different ways to derive the linear Kalman filter for this system according to different optimization criteria. We give the derivation based on the minimum mean square error (MMSE). Under the MMSE criterion, the conditional mean is the optimal estimate (Bar-Shalom et al. 2001), which is true not only for linear systems. We will now examine the conditional mean and covariance of the estimates.

We define two types of estimates based on the information upon which they are conditioned. The first one is the conditional mean

$$\hat{\mathbf{x}}_{k|k} \triangleq E\left[\mathbf{x}_k \mid \mathbf{y}_{1:k}\right] \tag{2.5}$$

based on $\mathbf{y}_{1:k} \triangleq \left[\mathbf{y}_1, \cdots, \mathbf{y}_k\right]$, the measurement history up to time k. The estimate $\hat{\mathbf{x}}_{k|k}$ is usually called the *a posteriori* estimate.

The second estimate is the propagated or predicted estimate through the given dynamics

$$\hat{\mathbf{x}}_{k|k-1} \triangleq E\left[\mathbf{x}_k \mid \mathbf{y}_{1:k-1}\right] \tag{2.6}$$

which describes how the state evolves between measurements. It is usually called *a priori* estimate. The *a priori* covariance $\mathbf{P}_{k|k-1}$ and *a posteriori* covariance $\mathbf{P}_{k|k}$ can be similarly defined.

The Kalman filtering includes the propagation/prediction step and the measurement update step. Assume that the conditional mean estimate $\hat{\mathbf{x}}_{k-1|k-1}$ and the conditional covariance $\mathbf{P}_{k-1|k-1}$ at the time $k-1$ are available. The predicted estimate can be derived as follows:

$$\begin{aligned}\hat{\mathbf{x}}_{k|k-1} &\triangleq E\left[\mathbf{x}_k \mid \mathbf{y}_{1:k-1}\right] \\ &= E\left[\mathbf{F}_{k-1}\mathbf{x}_{k-1} + \mathbf{v}_{k-1} \mid \mathbf{y}_{1:k-1}\right] \\ &= \mathbf{F}_{k-1}E\left[\mathbf{x}_{k-1} \mid \mathbf{y}_{1:k-1}\right] + E\left[\mathbf{v}_{k-1} \mid \mathbf{y}_{1:k-1}\right] \\ &= \mathbf{F}_{k-1}\hat{\mathbf{x}}_{k-1|k-1} + 0\end{aligned} \tag{2.7}$$

Therefore,

$$\hat{\mathbf{x}}_{k|k-1} = \mathbf{F}_{k-1}\hat{\mathbf{x}}_{k-1|k-1} \tag{2.8}$$

The predicted covariance becomes

$$\begin{aligned}\mathbf{P}_{k|k-1} &= E\left[\left(\mathbf{x}_k - \hat{\mathbf{x}}_{k|k-1}\right)\left(\mathbf{x}_k - \hat{\mathbf{x}}_{k|k-1}\right)^T\right] \\ &= E\left[\left(\mathbf{F}_{k-1}\mathbf{x}_{k-1} + \mathbf{v}_{k-1} - \mathbf{F}_{k-1}\hat{\mathbf{x}}_{k-1|k-1}\right)\left(\mathbf{F}_{k-1}\mathbf{x}_{k-1} + \mathbf{v}_{k-1} - \mathbf{F}_{k-1}\hat{\mathbf{x}}_{k-1|k-1}\right)^T\right] \\ &= \mathbf{F}_{k-1}\mathbf{P}_{k-1|k-1}\left(\mathbf{F}_{k-1}\right)^T + \mathbf{Q}_{k-1}\end{aligned} \tag{2.9}$$

where $\mathbf{P}_{k-1|k-1} = E\left[\left(\mathbf{x}_{k-1} - \hat{\mathbf{x}}_{k-1|k-1}\right)\left(\mathbf{x}_{k-1} - \hat{\mathbf{x}}_{k-1|k-1}\right)^T\right]$. Note that the last equality is obtained because $\left(\mathbf{x}_{k-1} - \hat{\mathbf{x}}_{k|k-1}\right)$ and $\mathbf{v}_{k-1}$ are uncorrelated, i.e.,

$$E\left[\mathbf{F}_{k-1}\left(\mathbf{x}_{k-1} - \hat{\mathbf{x}}_{k-1|k-1}\right)\left(\mathbf{v}_{k-1}\right)^T\right] = E\left[\mathbf{v}_{k-1}\left(\mathbf{x}_{k-1} - \hat{\mathbf{x}}_{k-1|k-1}\right)^T\left(\mathbf{F}_{k-1}\right)^T\right] = 0. \tag{2.10}$$

The estimate and covariance have been propagated from stage k 1 to k. Next, at stage k, the conditional mean and covariance are to be updated by incorporating the current measurement $\mathbf{y}_k$. To this end, define the innovation

$$\tilde{\mathbf{y}}_k = \mathbf{y}_k - \mathbf{H}_k\hat{\mathbf{x}}_{k|k\ 1} \tag{2.11}$$

The linear update equation is given by

$$\hat{\mathbf{x}}_{k|k} = \hat{\mathbf{x}}_{k|k\ 1} + \mathbf{K}_k\mathbf{y}_k. \tag{2.12}$$

Note that this linear relationship between the estimate and measurement is a natural result of the Gaussian assumption and the choice of the conditional mean as the estimate (Bar-Shalom 2001). $\mathbf{K}_k$ is called the Kalman gain and can be derived from the MMSE criterion as follows.

The estimation error is given by

$$
\begin{aligned}
\mathbf{e}_k &\triangleq \mathbf{x}_k - \hat{\mathbf{x}}_{k|k} \\
&= \mathbf{F}_{k-1}\mathbf{e}_{k-1} - \mathbf{K}_k\mathbf{H}_k\mathbf{F}_{k-1}\mathbf{e}_{k-1} + \mathbf{v}_{k-1} - \mathbf{K}_k\mathbf{H}_k\mathbf{v}_{k-1} - \mathbf{K}_k\mathbf{n}_k. \\
&= \left(\mathbf{I} - \mathbf{K}_k\mathbf{H}_k\right)\left(\mathbf{F}_{k-1}\mathbf{e}_{k-1} + \mathbf{v}_{k-1}\right) - \mathbf{K}_k\mathbf{n}_k
\end{aligned}
\tag{2.13}
$$

Next, the covariance matrix becomes

$$
\begin{aligned}
\mathbf{P}_{k|k} &= E\left[\mathbf{e}_k\mathbf{e}_k^T\right] \\
&= E\left[\begin{array}{l} \left(\mathbf{I} - \mathbf{K}_k\mathbf{H}_k\right)\mathbf{F}_{k-1}\mathbf{e}_{k-1}\left(\left(\mathbf{I} - \mathbf{K}_k\mathbf{H}_k\right)\mathbf{F}_{k-1}\mathbf{e}_{k-1}\right)^T \\ +\left(\mathbf{I} - \mathbf{K}_k\mathbf{H}_k\right)\mathbf{v}_{k-1}\left(\left(\mathbf{I} - \mathbf{K}_k\mathbf{H}_k\right)\mathbf{v}_{k-1}\right)^T + \mathbf{K}_k\mathbf{n}_k\left(\mathbf{K}_k\mathbf{n}_k\right)^T \end{array}\right] \\
&= \left(\mathbf{I} - \mathbf{K}_k\mathbf{H}_k\right)\mathbf{P}_{k|k-1}\left(\mathbf{I} - \mathbf{K}_k\mathbf{H}_k\right)^T + \mathbf{K}_k\mathbf{R}_k\mathbf{K}_k^T \\
&= \mathbf{P}_{k|k-1} - \mathbf{K}_k\mathbf{H}_k\mathbf{P}_{k|k-1} - \mathbf{P}_{k|k-1}\mathbf{H}_k^T\ \mathbf{K}_k^T + \mathbf{K}_k\mathbf{H}_k\mathbf{P}_{k|k-1}\mathbf{H}_k^T\mathbf{K}_k^T + \mathbf{K}_k\mathbf{R}_k\mathbf{K}_k^T
\end{aligned}
\tag{2.14}
$$

The optimal Kalman gain can be found by letting $\dfrac{\partial \mathrm{Tr}\left(\mathbf{P}_{k|k}\right)}{\partial \mathbf{K}_k} = 0$.
Note that

$$
\frac{\partial \mathrm{Tr}\left(\mathbf{P}_{k|k}\right)}{\partial \mathbf{K}_k} = -\left(\mathbf{H}_k\mathbf{P}_{k|k-1}\right)^T - \mathbf{P}_{k|k-1}\mathbf{H}_k^T + 2\mathbf{K}_k\mathbf{H}_k\mathbf{P}_{k|k-1}\mathbf{H}_k^T + 2\mathbf{K}_k\mathbf{R}_k, \tag{2.15}
$$

We have,

$$
\begin{aligned}
\mathbf{K}_k &= \left[\mathbf{P}_{k|k-1}^T\mathbf{H}_k^T + \mathbf{P}_{k|k-1}\mathbf{H}_k^T\right]\left[\mathbf{H}_k\mathbf{P}_{k|k-1}\mathbf{H}_k^T + \mathbf{R}_k\right]^{-1} / 2 \\
&= \mathbf{P}_{k|k-1}\mathbf{H}_k^T\left[\mathbf{H}_k\mathbf{P}_{k|k-1}\mathbf{H}_k^T + \mathbf{R}_k\right]^{-1}
\end{aligned}
\tag{2.16}
$$

Using (2.16) in (2.14), the covariance matrix $\mathrm{P}_{k|k}$ becomes

$$
\begin{aligned}
\mathbf{P}_{k|k} &= \mathbf{P}_{k|k-1} - \mathbf{K}_k\mathbf{H}_k\mathbf{P}_{k|k-1} - \mathbf{P}_{k|k-1}\mathbf{H}_k^T\mathbf{K}_k^T + \mathbf{K}_k\left(\mathbf{H}_k\mathbf{P}_{k|k-1}\mathbf{H}_k^T + \mathbf{R}_k\right)\mathbf{K}_k^T \\
&= \mathbf{P}_{k|k-1} - \mathbf{K}_k\mathbf{H}_k\mathbf{P}_{k|k-1} - \mathbf{P}_{k|k-1}\mathbf{H}_k^T\mathbf{K}_k^T + \mathbf{P}_{k|k-1}\mathbf{H}_k^T\mathbf{K}_k^T \\
&= \mathbf{P}_{k|k-1} - \mathbf{K}_k\mathbf{H}_k\mathbf{P}_{k|k-1}
\end{aligned}
\tag{2.17}
$$

The linear discrete-time Kalman filter can be summarized in Table 2.1

2.2 Information Kalman Filter

The Kalman filter equations can be presented in many different forms by algebraic manipulations. One of the alternative forms is called the information filter. It is computationally more efficient to handle the measurement update

Table 2.1 Discrete-time Kalman filter.

Discrete-time linear dynamics and measurement model:

$$\mathbf{x}_k = \mathbf{F}_{k-1}\mathbf{x}_{k-1} + \mathbf{v}_{k-1} \tag{2.1}$$

$$\mathbf{y}_k = \mathbf{H}_k\mathbf{x}_k + \mathbf{n}_k \tag{2.2}$$

<u>Assumptions</u>: $\mathbf{v}_{k-1}$ and $\mathbf{n}_k$ are white Gaussian process and measurement noise, respectively. $E\left[\mathbf{v}_{k-1}\mathbf{v}_{k-1}^T\right] = \mathbf{Q}_{k-1}$, $E\left[\mathbf{n}_k\mathbf{n}_k^T\right] = \mathbf{R}_k$. $\mathbf{v}_{k-1}$ and $\mathbf{n}_k$ are uncorrelated with $\mathbf{x}_0$ and with each other.

<u>Initialization</u>: $\hat{\mathbf{x}}_{0|0}$ and $P_{0|0}$

<u>Prediction</u>:

$$\hat{\mathbf{x}}_{k|k-1} = \mathbf{F}_{k-1}\hat{\mathbf{x}}_{k-1|k-1} \tag{2.8}$$

$$\mathbf{P}_{k|k-1} = \mathbf{F}_{k-1}\mathbf{P}_{k-1|k-1}\mathbf{F}_{k-1}^T + \mathbf{Q}_{k-1} \tag{2.9}$$

<u>Update</u>:

$$\hat{\mathbf{x}}_{k|k} = \hat{\mathbf{x}}_{k|k-1} + \mathbf{K}_k\left(\mathbf{y}_k - \mathbf{H}_k\hat{\mathbf{x}}_{k|k-1}\right). \tag{2.12}$$

$$\mathbf{P}_{k|k} = \mathbf{P}_{k|k-1} - \mathbf{K}_k\mathbf{H}_k\mathbf{P}_{k|k-1} \tag{2.17}$$

where

$$\mathbf{K}_k = \mathbf{P}_{k|k-1}\mathbf{H}_k^T\left[\mathbf{H}_k\mathbf{P}_{k|k-1}\mathbf{H}_k^T + \mathbf{R}_k\right]^{-1} \tag{2.16}$$

---

when the measurement sets are much larger than the state-space dimension because it can reduce the matrix inversion complexity in the Kalman filter algorithm.

Recall the Kalman gain

$$\mathbf{K}_k - \mathbf{P}_{k|k-1}\mathbf{H}_k^T\left[\mathbf{H}_k\mathbf{P}_{k|k-1}\mathbf{H}_k^T + \mathbf{R}_k\right]^{-1}$$

Post-multiplying this equation by $\left[\mathbf{H}_k\mathbf{P}_{k|k-1}\mathbf{H}_k^T + \mathbf{R}_k\right]$ yields

$$\mathbf{K}_k\mathbf{H}_k\mathbf{P}_{k|k-1}\mathbf{H}_k^T + \mathbf{K}_k\mathbf{R}_k = \mathbf{P}_{k|k-1}\mathbf{H}_k^T \tag{2.18}$$

Recall the covariance update equation:

$$\mathbf{P}_{k|k} = \mathbf{P}_{k|k-1} - \mathbf{K}_k\mathbf{H}_k\mathbf{P}_{k|k-1}$$

It can be rewritten as

$$\mathbf{K}_k\mathbf{H}_k\mathbf{P}_{k|k-1} = \mathbf{P}_{k|k-1} - \mathbf{P}_{k|k} \tag{2.19}$$

Substituting this equation into (2.18) yields

$$\left(\mathbf{P}_{k|k-1} - \mathbf{P}_{k|k}\right)\mathbf{H}_k^T + \mathbf{K}_k\mathbf{R}_k = \mathbf{P}_{k|k-1}\mathbf{H}_k^T \tag{2.20}$$

It results in another Kalman gain expression

$$\mathbf{K}_k = \mathbf{P}_{k|k}\mathbf{H}_k^T\mathbf{R}_k^{-1} \tag{2.21}$$

Substituting this equation into the covariance update Eq. (2.17) and post-multiplying the result by $\mathbf{P}_{k|k-1}^{-1}$ yields

$$\mathbf{I} - \mathbf{P}_{k|k}\mathbf{H}_k^T\mathbf{R}_k^{-1}\mathbf{H}_k = \mathbf{P}_{k|k}\mathbf{P}_{k|k-1}^{-1} \tag{2.22}$$

Rearranging this equation gives

$$\mathbf{P}_{k|k}\mathbf{P}_{k|k-1}^{-1} + \mathbf{P}_{k|k}\mathbf{H}_k^T\mathbf{R}_k^{-1}\mathbf{H}_k = \mathbf{I} = \mathbf{P}_{k|k}\mathbf{P}_{k|k}^{-1} \tag{2.23}$$

Factoring out the common factor $\mathbf{P}_{k|k}$ leads to

$$\mathbf{P}_{k|k}^{-1} = \mathbf{P}_{k|k-1}^{-1} + \mathbf{H}_k^T\mathbf{R}_k^{-1}\mathbf{H}_k \tag{2.24}$$

The inverse of the covariance matrix in (2.24) is called the information matrix. Let's define the information matrix as $\breve{\mathbf{Y}} = \mathbf{P}^{-1}$. Then, the information update Eq. (2.24) becomes

$$\breve{\mathbf{Y}}_{k|k} = \breve{\mathbf{Y}}_{k|k-1} + \mathbf{H}_k^T\mathbf{R}_k^{-1}\mathbf{H}_k \tag{2.25}$$

For the covariance prediction, recall

$$\mathbf{P}_{k|k-1} = \mathbf{F}_{k-1}\mathbf{P}_{k-1|k-1}\mathbf{F}_{k-1}^T + \mathbf{Q}_{k-1}$$

which can be rewritten as

$$\mathbf{P}_{k|k-1}^{-1} = \left(\mathbf{F}_{k-1}\mathbf{P}_{k-1|k-1}\mathbf{F}_{k-1}^T + \mathbf{Q}_{k-1}\right)^{-1} \tag{2.26}$$

Using the matrix inverse lemma

$$(\mathbf{M}+\mathbf{N})^{-1} = \mathbf{M}^{-1} - \mathbf{M}^{-1}\mathbf{N}\left(\mathbf{I}+\mathbf{M}^{-1}\mathbf{N}\right)^{-1}\mathbf{M}^{-1} \tag{2.27}$$

with $\mathbf{M} = \mathbf{F}_{k-1}\mathbf{P}_{k-1|k-1}\mathbf{F}_{k-1}^T$ and $\mathbf{N} = \mathbf{Q}_{k-1}$, and defining

$$\mathbf{\Gamma}_{k-1} = \left(\mathbf{F}_{k-1}\mathbf{P}_{k-1|k-1}\mathbf{F}_{k-1}^T\right)^{-1} = \mathbf{F}_{k-1}^{-T}\mathbf{P}_{k-1|k-1}^{-1}\mathbf{F}_{k-1}^{-1} \tag{2.28}$$

Equation (2.26) becomes

$$\breve{\mathbf{Y}}_{k|k-1} = \mathbf{P}_{k|k-1}^{-1} = \mathbf{\Gamma}_{k-1} - \mathbf{\Gamma}_{k-1}(\mathbf{\Gamma}_{k-1} + \mathbf{Q}_{k-1}^{-1})^{-1}\mathbf{\Gamma}_{k-1} \tag{2.29}$$

The state update and prediction equations in the information form can be also derived. Recall

$$\hat{\mathbf{x}}_{k|k} = \hat{\mathbf{x}}_{k|k-1} + \mathbf{K}_k\left(\mathbf{y}_k - \mathbf{H}_k\hat{\mathbf{x}}_{k|k-1}\right)$$

Pre-multiplying this equation by $\mathbf{P}_{k|k}^{-1}$ yields

$$\mathbf{P}_{k|k}^{-1}\hat{\mathbf{x}}_{k|k} = \mathbf{P}_{k|k}^{-1}\hat{\mathbf{x}}_{k|k-1} + \mathbf{P}_{k|k}^{-1}\mathbf{K}_k\left(\mathbf{y}_k - \mathbf{H}_k\hat{\mathbf{x}}_{k|k-1}\right) \tag{2.30}$$

Defining an information state vector $\breve{\mathbf{y}} = \mathbf{P}^{-1}\hat{\mathbf{x}}$, and substituting Eq. (2.21) into Eq. (2.30) lead to

$$\begin{aligned} \breve{\mathbf{y}}_{k|k} &= \mathbf{P}_{k|k}^{-1}\hat{\mathbf{x}}_{k|k-1} + \mathbf{H}_k^T\mathbf{R}_k^{-1}\left(\mathbf{y}_k - \mathbf{H}_k\hat{\mathbf{x}}_{k|k-1}\right) \\ &= \left(\mathbf{P}_{k|k}^{-1} - \mathbf{H}_k^T\mathbf{R}_k^{-1}\mathbf{H}_k\right)\hat{\mathbf{x}}_{k|k-1} + \mathbf{H}_k^T\mathbf{R}_k^{-1}\mathbf{y}_k \\ &= \mathbf{P}_{k|k-1}^{-1}\hat{\mathbf{x}}_{k|k-1} + \mathbf{H}_k^T\mathbf{R}_k^{-1}\mathbf{y}_k \\ &= \breve{\mathbf{y}}_{k|k-1} + \mathbf{H}_k^T\mathbf{R}_k^{-1}\mathbf{y}_k \end{aligned} \tag{2.31}$$

This is the prediction equation for the information state.

Pre-multiplying the state prediction equation $\hat{\mathbf{x}}_{k|k-1} = \mathbf{F}_{k-1}\hat{\mathbf{x}}_{k-1|k-1}$ by $\mathbf{P}_{k|k-1}^{-1}$ yields

$$\mathbf{P}_{k|k-1}^{-1}\hat{\mathbf{x}}_{k|k-1} = \mathbf{P}_{k|k-1}^{-1}\mathbf{F}_{k-1}\hat{\mathbf{x}}_{k-1|k-1} \tag{2.32}$$

Using the definition of the information state and the information matrix prediction Eq. (2.29) into Eq. (2.32) lead to

$$\begin{aligned} \breve{\mathbf{y}}_{k|k-1} &= \left[\boldsymbol{\Gamma}_{k-1} - \boldsymbol{\Gamma}_{k-1}\left(\boldsymbol{\Gamma}_{k-1} + \mathbf{Q}_{k-1}^{-1}\right)^{-1}\boldsymbol{\Gamma}_{k-1}\right]\mathbf{F}_{k-1}\hat{\mathbf{x}}_{k-1|k-1} \\ &= \left[\mathbf{I} - \boldsymbol{\Gamma}_{k-1}\left(\boldsymbol{\Gamma}_{k-1} + \mathbf{Q}_{k-1}^{-1}\right)^{-1}\right]\boldsymbol{\Gamma}_{k-1}\mathbf{F}_{k-1}\hat{\mathbf{x}}_{k-1|k-1} \\ &= \left[\mathbf{I} - \boldsymbol{\Gamma}_{k-1}\left(\boldsymbol{\Gamma}_{k-1} + \mathbf{Q}_{k-1}^{-1}\right)^{-1}\right]\mathbf{F}_{k-1}^{-T}\mathbf{P}_{k-1|k-1}\mathbf{F}_{k-1}^{-1}\mathbf{F}_{k-1}\hat{\mathbf{x}}_{k-1|k-1} \\ &= \left[\mathbf{I} - \boldsymbol{\Gamma}_{k-1}\left(\boldsymbol{\Gamma}_{k-1} + \mathbf{Q}_{k-1}^{-1}\right)^{-1}\right]\mathbf{F}_{k-1}^{-T}\mathbf{P}_{k-1|k-1}^{-1}\hat{\mathbf{x}}_{k-1|k-1} \\ &= \left[\mathbf{I} - \boldsymbol{\Gamma}_{k-1}\left(\boldsymbol{\Gamma}_{k-1} + \mathbf{Q}_{k-1}^{-1}\right)^{-1}\right]\mathbf{F}_{k-1}^{-T}\breve{\mathbf{y}}_{k-1|k-1} \end{aligned} \tag{2.33}$$

This is the information state prediction equation. The information form of the Kalman filter can be summarized in Table 2.2.

Both the discrete-time Kalman filter and the information filter generate identical results. When there are much more measurement sets than the states, the information form can avoid the large matrix inverse operation as needed in $\left[\mathbf{H}_k\mathbf{P}_{k|k-1}\mathbf{H}_k^T + \mathbf{R}_k\right]^{-1}$ for the Kalman gain calculation. Suppose there are m measurement sets and the $\mathbf{R}_k$ matrix is block diagonal. Equation (2.25) can be rewritten as

Table 2.2 Information filter.

Discrete-time linear dynamics and measurement model:

$$\mathbf{x}_k = \mathbf{F}_{k-1}\mathbf{x}_{k-1} + \mathbf{v}_{k-1} \tag{2.1}$$

$$\mathbf{y}_k = \mathbf{H}_k\mathbf{x}_k + \mathbf{n}_k \tag{2.2}$$

Assumptions: $\mathbf{v}_{k-1}$ and $\mathbf{n}_k$ are white Gaussian process and measurement noise, respectively. $E\left[\mathbf{v}_{k-1}\mathbf{v}_{k-1}^T\right] = \mathbf{Q}_{k-1}$, $E\left[\mathbf{n}_k\mathbf{n}_k^T\right] = \mathbf{R}_k$. $\mathbf{v}_{k-1}$ and $\mathbf{n}_k$ are uncorrelated with $\mathbf{x}_0$ and with each other.

Initialization: $\hat{\mathbf{x}}_{0|0}$ and $\mathbf{P}_{0|0}$, and the information covariance $\breve{\mathbf{Y}}_{0|0} = \mathbf{P}_{0|0}^{-1}$ and the information state vector $\breve{\mathbf{y}}_{0|0} = \breve{\mathbf{Y}}_{0|0}\hat{\mathbf{x}}_{0|0}$.

Prediction:

$$\breve{\mathbf{y}}_{k|k-1} = \left[\mathbf{I} - \mathbf{\Gamma}_{k-1}\left(\mathbf{\Gamma}_{k-1} + \mathbf{Q}_{k-1}^{-1}\right)^{-1}\right]\mathbf{F}_{k-1}^{-T}\breve{\mathbf{y}}_{k-1|k-1} \tag{2.33}$$

$$\breve{\mathbf{Y}}_{k|k-1} = \mathbf{\Gamma}_{k-1} - \mathbf{\Gamma}_{k-1}\left(\mathbf{\Gamma}_{k-1} + \mathbf{Q}_{k-1}^{-1}\right)^{-1}\mathbf{\Gamma}_{k-1} \tag{2.29}$$

where $\mathbf{\Gamma}_{k-1} = \mathbf{F}_{k-1}^{-T}\mathbf{P}_{k-1|k-1}^{-1}\mathbf{F}_{k-1}^{-1}$

Update:

$$\breve{\mathbf{y}}_{k|k} = \breve{\mathbf{y}}_{k|k-1} + \mathbf{H}_k^T\mathbf{R}_k^{-1}\mathbf{y}_k. \tag{2.31}$$

$$\breve{\mathbf{Y}}_{k|k} = \breve{\mathbf{Y}}_{k|k-1} + \mathbf{H}_k^T\mathbf{R}_k^{-1}\mathbf{H}_k \tag{2.25}$$

$$\begin{aligned}\breve{\mathbf{Y}}_{k|k} &= \breve{\mathbf{Y}}_{k|k-1} + \mathbf{H}_k^T\mathbf{R}_k^{-1}\mathbf{H}_k \\ &= \breve{\mathbf{Y}}_{k|k-1} + \left[\mathbf{H}_{k,1}^T \vdots \mathbf{H}_{k,2}^T \vdots \cdots \mathbf{H}_{k,m}^T\right]\begin{bmatrix}\mathbf{R}_{k,1}^{-1} & \mathbf{0} & \mathbf{0} & \mathbf{0} \\ \mathbf{0} & \mathbf{R}_{k,2}^{-1} & \mathbf{0} & \mathbf{0} \\ \mathbf{0} & \mathbf{0} & \ddots & \mathbf{0} \\ \mathbf{0} & \mathbf{0} & \mathbf{0} & \mathbf{R}_{k,m}^{-1}\end{bmatrix}\begin{bmatrix}\mathbf{H}_{k,1} \\ \mathbf{H}_{k,1} \\ \vdots \\ \mathbf{H}_{k,m}\end{bmatrix}\end{aligned} \tag{2.34}$$

The block-diagonal $\mathbf{R}_k$ means that each measurement set $\mathbf{y}_{k,i}\ (i = 1, \cdots, m)$ at time step k is uncorrelated with other measurement sets. Note that if the original $\mathbf{R}_k$ is not in this form, we can always perform the decorrelating transformation or stochastic decomposition mentioned in Chapter 1.2 to make $\mathbf{R}_k$ block diagonal (Kailath et al. 2000). Equation (2.34) can be rewritten as

$$\breve{\mathbf{Y}}_{k|k} = \breve{\mathbf{Y}}_{k|k-1} + \sum_{i=1}^{m}\mathbf{H}_{k,i}^T\mathbf{R}_{k,i}^{-1}\mathbf{H}_{k,i} \tag{2.35}$$

This form gives an efficient way to update the information matrix by processing the measurement one at a time until all the measurement sets at time k are assimilated. The results are the same as if we process all measurement sets all at once as a large measurement vector. It can be applied to the information state update Eq. (2.31) in the same way. After this one-at-a-time measurement update, the prediction step can proceed as usual.

There is another form of the information filter given by (Mutambara 1998) for measurement update:

$$\breve{\mathbf{y}}_{k|k} = \breve{\mathbf{y}}_{k|k-1} + \mathbf{i}_k \tag{2.36}$$

$$\breve{\mathbf{Y}}_{k|k} = \breve{\mathbf{Y}}_{k|k-1} + \mathbf{I}_k \tag{2.37}$$

where $\mathbf{i}_k = \mathbf{H}_k^T \mathbf{R}_k^{-1} \mathbf{y}_k$, $\mathbf{I}_k = \mathbf{H}_k^T \mathbf{R}_k^{-1} \mathbf{H}_k$ are called the information state contribution and information matrix contribution, respectively.

The information filter is especially useful when the measurements come from multiple sensors, which will be revisited in Chapter 6 for multiple sensor estimation application.

2.3 The Relation Between the Bayesian Estimation and Kalman Filter

In Section 2.1, the linear discrete Kalman filter is derived from minimizing the MSE. In this section, we will show the relationship between this filter and the Bayesian estimation in the prediction and update steps (Šimandl 2006).

Prediction

Based on the Chapman-Kolmogorov Eq. (1.44), we have

$$\begin{aligned} p(\mathbf{x}_k \mid \mathbf{x}_{k-1}) &= \int p(\mathbf{x}_k \mid \mathbf{x}_{k-1}) p(\mathbf{x}_{k-1} \mid \mathbf{y}_{k-1}) d\mathbf{x}_{k-1} \\ &= \int N(\mathbf{x}_k ; \mathbf{F}_{k-1}\mathbf{x}_{k-1}, \mathbf{Q}_{k-1}) N(\mathbf{x}_{k-1} ; \hat{\mathbf{x}}_{k-1|k-1}, \mathbf{P}_{k-1|k-1}) d\mathbf{x}_{k-1} \\ &= \int a \exp\left(-\frac{1}{2} b\right) d\mathbf{x}_{k-1} \end{aligned} \tag{2.38}$$

where
$$a = \frac{1}{(2\pi)^{n_x/2} \det(\mathbf{Q}_{k-1})^{1/2} (2\pi)^{n_x/2} \det(\mathbf{P}_{k-1|k-1})^{1/2}} \tag{2.39}$$

$$\begin{aligned} b &= (\mathbf{x}_k - \mathbf{F}_{k-1}\mathbf{x}_{k-1})^T \mathbf{Q}_{k-1}^{-1} (\mathbf{x}_k - \mathbf{F}_{k-1}\mathbf{x}_{k-1}) + (\mathbf{x}_{k-1} - \hat{\mathbf{x}}_{k-1|k-1})^T \mathbf{P}_{k-1|k-1}^{-1} (\mathbf{x}_{k-1} - \hat{\mathbf{x}}_{k-1|k-1}) \\ &= \mathbf{x}_{k-1}^T \left[\mathbf{F}_{k-1}^T \mathbf{Q}_{k-1}^{-1} \mathbf{F}_{k-1} + \mathbf{P}_{k-1|k-1}^{-1}\right] \mathbf{x}_{k-1} - \mathbf{x}_{k-1}^T \left[\mathbf{F}_{k-1}^T \mathbf{Q}_{k-1}^{-1} \mathbf{x}_k + \mathbf{P}_{k-1|k-1}^{-1} \hat{\mathbf{x}}_{k-1|k-1}\right] \\ &\quad - \left[\mathbf{x}_k^T \mathbf{Q}_{k-1}^{-1} \mathbf{F}_{k-1} + \hat{\mathbf{x}}_{k-1|k-1}^T \mathbf{P}_{k-1|k-1}^{-1}\right] \mathbf{x}_{k-1} \\ &\quad + \mathbf{x}_k^T \mathbf{Q}_{k-1}^{-1} \mathbf{x}_k + \hat{\mathbf{x}}_{k-1|k-1}^T \mathbf{P}_{k-1|k-1}^{-1} \hat{\mathbf{x}}_{k-1|k-1} \\ &= \mathbf{x}_{k-1}^T \tilde{\mathbf{A}} \mathbf{x}_{k-1} - (\mathbf{x}_{k-1})^T \tilde{\mathbf{b}} - \tilde{\mathbf{b}}^T \mathbf{x}_{k-1} + c \end{aligned} \tag{2.40}$$

where $\tilde{\mathbf{A}} = \mathbf{F}_{k-1}^T \mathbf{Q}_{k-1}^{-1} \mathbf{F}_{k-1} + \mathbf{P}_{k-1|k-1}^{-1}$, $\tilde{\mathbf{b}} = \mathbf{F}_{k-1}^T \mathbf{Q}_{k-1}^{-1} \mathbf{x}_k + \mathbf{P}_{k-1|k-1}^{-1} \hat{\mathbf{x}}_{k-1|k-1}$, and $c = \mathbf{x}_k^T \mathbf{Q}_{k-1}^{-1} \mathbf{x}_k + \hat{\mathbf{x}}_{k-1|k-1}^T \mathbf{P}_{k-1|k-1}^{-1} \hat{\mathbf{x}}_{k-1|k-1}$.

Equation (2.40) can be rewritten as

$$b = \left(\mathbf{x}_{k-1} - \tilde{\mathbf{A}}^{-1}\tilde{\mathbf{b}}\right)^T \tilde{\mathbf{A}}\left(\mathbf{x}_{k-1} - \tilde{\mathbf{A}}^{-1}\tilde{\mathbf{b}}\right) - \tilde{\mathbf{b}}^T \tilde{\mathbf{A}}^{-1}\tilde{\mathbf{b}} + c \tag{2.41}$$

and Eq. (2.38) can then be rewritten as

$$\begin{aligned}
& p\left(\mathbf{x}_k \mid \mathbf{x}_{k-1}\right) \\
&= \int a \cdot exp\left(-\frac{1}{2} b\right) d\mathbf{x}_{k-1} \\
&= a \int exp\left(-\frac{1}{2}\left(\left(\mathbf{x}_{k-1} - \tilde{\mathbf{A}}^{-1}\tilde{\mathbf{b}}\right)^T \tilde{\mathbf{A}}\left(\mathbf{x}_{k-1} - \tilde{\mathbf{A}}^{-1}\tilde{\mathbf{b}}\right) - \tilde{\mathbf{b}}^T \tilde{\mathbf{A}}^{-1}\tilde{\mathbf{b}} + c\right)\right) d\mathbf{x}_{k-1} \\
&= a \cdot exp\left(-\frac{1}{2}\left(-\tilde{\mathbf{b}}^T \tilde{\mathbf{A}}^{-1}\tilde{\mathbf{b}} + c\right)\right)\left((2\pi)^{\frac{n_x}{2}} \det\left(\tilde{\mathbf{A}}^{-1}\right)^{\frac{1}{2}}\right) \\
&\quad \cdot \int exp\left(-\frac{1}{2}\left(\left(\mathbf{x}_{k-1} - \tilde{\mathbf{A}}^{-1}\tilde{\mathbf{b}}\right)^T \tilde{\mathbf{A}}\left(\mathbf{x}_{k-1} - \tilde{\mathbf{A}}^{-1}\tilde{\mathbf{b}}\right)\right)\right) \Big/ \left((2\pi)^{\frac{n_x}{2}} \det\left(\tilde{\mathbf{A}}^{-1}\right)^{\frac{1}{2}}\right) d\mathbf{x}_{k-1} \\
&= a(2\pi)^{\frac{n_x}{2}} \det\left(\tilde{\mathbf{A}}^{-1}\right)^{\frac{1}{2}} exp\left(-\frac{1}{2}\left(-\tilde{\mathbf{b}}^T \tilde{\mathbf{A}}^{-1}\tilde{\mathbf{b}} + c\right)\right)
\end{aligned} \tag{2.42}$$

Re-organizing $-\tilde{\mathbf{b}}^T \tilde{\mathbf{A}}^{-1}\tilde{\mathbf{b}} + c$ yields

$$\begin{aligned}
-\tilde{\mathbf{b}}^T \tilde{\mathbf{A}}^{-1}\tilde{\mathbf{b}} + c &= -\left[\mathbf{F}_{k-1}^T \mathbf{Q}_{k-1}^{-1} \mathbf{x}_k + \mathbf{P}_{k-1|k-1}^{-1} \hat{\mathbf{x}}_{k-1|k-1}\right]^T \tilde{\mathbf{A}}^{-1}\left[\mathbf{F}_{k-1}^T \mathbf{Q}_{k-1}^{-1} \mathbf{x}_k + \mathbf{P}_{k-1|k-1}^{-1} \hat{\mathbf{x}}_{k-1|k-1}\right] \\
&\quad + \mathbf{x}_k^T \mathbf{Q}_{k-1}^{-1} \mathbf{x}_k + \hat{\mathbf{x}}_{k-1|k-1}^T \mathbf{P}_{k-1|k-1}^{-1} \hat{\mathbf{x}}_{k-1|k-1} \\
&= -\mathbf{x}_k^T \left[\mathbf{Q}_{k-1}^{-1} \mathbf{F}_{k-1} \tilde{\mathbf{A}}^{-1} \mathbf{F}_{k-1}^T \mathbf{Q}_{k-1}^{-1} - \mathbf{Q}_{k-1}^{-1}\right] \mathbf{x}_k \\
&\quad - \mathbf{x}_k^T \mathbf{Q}_{k-1}^{-1} \mathbf{F}_{k-1} \tilde{\mathbf{A}}^{-1} \mathbf{P}_{k-1|k-1}^{-1} \hat{\mathbf{x}}_{k-1|k-1} \\
&\quad - \hat{\mathbf{x}}_{k-1|k-1}^T \mathbf{P}_{k-1|k-1}^{-1} \tilde{\mathbf{A}}^{-1} \mathbf{F}_{k-1}^T \mathbf{Q}_{k-1}^{-1} \mathbf{x}_k \\
&\quad - \hat{\mathbf{x}}_{k-1|k-1}^T \mathbf{P}_{k-1|k-1}^{-1} \tilde{\mathbf{A}}^{-1} \mathbf{P}_{k-1|k-1}^{-1} \hat{\mathbf{x}}_{k-1|k-1} + \hat{\mathbf{x}}_{k-1|k-1}^T \mathbf{P}_{k-1|k-1}^{-1} \hat{\mathbf{x}}_{k-1|k-1}
\end{aligned} \tag{2.43}$$

Because $p\left(\mathbf{x}_k \mid \mathbf{x}_{k-1}\right)$ is also a Gaussian distribution, it can be defined as

$$\begin{aligned}
& p\left(\mathbf{x}_k \mid \mathbf{x}_{k-1}\right) \sim N\left(\mathbf{x}_k ; \hat{\mathbf{x}}_{k|k-1}, \mathbf{P}_{k|k-1}\right) \\
&= \frac{1}{(2\pi)^{\frac{n_x}{2}} \det\left(\mathbf{P}_{k|k-1}\right)^{\frac{1}{2}}} exp\left(-\frac{1}{2}\left(\left(\mathbf{x}_k - \hat{\mathbf{x}}_{k|k-1}\right)^T \mathbf{P}_{k|k-1}^{-1}\left(\mathbf{x}_k - \hat{\mathbf{x}}_{k|k-1}\right)\right)\right).
\end{aligned} \tag{2.44}$$

Comparing the first quadratic term in Eq. (2.43) with the exponential term in (2.44), we have,

$$-\mathbf{Q}_{k-1}^{-1}\mathbf{F}_{k-1}\tilde{\mathbf{A}}^{-1}\mathbf{F}_{k-1}^{T}\mathbf{Q}_{k-1}^{-1}+\mathbf{Q}_{k-1}^{-1}=\mathbf{P}_{k|k-1}^{-1} \tag{2.45}$$

Using the expression of $\tilde{\mathbf{A}}$ into the left-hand side of Eq. (2.45) gives

$$-\mathbf{Q}_{k-1}^{-1}\mathbf{F}_{k-1}\tilde{\mathbf{A}}^{-1}\mathbf{F}_{k-1}^{T}\mathbf{Q}_{k-1}^{-1}+\mathbf{Q}_{k-1}^{-1}=-\mathbf{Q}_{k-1}^{-1}\mathbf{F}_{k-1}\left[\mathbf{F}_{k-1}^{T}\mathbf{Q}_{k-1}^{-1}\mathbf{F}_{k-1}+\mathbf{P}_{k-1|k-1}^{-1}\right]^{-1}\mathbf{F}_{k-1}^{T}\mathbf{Q}_{k-1}^{-1}+\mathbf{Q}_{k-1}^{-1} \tag{2.46}$$

Using the matrix inversion lemma (Kailath et al. 2000),

$$\left(\mathbf{BCD}+\mathbf{A}\right)^{-1}=\mathbf{A}^{-1}-\mathbf{A}^{-1}\mathbf{B}\left(\mathbf{DA}^{-1}\mathbf{B}+\mathbf{C}^{-1}\right)^{-1}\mathbf{DA}^{-1} \tag{2.47}$$

Equation (2.46) becomes

$$-\mathbf{Q}_{k-1}^{-1}\mathbf{F}_{k-1}\tilde{\mathbf{A}}^{-1}\mathbf{F}_{k-1}^{T}\mathbf{Q}_{k-1}^{-1}+\mathbf{Q}_{k-1}^{-1}=\left(\mathbf{Q}_{k-1}+\mathbf{F}_{k-1}\mathbf{P}_{k-1|k-1}\mathbf{F}_{k-1}^{T}\right)^{-1}. \tag{2.48}$$

Hence, we have

$$\mathbf{P}_{k|k-1}=\mathbf{F}_{k-1}\mathbf{P}_{k-1|k-1}\mathbf{F}_{k-1}^{T}+\mathbf{Q}_{k-1} \tag{2.49}$$

which is the Eq. (2.9), the covariance prediction of the discrete-time Kalman filter.

Matching the second term in Eq. (2.43) with the corresponding term in (2.44) yields

$$-\mathbf{Q}_{k-1}^{-1}\mathbf{F}_{k-1}\tilde{\mathbf{A}}^{-1}\mathbf{P}_{k-1|k-1}^{-1}\hat{\mathbf{x}}_{k-1|k-1}=-\mathbf{P}_{k|k-1}^{-1}\hat{\mathbf{x}}_{k|k-1} \tag{2.50}$$

Substituting the expression of $\tilde{\mathbf{A}}$ into the left-hand side of Eq. (2.50) leads to

$$\begin{aligned}
&-\mathbf{Q}_{k-1}^{-1}\mathbf{F}_{k-1}\left[\mathbf{F}_{k-1}^{T}\mathbf{Q}_{k-1}^{-1}\mathbf{F}_{k-1}+\mathbf{P}_{k-1|k-1}^{-1}\right]^{-1}\mathbf{P}_{k-1|k-1}^{-1}\hat{\mathbf{x}}_{k-1|k-1}\\
&=-\mathbf{Q}_{k-1}^{-1}\mathbf{F}_{k-1}\left[\mathbf{P}_{k-1|k-1}-\mathbf{P}_{k-1|k-1}\mathbf{F}_{k-1}^{T}\left(\mathbf{F}_{k-1}\mathbf{P}_{k-1|k-1}\mathbf{F}_{k-1}^{T}+\mathbf{Q}_{k-1}\right)^{-1}\right.\\
&\quad\mathbf{P}_{k-1|k-1}^{-1}\hat{\mathbf{x}}_{k-1|k-1}\\
&=-\mathbf{Q}_{k-1}^{-1}\left[\mathbf{I}-\mathbf{F}_{k-1}\mathbf{P}_{k-1|k-1}\mathbf{F}_{k-1}^{T}\left(\mathbf{F}_{k-1}\mathbf{P}_{k-1|k-1}\mathbf{F}_{k-1}^{T}+\mathbf{Q}_{k-1}\right)^{-1}\right]\mathbf{F}_{k-1}\hat{\mathbf{x}}_{k-1|k-1}\\
&=-\mathbf{Q}_{k-1}^{-1}\left[\left(\mathbf{F}_{k-1}\mathbf{P}_{k-1|k-1}\mathbf{F}_{k-1}^{T}+\mathbf{Q}_{k-1}-\mathbf{F}_{k-1}\mathbf{P}_{k-1|k-1}\mathbf{F}_{k-1}^{T}\right)\left(\mathbf{F}_{k-1}\mathbf{P}_{k-1|k-1}\mathbf{F}_{k-1}^{T}+\mathbf{Q}_{k-1}\right)^{-1}\right]\\
&\quad\mathbf{F}_{k-1}\hat{\mathbf{x}}_{k-1|k-1}\\
&=-\left(\mathbf{F}_{k-1}\mathbf{P}_{k-1|k-1}\mathbf{F}_{k-1}^{T}+\mathbf{Q}_{k-1}\right)^{-1}\mathbf{F}_{k-1}\hat{\mathbf{x}}_{k-1|k-1}\\
&=-\mathbf{P}_{k|k-1}^{-1}\mathbf{F}_{k-1}\hat{\mathbf{x}}_{k-1|k-1}
\end{aligned} \tag{2.51}$$

By comparing Eq. (2.50) with Eq. (2.51), we have

$$\hat{\mathbf{x}}_{k|k-1} = \mathbf{F}_{k-1}\hat{\mathbf{x}}_{k-1|k-1} \tag{2.52}$$

which is the Eq. (2.8), the state prediction of the discrete-time Kalman filter. The summation of the last two terms in Eq. (2.43) is given by

$$\begin{aligned}
&-\hat{\mathbf{x}}_{k-1|k-1}^T \mathbf{P}_{k-1|k-1}^{-1} \tilde{\mathbf{A}}^{-1} \mathbf{P}_{k-1|k-1}^{-1} \hat{\mathbf{x}}_{k-1|k-1} + \hat{\mathbf{x}}_{k-1|k-1}^T \mathbf{P}_{k-1|k-1}^{-1} \hat{\mathbf{x}}_{k-1|k-1} \\
&= -\hat{\mathbf{x}}_{k-1|k-1}^T \mathbf{P}_{k-1|k-1}^{-1} \left[\mathbf{F}_{k-1}^T \mathbf{Q}_{k-1}^{-1} \mathbf{F}_{k-1} + \mathbf{P}_{k-1|k-1}^{-1}\right]^{-1} \mathbf{P}_{k-1|k-1}^{-1} \hat{\mathbf{x}}_{k-1|k-1} \\
&\quad + \hat{\mathbf{x}}_{k-1|k-1}^T \mathbf{P}_{k-1|k-1}^{-1} \hat{\mathbf{x}}_{k-1|k-1} \\
&= -\hat{\mathbf{x}}_{k-1|k-1}^T \mathbf{P}_{k-1|k-1}^{-1} \left[\mathbf{P}_{k-1|k-1} - \mathbf{P}_{k-1|k-1} \mathbf{F}_{k-1}^T \left(\mathbf{F}_{k-1} \mathbf{P}_{k-1|k-1} \mathbf{F}_{k-1}^T + \mathbf{Q}_{k-1}\right)^{-1} \mathbf{F}_{k-1} \mathbf{P}_{k-1|k-1}\right] \\
&\quad \mathbf{P}_{k-1|k-1}^{-1} \hat{\mathbf{x}}_{k-1|k-1} + \hat{\mathbf{x}}_{k-1|k-1}^T \mathbf{P}_{k-1|k-1}^{-1} \hat{\mathbf{x}}_{k-1|k-1} \\
&= -\hat{\mathbf{x}}_{k-1|k-1}^T \mathbf{P}_{k-1|k-1}^{-1} \hat{\mathbf{x}}_{k-1|k-1} + \hat{\mathbf{x}}_{k-1|k-1}^T \mathbf{F}_{k-1}^T \left(\mathbf{F}_{k-1} \mathbf{P}_{k-1|k-1} \mathbf{F}_{k-1}^T + \mathbf{Q}_{k-1}\right)^{-1} \mathbf{F}_{k-1} \hat{\mathbf{x}}_{k-1|k-1} \\
&\quad + \hat{\mathbf{x}}_{k-1|k-1}^T \mathbf{P}_{k-1|k-1}^{-1} \hat{\mathbf{x}}_{k-1|k-1} \\
&= \hat{\mathbf{x}}_{k-1|k-1}^T \mathbf{F}_{k-1}^T \left(\mathbf{F}_{k-1} \mathbf{P}_{k-1|k-1} \mathbf{F}_{k-1}^T + \mathbf{Q}_{k-1}\right)^{-1} \mathbf{F}_{k-1} \hat{\mathbf{x}}_{k-1|k-1} \\
&= \hat{\mathbf{x}}_{k|k-1}^T \mathbf{P}_{k|k-1}^{-1} \hat{\mathbf{x}}_{k|k-1}
\end{aligned} \tag{2.53}$$

Considering Eqs. (2.49)–(2.53), the exponential terms in Eq. (2.42) and Eq. (2.44) are the same. The remaining thing is to prove that the constants in Eqs. (2.42) and (2.44) are the same.

$$\begin{aligned}
&a(2\pi)^{\frac{n_x}{2}} \det\left(\tilde{\mathbf{A}}^{-1}\right)^{\frac{1}{2}} \\
&= \frac{(2\pi)^{n_x/2} \det\left(\left[\mathbf{F}_{k-1}^T \mathbf{Q}_{k-1}^{-1} \mathbf{F}_{k-1} + \mathbf{P}_{k-1|k-1}^{-1}\right]^{-1}\right)^{1/2}}{(2\pi)^{n_x/2} \det\left(\mathbf{Q}_{k-1}\right)^{1/2} (2\pi)^{n_x/2} \det\left(\mathbf{P}_{k-1|k-1}\right)^{1/2}} \\
&= \frac{1}{(2\pi)^{n_x/2} \det\left(\mathbf{Q}_{k-1}\right)^{1/2} \det\left(\mathbf{P}_{k-1|k-1}\right)^{1/2} \det\left(\left[\mathbf{F}_{k-1}^T \mathbf{Q}_{k-1}^{-1} \mathbf{F}_{k-1} + \mathbf{P}_{k-1|k-1}^{-1}\right]\right)^{1/2}} \\
&= \frac{1}{(2\pi)^{n_x/2} \det\left(\mathbf{Q}_{k-1}\right)^{1/2} \det\left(\mathbf{P}_{k-1|k-1} \mathbf{F}_{k-1}^T \mathbf{Q}_{k-1}^{-1} \mathbf{F}_{k-1} + \mathbf{I}\right)^{1/2}}
\end{aligned} \tag{2.54}$$

By using the identity,

$$\det(\mathbf{I} + \mathbf{AB}) = \det(\mathbf{I} + \mathbf{BA}) \tag{2.55}$$

Equation (2.54) becomes

$$
\begin{aligned}
& a(2\pi)^{\frac{n_x}{2}} \det\left(\tilde{\mathbf{A}}^{-1}\right)^{\frac{1}{2}} \\
&= \frac{1}{(2\pi)^{n_x/2} \det\left(\mathbf{F}_{k-1}\mathbf{P}_{k-1|k-1}\mathbf{F}_{k-1}^T\mathbf{Q}_{k-1}^{-1} + \mathbf{I}\right)^{1/2} \det\left(\mathbf{Q}_{k-1}\right)^{1/2}} \\
&= \frac{1}{(2\pi)^{n_x/2} \det\left(\mathbf{F}_{k-1}\mathbf{P}_{k-1|k-1}\mathbf{F}_{k-1}^T + \mathbf{Q}_{k-1}\right)^{1/2}} \\
&= \frac{1}{(2\pi)^{n_x/2} \det\left(\mathbf{P}_{k|k-1}\right)^{1/2}}
\end{aligned}
\tag{2.56}
$$

Hence, the constant parts in Eqs. (2.42) and (2.44) are the same. Since the exponential term and the constants in Eqs. (2.42) and (2.44) are both the same, Eqs. (2.42) and (2.44) are equivalent.

Update

Using the Bayesian equation, we have

$$
\begin{aligned}
p(\mathbf{x}_k \mid \mathbf{y}_k) &= \frac{p(\mathbf{y}_k \mid \mathbf{x}_k) p(\mathbf{x}_k \mid \mathbf{x}_{k-1})}{p(\mathbf{y}_k)} \\
&= \frac{N(\mathbf{y}_k; \mathbf{H}_k\mathbf{x}_k, \mathbf{R}_k) N(\mathbf{x}_k; \hat{\mathbf{x}}_{k|k-1}, \mathbf{P}_{k|k-1})}{N(\mathbf{y}_k; \mathbf{H}_k\hat{\mathbf{x}}_{k|k-1}, \mathbf{H}_k\mathbf{P}_{k|k-1}\mathbf{H}_k^T + \mathbf{R}_k)} \\
&= \frac{(2\pi)^{n_z/2} \det\left(\mathbf{H}_k\mathbf{P}_{k|k-1}\mathbf{H}_k^T + \mathbf{R}_k\right)^{1/2}}{(2\pi)^{n_z/2} \det(\mathbf{R}_k)^{1/2} (2\pi)^{n_x/2} \det\left(\mathbf{P}_{k|k-1}\right)^{1/2}} \\
&\quad \times exp\left(-\frac{1}{2}(\mathbf{y}_k - \mathbf{H}_k\mathbf{x}_k)^T \mathbf{R}_k^{-1} (\mathbf{y}_k - \mathbf{H}_k\mathbf{x}_k)\right) \\
&\quad \times exp\left(-\frac{1}{2}\left(\mathbf{x}_k - \hat{\mathbf{x}}_{k|k-1}\right)^T \mathbf{P}_{k|k-1}^{-1} \left(\mathbf{x}_k - \hat{\mathbf{x}}_{k|k-1}\right)\right) \\
&\quad \times exp\left(\frac{1}{2}\left(\mathbf{y}_k - \mathbf{H}_k\hat{\mathbf{x}}_{k|k-1}\right)^T \left(\mathbf{H}_k\mathbf{P}_{k|k-1}\mathbf{H}_k^T + \mathbf{R}_k\right)^{-1} \left(\mathbf{y}_k - \mathbf{H}_k\hat{\mathbf{x}}_{k|k-1}\right)\right)
\end{aligned}
\tag{2.57}
$$

It can be seen that the $\mathbf{p}(\mathbf{x}_k \mid \mathbf{y}_k)$ also satisfies the normal distribution. Hence, we assume,

$$
p(\mathbf{x}_k \mid \mathbf{y}_k) \triangleq N\left(\mathbf{x}_k; \hat{\mathbf{x}}_{k|k}, \mathbf{P}_{k|k}\right) = \frac{exp\left(-\frac{1}{2}(\mathbf{x}_k - \hat{\mathbf{x}}_k)^T \mathbf{P}_{k|k}^{-1} (\mathbf{x}_k - \hat{\mathbf{x}}_k)\right)}{(2\pi)^{n_x/2} \det\left(\mathbf{P}_{k|k}\right)^{1/2}}
\tag{2.58}
$$

Equation (2.57) can be rewritten as

$$p(\mathbf{x}_k \mid \mathbf{y}_k) = C \cdot e^{-\frac{1}{2}D} \tag{2.59}$$

where

$$C = \frac{(2\pi)^{n_z/2} \det\left(\mathbf{H}_k \mathbf{P}_{k|k-1} \mathbf{H}_k^T + \mathbf{R}_k\right)^{1/2}}{(2\pi)^{n_z/2} \det(\mathbf{R}_k)^{1/2} (2\pi)^{n_x/2} \det\left(\mathbf{P}_{k|k-1}\right)^{1/2}} \tag{2.60}$$

$$\begin{aligned}
D &= (\mathbf{y}_k - \mathbf{H}_k\mathbf{x}_k)^T \mathbf{R}_k^{-1} (\mathbf{y}_k - \mathbf{H}_k\mathbf{x}_k) + \left(\mathbf{x}_k - \hat{\mathbf{x}}_{k|k-1}\right)^T \mathbf{P}_{k|k-1}^{-1} \left(\mathbf{x}_k - \hat{\mathbf{x}}_{k|k-1}\right) \\
&\quad - \left(\mathbf{y}_k - \mathbf{H}_k \hat{\mathbf{x}}_{k|k-1}\right)^T \left(\mathbf{H}_k \mathbf{P}_{k|k-1} \mathbf{H}_k^T + \mathbf{R}_k\right)^{-1} \left(\mathbf{y}_k - \mathbf{H}_k \hat{\mathbf{x}}_{k|k-1}\right) \\
&= \mathbf{x}_k^T \left[\mathbf{H}_k^T \mathbf{R}_k^{-1} \mathbf{H}_k + \mathbf{P}_{k|k-1}^{-1}\right] \mathbf{x}_k - \mathbf{x}_k^T \left[\mathbf{H}_k^T \mathbf{R}_k^{-1} \mathbf{y}_k + \mathbf{P}_{k|k-1}^{-1} \hat{\mathbf{x}}_{k|k-1}\right] \\
&\quad - \left[\mathbf{y}_k^T \mathbf{R}_k^{-1} \mathbf{H}_k + \hat{\mathbf{x}}_{k|k-1}^T \mathbf{P}_{k|k-1}^{-1}\right] \mathbf{x}_k \\
&\quad + \mathbf{y}_k^T \left[\mathbf{R}_k^{-1} - \left(\mathbf{H}_k \mathbf{P}_{k|k-1} \mathbf{H}_k^T + \mathbf{R}_k\right)^{-1}\right] \mathbf{y}_k + \mathbf{y}_k^T \left(\mathbf{H}_k \mathbf{P}_{k|k-1} \mathbf{H}_k^T + \mathbf{R}_k\right)^{-1} \mathbf{H}_k \hat{\mathbf{x}}_{k|k-1} \\
&\quad + \hat{\mathbf{x}}_{k|k-1}^T \mathbf{H}_k^T \left(\mathbf{H}_k \mathbf{P}_{k|k-1} \mathbf{H}_k^T + \mathbf{R}_k\right)^{-1} \mathbf{y}_k \\
&\quad - \hat{\mathbf{x}}_{k|k-1}^T \left[\mathbf{H}_k^T \left(\mathbf{H}_k \mathbf{P}_{k|k-1} \mathbf{H}_k^T + \mathbf{R}_k\right)^{-1} \mathbf{H}_k - \mathbf{P}_{k|k-1}^{-1}\right] \hat{\mathbf{x}}_{k|k-1}
\end{aligned} \tag{2.61}$$

Similarly, Eq. (2.58) can be rewritten as,

$$p(\mathbf{x}_k \mid \mathbf{y}_k) = \tilde{C} \cdot e^{-\frac{1}{2}\tilde{D}} \tag{2.62}$$

where

$$\tilde{C} = \frac{1}{(2\pi)^{n_x/2} \det\left(\mathbf{P}_{k|k}\right)^{1/2}} \tag{2.63}$$

$$\begin{aligned}
\tilde{D} &= (\mathbf{x}_k - \hat{\mathbf{x}}_k)^T \mathbf{P}_{k|k}^{-1} (\mathbf{x}_k - \hat{\mathbf{x}}_k) \\
&= \mathbf{x}_k^T \mathbf{P}_{k|k}^{-1} \mathbf{x}_k - \hat{\mathbf{x}}_k^T \mathbf{P}_{k|k}^{-1} \mathbf{x}_k - \mathbf{x}_k^T \mathbf{P}_{k|k}^{-1} \hat{\mathbf{x}}_k + \hat{\mathbf{x}}_k^T \mathbf{P}_{k|k}^{-1} \hat{\mathbf{x}}_k
\end{aligned} \tag{2.64}$$

To show the equivalence of Eq. (2.57) and Eq. (2.58), we first show that $D = \tilde{D}$. By comparing the first term of Eq. (2.61) and Eq. (2.64), we have,

$$\mathbf{P}_{k|k}^{-1} = \mathbf{H}_k^T \mathbf{R}_k^{-1} \mathbf{H}_k + \mathbf{P}_{k|k-1}^{-1} \tag{2.65}$$

which is true because it is the information update Eq. (2.24).

By comparing the second term of Eq. (2.61) with the third term of Eq. (2.64), we have,

$$\mathbf{P}_{k|k}^{-1} \hat{\mathbf{x}}_k = \mathbf{H}_k^T \mathbf{R}_k^{-1} \mathbf{y}_k + \mathbf{P}_{k|k-1}^{-1} \hat{\mathbf{x}}_{k|k-1} \tag{2.66}$$

which is true as well because it is the information state update Eq. (2.31).

Note that in both Eq. (2.61) and Eq. (2.64), the third term and the second term are the same. Thus, the next thing is to prove that the summation of the last four terms in Eq. (2.61) is equivalent to the last term of Eq. (2.64), i.e.,

$$\begin{aligned}
&\mathbf{y}_k^T\left[\mathbf{R}_k^{-1}-\left(\mathbf{H}_k\mathbf{P}_{k|k-1}\mathbf{H}_k^T+\mathbf{R}_k\right)^{-1}\right]\mathbf{y}_k+\mathbf{y}_k^T\left(\mathbf{H}_k\mathbf{P}_{k|k-1}\mathbf{H}_k^T+\mathbf{R}_k\right)^{-1}\mathbf{H}_k\hat{\mathbf{x}}_{k|k-1}\\
&+\hat{\mathbf{x}}_{k|k-1}^T\mathbf{H}_k^T\left(\mathbf{H}_k\mathbf{P}_{k|k-1}\mathbf{H}_k^T+\mathbf{R}_k\right)^{-1}\mathbf{y}_k\\
&-\hat{\mathbf{x}}_{k|k-1}^T\left[\mathbf{H}_k^T\left(\mathbf{H}_k\mathbf{P}_{k|k-1}\mathbf{H}_k^T+\mathbf{R}_k\right)^{-1}\mathbf{H}_k-\mathbf{P}_{k|k-1}^{-1}\right]\hat{\mathbf{x}}_{k|k-1}=\hat{\mathbf{x}}_k^T\mathbf{P}_{k|k}^{-1}\hat{\mathbf{x}}_k
\end{aligned} \tag{2.67}$$

By using Eq. (2.66), we have

$$\begin{aligned}
\hat{\mathbf{x}}_k&=\left(\mathbf{P}_{k|k}\right)\left(\mathbf{H}_k^T\mathbf{R}_k^{-1}\mathbf{y}_k+\mathbf{P}_{k|k-1}^{-1}\hat{\mathbf{x}}_{k|k-1}\right)\\
&=\left(\mathbf{P}_{k|k}\right)\left(\mathbf{H}_k^T\mathbf{R}_k^{-1}\mathbf{y}_k+\left(\mathbf{P}_{k|k}^{-1}-\mathbf{H}_k^T\mathbf{R}_k^{-1}\mathbf{H}_k\right)\hat{\mathbf{x}}_{k|k-1}\right)\\
&=\mathbf{P}_{k|k}\mathbf{H}_k^T\mathbf{R}_k^{-1}\mathbf{y}_k+\left(\hat{\mathbf{x}}_{k|k-1}-\mathbf{P}_{k|k}\mathbf{H}_k^T\mathbf{R}_k^{-1}\mathbf{H}_k\hat{\mathbf{x}}_{k|k-1}\right)\\
&=\hat{\mathbf{x}}_{k|k-1}+\mathbf{P}_{k|k}\mathbf{H}_k^T\mathbf{R}_k^{-1}\left(\mathbf{y}_k-\mathbf{H}_k\hat{\mathbf{x}}_{k|k-1}\right)
\end{aligned} \tag{2.68}$$

Using Eq. (2.65) and the matrix inversion lemma, $\mathbf{P}_{k|k}$ can be rewritten as

$$\begin{aligned}
\mathbf{P}_{k|k}&=\left(\mathbf{H}_k^T\mathbf{R}_k^{-1}\mathbf{H}_k+\mathbf{P}_{k|k-1}^{-1}\right)^{-1}\\
&=\mathbf{P}_{k|k-1}-\mathbf{P}_{k|k-1}\mathbf{H}_k^T\left(\mathbf{H}_k\mathbf{P}_{k|k-1}\mathbf{H}_k^T+\mathbf{R}_k\right)^{-1}\mathbf{H}_k\mathbf{P}_{k|k-1}
\end{aligned} \tag{2.69}$$

Using Eq. (2.69), Eq. (2.68) can be rewritten as

$$\begin{aligned}
&\hat{\mathbf{x}}_k=\hat{\mathbf{x}}_{k|k-1}+\mathbf{P}_{k|k}\mathbf{H}_k^T\mathbf{R}_k^{-1}\left(\mathbf{y}_k-\mathbf{H}_k\hat{\mathbf{x}}_{k|k-1}\right)\\
&=\hat{\mathbf{x}}_{k|k-1}+\left(\mathbf{P}_{k|k-1}-\mathbf{P}_{k|k-1}\mathbf{H}_k^T\left(\mathbf{H}_k\mathbf{P}_{k|k-1}\mathbf{H}_k^T+\mathbf{R}_k\right)^{-1}\mathbf{H}_k\mathbf{P}_{k|k-1}\right)\mathbf{H}_k^T\mathbf{R}_k^{-1}\left(\mathbf{y}_k-\mathbf{H}_k\hat{\mathbf{x}}_{k|k-1}\right)\\
&=\hat{\mathbf{x}}_{k|k-1}+\left(\mathbf{P}_{k|k-1}\mathbf{H}_k^T\mathbf{R}_k^{-1}-\mathbf{P}_{k|k-1}\mathbf{H}_k^T\left(\mathbf{H}_k\mathbf{P}_{k|k-1}\mathbf{H}_k^T+\mathbf{R}_k\right)^{-1}\mathbf{H}_k\mathbf{P}_{k|k-1}\mathbf{H}_k^T\mathbf{R}_k^{-1}\right)\\
&\quad\left(\mathbf{y}_k-\mathbf{H}_k\hat{\mathbf{x}}_{k|k-1}\right)\\
&=\hat{\mathbf{x}}_{k|k-1}+\left(\mathbf{P}_{k|k-1}\mathbf{H}_k^T\left(\mathbf{I}-\left(\mathbf{H}_k\mathbf{P}_{k|k-1}\mathbf{H}_k^T+\mathbf{R}_k\right)^{-1}\mathbf{H}_k\mathbf{P}_{k|k-1}\mathbf{H}_k^T\right)\mathbf{R}_k^{-1}\right)\left(\mathbf{y}_k-\mathbf{H}_k\hat{\mathbf{x}}_{k|k-1}\right)\\
&=\hat{\mathbf{x}}_{k|k-1}+\left(\mathbf{P}_{k|k-1}\mathbf{H}_k^T\left(\left(\mathbf{H}_k\mathbf{P}_{k|k-1}\mathbf{H}_k^T+\mathbf{R}_k\right)^{-1}\left(\mathbf{H}_k\mathbf{P}_{k|k-1}\mathbf{H}_k^T+\mathbf{R}_k-\mathbf{H}_k\mathbf{P}_{k|k-1}\mathbf{H}_k^T\right)\right)\mathbf{R}_k^{-1}\right)\\
&\quad\left(\mathbf{y}_k-\mathbf{H}_k\hat{\mathbf{x}}_{k|k-1}\right)\\
&=\hat{\mathbf{x}}_{k|k-1}+\left(\mathbf{P}_{k|k-1}\mathbf{H}_k^T\left(\left(\mathbf{H}_k\mathbf{P}_{k|k-1}\mathbf{H}_k^T+\mathbf{R}_k\right)^{-1}\mathbf{R}_k\right)\mathbf{R}_k^{-1}\right)\left(\mathbf{y}_k-\mathbf{H}_k\hat{\mathbf{x}}_{k|k-1}\right)\\
&=\hat{\mathbf{x}}_{k|k-1}+\mathbf{P}_{k|k-1}\mathbf{H}_k^T\left(\mathbf{H}_k\mathbf{P}_{k|k-1}\mathbf{H}_k^T+\mathbf{R}_k\right)^{-1}\left(\mathbf{y}_k-\mathbf{H}_k\hat{\mathbf{x}}_{k|k-1}\right)
\end{aligned} \tag{2.70}$$

Let's define $\mathbf{K}_k = \mathbf{P}_{k|k-1}\mathbf{H}_k^T\left(\mathbf{H}_k\mathbf{P}_{k|k-1}\mathbf{H}_k^T + \mathbf{R}_k\right)^{-1}$ and $\breve{\mathbf{Y}}_k = \mathbf{H}_k^T\mathbf{R}_k^{-1}\mathbf{H}_k + \mathbf{P}_{k|k-1}^{-1}$, then the last term of Eq. (2.64) becomes

$$\begin{aligned}
&\hat{\mathbf{x}}_k^T \mathbf{P}_{k|k}^{-1}\hat{\mathbf{x}}_k \\
&= \left(\hat{\mathbf{x}}_{k|k-1} + \mathbf{K}_k\left(\mathbf{y}_k - \mathbf{H}_k\hat{\mathbf{x}}_{k|k-1}\right)\right)^T \breve{\mathbf{Y}}_k\left(\hat{\mathbf{x}}_{k|k-1} + \mathbf{K}_k\left(\mathbf{y}_k - \mathbf{H}_k\hat{\mathbf{x}}_{k|k-1}\right)\right) \\
&= \mathbf{y}_k^T\mathbf{K}_k^T\breve{\mathbf{Y}}_k\mathbf{K}_k\mathbf{y}_k + \mathbf{y}_k^T\mathbf{K}_k^T\breve{\mathbf{Y}}_k\left(\mathbf{I} - \mathbf{K}_k\mathbf{H}_k\right)\hat{\mathbf{x}}_{k|k-1} + \hat{\mathbf{x}}_{k|k-1}^T\left(\mathbf{I} - \mathbf{K}_k\mathbf{H}_k\right)^T\breve{\mathbf{Y}}_k\mathbf{K}_k\mathbf{y}_k \\
&\quad + \hat{\mathbf{x}}_{k|k-1}^T\left(\mathbf{I} - \mathbf{K}_k\mathbf{H}_k\right)^T\breve{\mathbf{Y}}_k\left(\mathbf{I} - \mathbf{K}_k\mathbf{H}_k\right)\hat{\mathbf{x}}_{k|k-1}
\end{aligned} \tag{2.71}$$

Substitute Eq. (2.71) into Eq. (2.67) and the resultant equation is satisfied if the following three equations are fulfilled,

$$\mathbf{R}_k^{-1} - \left(\mathbf{H}_k\mathbf{P}_{k|k-1}\mathbf{H}_k^T + \mathbf{R}_k\right)^{-1} = \mathbf{K}_k^T\breve{\mathbf{Y}}_k\mathbf{K}_k \tag{2.72a}$$

$$\left(\mathbf{H}_k\mathbf{P}_{k|k-1}\mathbf{H}_k^T + \mathbf{R}_k\right)^{-1}\mathbf{H}_k = \mathbf{K}_k^T\breve{\mathbf{Y}}_k\left(\mathbf{I} - \mathbf{K}_k\mathbf{H}_k\right) \tag{2.72b}$$

$$-\mathbf{H}_k^T\left(\mathbf{H}_k\mathbf{P}_{k|k-1}\mathbf{H}_k^T + \mathbf{R}_k\right)^{-1}\mathbf{H}_k + \mathbf{P}_{k|k-1}^{-1} = \left(\mathbf{I} - \mathbf{K}_k\mathbf{H}_k\right)^T\breve{\mathbf{Y}}_k\left(\mathbf{I} - \mathbf{K}_k\mathbf{H}_k\right) \tag{2.72c}$$

To prove (2.72a), we have,

$$\begin{aligned}
&\mathbf{R}_k^{-1} - \left(\mathbf{H}_k\mathbf{P}_{k|k-1}\mathbf{H}_k^T + \mathbf{R}_k\right)^{-1} - \mathbf{K}_k^T\breve{\mathbf{Y}}_k\mathbf{K}_k \\
&= \mathbf{R}_k^{-1} - \left(\mathbf{H}_k\mathbf{P}_{k|k-1}\mathbf{H}_k^T + \mathbf{R}_k\right)^{-1} - \left(\mathbf{P}_{k|k-1}\mathbf{H}_k^T\left(\mathbf{H}_k\mathbf{P}_{k|k-1}\mathbf{H}_k^T + \mathbf{R}_k\right)^{-1}\right)^T \\
&\quad \left(\mathbf{H}_k^T\mathbf{R}_k^{-1}\mathbf{H}_k + \mathbf{P}_{k|k-1}^{-1}\right)\mathbf{P}_{k|k-1}\mathbf{H}_k^T\left(\mathbf{H}_k\mathbf{P}_{k|k-1}\mathbf{H}_k^T + \mathbf{R}_k\right)^{-1} \\
&= \left(\mathbf{H}_k\mathbf{P}_{k|k-1}\mathbf{H}_k^T + \mathbf{R}_k\right)^{-1}\left(\begin{matrix}\left(\mathbf{H}_k\mathbf{P}_{k|k-1}\mathbf{H}_k^T + \mathbf{R}_k\right)\mathbf{R}_k^{-1} - \mathbf{I} - \mathbf{H}_k\mathbf{P}_{k|k-1} \\ \times\left(\mathbf{H}_k^T\mathbf{R}_k^{-1}\mathbf{H}_k + \mathbf{P}_{k|k-1}^{-1}\right)\mathbf{P}_{k|k-1}\mathbf{H}_k^T\left(\mathbf{H}_k\mathbf{P}_{k|k-1}\mathbf{H}_k^T + \mathbf{R}_k\right)^{-1}\end{matrix}\right) \\
&= \left(\mathbf{H}_k\mathbf{P}_{k|k-1}\mathbf{H}_k^T + \mathbf{R}_k\right)^{-1} \\
&\quad \left(\begin{matrix}\left(\mathbf{H}_k\mathbf{P}_{k|k-1}\mathbf{H}_k^T + \mathbf{R}_k\right)\mathbf{R}_k^{-1}\left(\mathbf{H}_k\mathbf{P}_{k|k-1}\mathbf{H}_k^T + \mathbf{R}_k\right) \\ -\left(\mathbf{H}_k\mathbf{P}_{k|k-1}\mathbf{H}_k^T + \mathbf{R}_k\right) - \mathbf{H}_k\mathbf{P}_{k|k-1}\left(\mathbf{H}_k^T\mathbf{R}_k^{-1}\mathbf{H}_k + \mathbf{P}_{k|k-1}^{-1}\right)\mathbf{P}_{k|k-1}\mathbf{H}_k^T\end{matrix}\right) \\
&\quad \left(\mathbf{H}_k\mathbf{P}_{k|k-1}\mathbf{H}_k^T + \mathbf{R}_k\right)^{-1}
\end{aligned} \tag{2.73}$$

In Eq. (2.73), the term

$$
\begin{aligned}
&\left(\mathbf{H}_k\mathbf{P}_{k|k-1}\mathbf{H}_k^T+\mathbf{R}_k\right)\mathbf{R}_k^{-1}\left(\mathbf{H}_k\mathbf{P}_{k|k-1}\mathbf{H}_k^T+\mathbf{R}_k\right)-\left(\mathbf{H}_k\mathbf{P}_{k|k-1}\mathbf{H}_k^T+\mathbf{R}_k\right)\\
&\quad-\mathbf{H}_k\mathbf{P}_{k|k-1}\left(\mathbf{H}_k^T\mathbf{R}_k^{-1}\mathbf{H}_k+\mathbf{P}_{k|k-1}^{-1}\right)\mathbf{P}_{k|k-1}\mathbf{H}_k^T\\
&=\left(\mathbf{H}_k\mathbf{P}_{k|k-1}\mathbf{H}_k^T+\mathbf{R}_k\right)\mathbf{R}_k^{-1}\left(\mathbf{H}_k\mathbf{P}_{k|k-1}\mathbf{H}_k^T\right)+\left(\mathbf{H}_k\mathbf{P}_{k|k-1}\mathbf{H}_k^T+\mathbf{R}_k\right)\\
&\quad-\left(\mathbf{H}_k\mathbf{P}_{k|k-1}\mathbf{H}_k^T+\mathbf{R}_k\right)\\
&\quad-\mathbf{H}_k\mathbf{P}_{k|k-1}\mathbf{H}_k^T\mathbf{R}_k^{-1}\mathbf{H}_k\mathbf{P}_{k|k-1}\mathbf{H}_k^T-\mathbf{H}_k\mathbf{P}_{k|k-1}\mathbf{P}_{k|k-1}^{-1}\mathbf{P}_{k|k-1}\mathbf{H}_k^T\\
&=\mathbf{H}_k\mathbf{P}_{k|k-1}\mathbf{H}_k^T\mathbf{R}_k^{-1}\mathbf{H}_k\mathbf{P}_{k|k-1}\mathbf{H}_k^T+\mathbf{H}_k\mathbf{P}_{k|k-1}\mathbf{H}_k^T+\left(\mathbf{H}_k\mathbf{P}_{k|k-1}\mathbf{H}_k^T+\mathbf{R}_k\right)\\
&\quad-\left(\mathbf{H}_k\mathbf{P}_{k|k-1}\mathbf{H}_k^T+\mathbf{R}_k\right)\\
&\quad-\mathbf{H}_k\mathbf{P}_{k|k-1}\mathbf{H}_k^T\mathbf{R}_k^{-1}\mathbf{H}_k\mathbf{P}_{k|k-1}\mathbf{H}_k^T-\mathbf{H}_k\mathbf{P}_{k|k-1}\mathbf{H}_k^T\\
&=\mathbf{0}
\end{aligned}
\tag{2.74}
$$

Hence, Eq. (2.72a) is verified. Next, we will verify (2.72b)

$$
\begin{aligned}
&\mathbf{K}_k^T\breve{\mathbf{Y}}_k\left(\mathbf{I}-\mathbf{K}_k\mathbf{H}_k\right)\\
&=\left(\mathbf{P}_{k|k-1}\mathbf{H}_k^T\left(\mathbf{H}_k\mathbf{P}_{k|k-1}\mathbf{H}_k^T+\mathbf{R}_k\right)^{-1}\right)^T\left(\mathbf{H}_k^T\,\mathbf{R}_k^{-1}\,\mathbf{H}_k+\mathbf{P}_{k|k-1}^{-1}\right)\\
&\quad\left(\mathbf{I}-\mathbf{P}_{k|k-1}\mathbf{H}_k^T\left(\mathbf{H}_k\mathbf{P}_{k|k-1}\mathbf{H}_k^T+\mathbf{R}_k\right)^{-1}\mathbf{H}_k\right)\\
&=\left(\mathbf{H}_k\mathbf{P}_{k|k-1}\mathbf{H}_k^T+\mathbf{R}_k\right)^{-1}\mathbf{H}_k\mathbf{P}_{k|k-1}^T\left(\mathbf{H}_k^T\,\mathbf{R}_k^{-1}\,\mathbf{H}_k+\mathbf{P}_{k|k-1}^{-1}\right)\\
&\quad\left(\mathbf{I}-\mathbf{P}_{k|k-1}\mathbf{H}_k^T\left(\mathbf{H}_k\mathbf{P}_{k|k-1}\mathbf{H}_k^T+\mathbf{R}_k\right)^{-1}\mathbf{H}_k\right)\\
&=\left(\mathbf{H}_k\mathbf{P}_{k|k-1}\mathbf{H}_k^T+\mathbf{R}_k\right)^{-1}\left(\mathbf{H}_k\mathbf{P}_{k|k-1}^T\mathbf{H}_k^T\mathbf{R}_k^{-1}\mathbf{H}_k+\mathbf{H}_k\right)\\
&\quad\left(\mathbf{I}-\mathbf{P}_{k|k-1}\mathbf{H}_k^T\left(\mathbf{H}_k\mathbf{P}_{k|k-1}\mathbf{H}_k^T+\mathbf{R}_k\right)^{-1}\mathbf{H}_k\right)\\
&=\left(\mathbf{H}_k\mathbf{P}_{k|k-1}\mathbf{H}_k^T+\mathbf{R}_k\right)^{-1}\begin{pmatrix}\mathbf{H}_k\mathbf{P}_{k|k-1}^T\mathbf{H}_k^T\mathbf{R}_k^{-1}\mathbf{H}_k+\mathbf{H}_k\\ -\mathbf{H}_k\mathbf{P}_{k|k-1}^T\mathbf{H}_k^T\mathbf{R}_k^{-1}\mathbf{H}_k\mathbf{P}_{k|k-1}\mathbf{H}_k^T\left(\mathbf{H}_k\mathbf{P}_{k|k-1}\mathbf{H}_k^T+\mathbf{R}_k\right)^{-1}\mathbf{H}_k\\ -\mathbf{H}_k\mathbf{P}_{k|k-1}\mathbf{H}_k^T\left(\mathbf{H}_k\mathbf{P}_{k|k-1}\mathbf{H}_k^T+\mathbf{R}_k\right)^{-1}\mathbf{H}_k\end{pmatrix}
\end{aligned}
\tag{2.75}
$$

To prove Eq. (2.72b), we have,

$$
\begin{aligned}
&\left(\mathbf{H}_k\mathbf{P}_{k|k-1}\mathbf{H}_k^T+\mathbf{R}_k\right)^{-1}\mathbf{H}_k-\mathbf{K}_k^T\breve{\mathbf{Y}}_k\left(\mathbf{I}-\mathbf{K}_k\mathbf{H}_k\right)\\
&=\left(\mathbf{H}_k\mathbf{P}_{k|k-1}\mathbf{H}_k^T+\mathbf{R}_k\right)^{-1}\mathbf{H}_k\\
&-\left(\mathbf{H}_k\mathbf{P}_{k|k-1}\mathbf{H}_k^T+\mathbf{R}_k\right)^{-1}\begin{pmatrix}\mathbf{H}_k\mathbf{P}_{k|k-1}^T\mathbf{H}_k^T\mathbf{R}_k^{-1}\mathbf{H}_k+\mathbf{H}_k\\
-\mathbf{H}_k\mathbf{P}_{k|k-1}^T\mathbf{H}_k^T\mathbf{R}_k^{-1}\mathbf{H}_k\mathbf{P}_{k|k-1}\mathbf{H}_k^T\left(\mathbf{H}_k\mathbf{P}_{k|k-1}\mathbf{H}_k^T+\mathbf{R}_k\right)^{-1}\mathbf{H}_k\\
-\mathbf{H}_k\mathbf{P}_{k|k-1}\mathbf{H}_k^T\left(\mathbf{H}_k\mathbf{P}_{k|k-1}\mathbf{H}_k^T+\mathbf{R}_k\right)^{-1}\mathbf{H}_k\end{pmatrix}\\
&=\left(\mathbf{H}_k\mathbf{P}_{k|k-1}\mathbf{H}_k^T+\mathbf{R}_k\right)^{-1}\begin{pmatrix}-\mathbf{H}_k\mathbf{P}_{k|k-1}^T\mathbf{H}_k^T\mathbf{R}_k^{-1}\\
+\mathbf{H}_k\mathbf{P}_{k|k-1}^T\mathbf{H}_k^T\mathbf{R}_k^{-1}\mathbf{H}_k\mathbf{P}_{k|k-1}\mathbf{H}_k^T\left(\mathbf{H}_k\mathbf{P}_{k|k-1}\mathbf{H}_k^T+\mathbf{R}_k\right)^{-1}\\
+\mathbf{H}_k\mathbf{P}_{k|k-1}\mathbf{H}_k^T\left(\mathbf{H}_k\mathbf{P}_{k|k-1}\mathbf{H}_k^T+\mathbf{R}_k\right)^{-1}\end{pmatrix}\mathbf{H}_k
\end{aligned}
\tag{2.76}
$$

The term

$$
\begin{aligned}
&-\mathbf{H}_k\mathbf{P}_{k|k-1}^T\mathbf{H}_k^T\mathbf{R}_k^{-1}+\mathbf{H}_k\mathbf{P}_{k|k-1}^T\mathbf{H}_k^T\mathbf{R}_k^{-1}\mathbf{H}_k\mathbf{P}_{k|k-1}\mathbf{H}_k^T\left(\mathbf{H}_k\mathbf{P}_{k|k-1}\mathbf{H}_k^T+\mathbf{R}_k\right)^{-1}\\
&+\mathbf{H}_k\mathbf{P}_{k|k-1}\mathbf{H}_k^T\left(\mathbf{H}_k\mathbf{P}_{k|k-1}\mathbf{H}_k^T+\mathbf{R}_k\right)^{-1}\\
&=\mathbf{H}_k\mathbf{P}_{k|k-1}^T\mathbf{H}_k^T\Big(-\mathbf{R}_k^{-1}+\mathbf{R}_k^{-1}\mathbf{H}_k\mathbf{P}_{k|k-1}\mathbf{H}_k^T\left(\mathbf{H}_k\mathbf{P}_{k|k-1}\mathbf{H}_k^T+\mathbf{R}_k\right)^{-1}\\
&+\left(\mathbf{H}_k\mathbf{P}_{k|k-1}\mathbf{H}_k^T+\mathbf{R}_k\right)^{-1}\Big)\\
&=\mathbf{H}_k\mathbf{P}_{k|k-1}^T\mathbf{H}_k^T\Big(-\mathbf{R}_k^{-1}+\mathbf{R}_k^{-1}\left(\mathbf{H}_k\mathbf{P}_{k|k-1}\mathbf{H}_k^T+\mathbf{R}_k-\mathbf{R}_k\right)\left(\mathbf{H}_k\mathbf{P}_{k|k-1}\mathbf{H}_k^T+\mathbf{R}_k\right)^{-1}\\
&+\left(\mathbf{H}_k\mathbf{P}_{k|k-1}\mathbf{H}_k^T+\mathbf{R}_k\right)^{-1}\Big)\\
&=\mathbf{H}_k\mathbf{P}_{k|k-1}^T\mathbf{H}_k^T\left(-\mathbf{R}_k^{-1}+\mathbf{R}_k^{-1}-\left(\mathbf{H}_k\mathbf{P}_{k|k-1}\mathbf{H}_k^T+\mathbf{R}_k\right)^{-1}+\left(\mathbf{H}_k\mathbf{P}_{k|k-1}\mathbf{H}_k^T+\mathbf{R}_k\right)^{-1}\right)\\
&=\mathbf{0}
\end{aligned}
\tag{2.77}
$$

Hence, Eq. (2.76) is equal to $\mathbf{0}$ and (2.72b) is verified.
By Eq. (2.72c), we have

$$
\begin{aligned}
&\left(\mathbf{I}-\mathbf{K}_k\mathbf{H}_k\right)^T \breve{\mathbf{Y}}_k \left(\mathbf{I}-\mathbf{K}_k\mathbf{H}_k\right)+\mathbf{H}_k^T\left(\mathbf{H}_k\mathbf{P}_{k|k-1}\mathbf{H}_k^T+\mathbf{R}_k\right)^{-1}\mathbf{H}_k-\mathbf{P}_{k|k-1}^{-1} \\
&=\breve{\mathbf{Y}}_k-\mathbf{H}_k^T\mathbf{K}_k^T\breve{\mathbf{Y}}_k-\breve{\mathbf{Y}}_k\mathbf{K}_k\mathbf{H}_k+\mathbf{H}_k^T\mathbf{K}_k^T\breve{\mathbf{Y}}_k\mathbf{K}_k\mathbf{H}_k+\mathbf{H}_k^T\left(\mathbf{H}_k\mathbf{P}_{k|k-1}\mathbf{H}_k^T+\mathbf{R}_k\right)^{-1}\mathbf{H}_k-\mathbf{P}_{k|k-1}^{-1} \\
&=\mathbf{H}_k^T\mathbf{R}_k^{-1}\mathbf{H}_k+\mathbf{P}_{k|k-1}^{-1}-\mathbf{H}_k^T\left(\mathbf{P}_{k|k-1}\mathbf{H}_k^T\left(\mathbf{H}_k\mathbf{P}_{k|k-1}\mathbf{H}_k^T+\mathbf{R}_k\right)^{-1}\right)^T\left(\mathbf{H}_k^T\mathbf{R}_k^{-1}\mathbf{H}_k+\mathbf{P}_{k|k-1}^{-1}\right) \\
&\quad-\left(\mathbf{H}_k^T\mathbf{R}_k^{-1}\mathbf{H}_k+\mathbf{P}_{k|k-1}^{-1}\right)\mathbf{P}_{k|k-1}\mathbf{H}_k^T\left(\mathbf{H}_k\mathbf{P}_{k|k-1}\mathbf{H}_k^T+\mathbf{R}_k\right)^{-1}\mathbf{H}_k \\
&\quad+\mathbf{H}_k^T\left(\mathbf{P}_{k|k-1}\mathbf{H}_k^T\left(\mathbf{H}_k\mathbf{P}_{k|k-1}\mathbf{H}_k^T+\mathbf{R}_k\right)^{-1}\right)^T \\
&\quad\left(\mathbf{H}_k^T\mathbf{R}_k^{-1}\mathbf{H}_k+\mathbf{P}_{k|k-1}^{-1}\right)\mathbf{P}_{k|k-1}\mathbf{H}_k^T\left(\mathbf{H}_k\mathbf{P}_{k|k-1}\mathbf{H}_k^T+\mathbf{R}_k\right)^{-1}\mathbf{H}_k \\
&\quad+\mathbf{H}_k^T\left(\mathbf{H}_k\mathbf{P}_{k|k-1}\mathbf{H}_k^T+\mathbf{R}_k\right)^{-1}\mathbf{H}_k-\mathbf{P}_{k|k-1}^{-1} \\
&=\mathbf{H}_k^T\mathbf{R}_k^{-1}\mathbf{H}_k+\mathbf{P}_{k|k-1}^{-1}-\mathbf{H}_k^T\left(\mathbf{H}_k\mathbf{P}_{k|k-1}\mathbf{H}_k^T+\mathbf{R}_k\right)^{-1} \\
&\quad\mathbf{H}_k\mathbf{P}_{k|k-1}^T\mathbf{H}_k^T\mathbf{R}_k^{-1}\mathbf{H}_k-\mathbf{H}_k^T\left(\mathbf{H}_k\mathbf{P}_{k|k-1}\mathbf{H}_k^T+\mathbf{R}_k\right)^{-1}\mathbf{H}_k \\
&\quad-\mathbf{H}_k^T\mathbf{R}_k^{-1}\mathbf{H}_k\mathbf{P}_{k|k-1}\mathbf{H}_k^T\left(\mathbf{H}_k\mathbf{P}_{k|k-1}\mathbf{H}_k^T+\mathbf{R}_k\right)^{-1}\mathbf{H}_k-\mathbf{H}_k^T\left(\mathbf{H}_k\mathbf{P}_{k|k-1}\mathbf{H}_k^T+\mathbf{R}_k\right)^{-1}\mathbf{H}_k \\
&\quad+\mathbf{H}_k^T\left(\mathbf{H}_k\mathbf{P}_{k|k-1}\mathbf{H}_k^T+\mathbf{R}_k\right)^{-1}\mathbf{H}_k\mathbf{P}_{k|k-1}^T\mathbf{H}_k^T\mathbf{R}_k^{-1}\mathbf{H}_k\mathbf{P}_{k|k-1}\mathbf{H}_k^T\left(\mathbf{H}_k\mathbf{P}_{k|k-1}\mathbf{H}_k^T+\mathbf{R}_k\right)^{-1}\mathbf{H}_k \\
&\quad+\mathbf{H}_k^T\left(\mathbf{H}_k\mathbf{P}_{k|k-1}\mathbf{H}_k^T+\mathbf{R}_k\right)^{-1}\mathbf{H}_k\mathbf{P}_{k|k-1}\mathbf{H}_k^T\left(\mathbf{H}_k\mathbf{P}_{k|k-1}\mathbf{H}_k^T+\mathbf{R}_k\right)^{-1}\mathbf{H}_k \\
&\quad+\mathbf{H}_k^T\left(\mathbf{H}_k\mathbf{P}_{k|k-1}\mathbf{H}_k^T+\mathbf{R}_k\right)^{-1}\mathbf{H}_k-\mathbf{P}_{k|k-1}^{-1} \\
&=\mathbf{H}_k^T\mathbf{R}_k^{-1}\mathbf{H}_k-\mathbf{H}_k^T\left(\mathbf{H}_k\mathbf{P}_{k|k-1}\mathbf{H}_k^T+\mathbf{R}_k\right)^{-1}\mathbf{H}_k\mathbf{P}_{k|k-1}^T\mathbf{H}_k^T\mathbf{R}_k^{-1}\mathbf{H}_k \\
&\quad-\mathbf{H}_k^T\mathbf{R}_k^{-1}\mathbf{H}_k\mathbf{P}_{k|k-1}\mathbf{H}_k^T\left(\mathbf{H}_k\mathbf{P}_{k|k-1}\mathbf{H}_k^T+\mathbf{R}_k\right)^{-1}\mathbf{H}_k-\mathbf{H}_k^T\left(\mathbf{H}_k\mathbf{P}_{k|k-1}\mathbf{H}_k^T+\mathbf{R}_k\right)^{-1}\mathbf{H}_k \\
&\quad+\mathbf{H}_k^T\left(\mathbf{H}_k\mathbf{P}_{k|k-1}\mathbf{H}_k^T+\mathbf{R}_k\right)^{-1}\mathbf{H}_k\mathbf{P}_{k|k-1}^T\mathbf{H}_k^T\mathbf{R}_k^{-1}\mathbf{H}_k\mathbf{P}_{k|k-1}\mathbf{H}_k^T\left(\mathbf{H}_k\mathbf{P}_{k|k-1}\mathbf{H}_k^T+\mathbf{R}_k\right)^{-1}\mathbf{H}_k \\
&\quad+\mathbf{H}_k^T\left(\mathbf{H}_k\mathbf{P}_{k|k-1}\mathbf{H}_k^T+\mathbf{R}_k\right)^{-1}\mathbf{H}_k\mathbf{P}_{k|k-1}\mathbf{H}_k^T\left(\mathbf{H}_k\mathbf{P}_{k|k-1}\mathbf{H}_k^T+\mathbf{R}_k\right)^{-1}\mathbf{H}_k \\
&=\mathbf{H}_k^T\left(\mathbf{H}_k\mathbf{P}_{k|k-1}\mathbf{H}_k^T+\mathbf{R}_k\right)^{-1}
\left\{\begin{array}{l}
\left(\mathbf{H}_k\mathbf{P}_{k|k-1}\mathbf{H}_k^T+\mathbf{R}_k\right)\mathbf{R}_k^{-1}\left(\mathbf{H}_k\mathbf{P}_{k|k-1}\mathbf{H}_k^T+\mathbf{R}_k\right) \\
-\mathbf{H}_k\mathbf{P}_{k|k-1}^T\mathbf{H}_k^T\mathbf{R}_k^{-1}\left(\mathbf{H}_k\mathbf{P}_{k|k-1}\mathbf{H}_k^T+\mathbf{R}_k\right) \\
-\left(\mathbf{H}_k\mathbf{P}_{k|k-1}\mathbf{H}_k^T+\mathbf{R}_k\right)\mathbf{R}_k^{-1}\mathbf{H}_k\mathbf{P}_{k|k-1}\mathbf{H}_k^T \\
-\left(\mathbf{H}_k\mathbf{P}_{k|k-1}\mathbf{H}_k^T+\mathbf{R}_k\right) \\
+\mathbf{H}_k\mathbf{P}_{k|k-1}^T\mathbf{H}_k^T\mathbf{R}_k^{-1}\mathbf{H}_k\mathbf{P}_{k|k-1}\mathbf{H}_k^T \\
+\mathbf{H}_k\mathbf{P}_{k|k-1}\mathbf{H}_k^T
\end{array}\right\} \qquad (2.78) \\
&\quad\left(\mathbf{H}_k\mathbf{P}_{k|k-1}\mathbf{H}_k^T+\mathbf{R}_k\right)^{-1}\mathbf{H}_k
\end{aligned}
$$

The term in the bracket is given by

$$
\begin{aligned}
&\left(\mathbf{H}_k\mathbf{P}_{k|k-1}\mathbf{H}_k^T+\mathbf{R}_k\right)\mathbf{R}_k^{-1}\left(\mathbf{H}_k\mathbf{P}_{k|k-1}\mathbf{H}_k^T+\mathbf{R}_k\right)-\mathbf{H}_k\mathbf{P}_{k|k-1}^T\mathbf{H}_k^T\mathbf{R}_k^{-1}\left(\mathbf{H}_k\mathbf{P}_{k|k-1}\mathbf{H}_k^T+\mathbf{R}_k\right)\\
&-\left(\mathbf{H}_k\mathbf{P}_{k|k-1}\mathbf{H}_k^T+\mathbf{R}_k\right)\mathbf{R}_k^{-1}\mathbf{H}_k\mathbf{P}_{k|k-1}\mathbf{H}_k^T-\left(\mathbf{H}_k\mathbf{P}_{k|k-1}\mathbf{H}_k^T+\mathbf{R}_k\right)\\
&+\mathbf{H}_k\mathbf{P}_{k|k-1}^T\mathbf{H}_k^T\mathbf{R}_k^{-1}\mathbf{H}_k\mathbf{P}_{k|k-1}\mathbf{H}_k^T+\mathbf{H}_k\mathbf{P}_{k|k-1}\mathbf{H}_k^T\\
&=\mathbf{H}_k\mathbf{P}_{k|k-1}\mathbf{H}_k^T\mathbf{R}_k^{-1}\left(\mathbf{H}_k\mathbf{P}_{k|k-1}\mathbf{H}_k^T+\mathbf{R}_k\right)+\left(\mathbf{H}_k\mathbf{P}_{k|k-1}\mathbf{H}_k^T+\mathbf{R}_k\right)\\
&-\mathbf{H}_k\mathbf{P}_{k|k-1}^T\mathbf{H}_k^T\mathbf{R}_k^{-1}\left(\mathbf{H}_k\mathbf{P}_{k|k-1}\mathbf{H}_k^T+\mathbf{R}_k\right)\\
&-\mathbf{H}_k\mathbf{P}_{k|k-1}\mathbf{H}_k^T\mathbf{R}_k^{-1}\mathbf{H}_k\mathbf{P}_{k|k-1}\mathbf{H}_k^T-\mathbf{H}_k\mathbf{P}_{k|k-1}\mathbf{H}_k^T-\left(\mathbf{H}_k\mathbf{P}_{k|k-1}\mathbf{H}_k^T+\mathbf{R}_k\right)\\
&+\mathbf{H}_k\mathbf{P}_{k|k-1}^T\mathbf{H}_k^T\mathbf{R}_k^{-1}\mathbf{H}_k\mathbf{P}_{k|k-1}\mathbf{H}_k^T+\mathbf{H}_k\mathbf{P}_{k|k-1}\mathbf{H}_k^T\\
&=\mathbf{0}
\end{aligned}
\tag{2.79}
$$

Hence, Eq. (2.78) is equal to $\mathbf{0}$, and (2.72c) is verified.
We have proved that $D=\widetilde{D}$. Next, we will show $C=\widetilde{C}$.

$$
\frac{(2\pi)^{n_z/2}\det\left(\mathbf{H}_k\mathbf{P}_{k|k-1}\mathbf{H}_k^T+\mathbf{R}_k\right)^{1/2}}{(2\pi)^{n_z/2}\det(\mathbf{R}_k)^{1/2}(2\pi)^{n_x/2}\det\left(\mathbf{P}_{k|k-1}\right)^{1/2}}=\frac{1}{(2\pi)^{n_x/2}\det\left(\mathbf{P}_{k|k}\right)^{1/2}}
\tag{2.80}
$$

It is equivalent to proving

$$
\det\left(\mathbf{P}_{k|k}\right)=\frac{\det(\mathbf{R}_k)\det\left(\mathbf{P}_{k|k-1}\right)}{\det\left(\mathbf{H}_k\mathbf{P}_{k|k-1}\mathbf{H}_k^T+\mathbf{R}_k\right)}.
\tag{2.81}
$$

Using Eq. (2.69), we have,

$$
\begin{aligned}
\det\left(\mathbf{P}_{k|k}\right)&=\det\left(\mathbf{P}_{k|k-1}\right)\det\left(\mathbf{I}-\mathbf{H}_k^T\left(\mathbf{H}_k\mathbf{P}_{k|k-1}\mathbf{H}_k^T+\mathbf{R}_k\right)^{-1}\mathbf{H}_k\mathbf{P}_{k|k-1}\right)\\
&=\det\left(\mathbf{P}_{k|k-1}\right)\det\left(\mathbf{I}-\mathbf{H}_k\mathbf{P}_{k|k-1}\mathbf{H}_k^T\left(\mathbf{H}_k\mathbf{P}_{k|k-1}\mathbf{H}_k^T+\mathbf{R}_k\right)^{-1}\right)\\
&=\det\left(\mathbf{P}_{k|k-1}\right)\det\begin{pmatrix}\left(\mathbf{H}_k\mathbf{P}_{k|k-1}\mathbf{H}_k^T+\mathbf{R}_k\right)\left(\mathbf{H}_k\mathbf{P}_{k|k-1}\mathbf{H}_k^T+\mathbf{R}_k\right)^{-1}\\-\mathbf{H}_k\mathbf{P}_{k|k-1}\mathbf{H}_k^T\left(\mathbf{H}_k\mathbf{P}_{k|k-1}\mathbf{H}_k^T+\mathbf{R}_k\right)^{-1}\end{pmatrix}\\
&=\det\left(\mathbf{P}_{k|k-1}\right)\det\left(\mathbf{R}_k\left(\mathbf{H}_k\mathbf{P}_{k|k-1}\mathbf{H}_k^T+\mathbf{R}_k\right)^{-1}\right)\\
&=\frac{\det\left(\mathbf{P}_{k|k-1}\right)\det(\mathbf{R}_k)}{\det\left(\mathbf{H}_k\mathbf{P}_{k|k-1}\mathbf{H}_k^T+\mathbf{R}_k\right)}
\end{aligned}
\tag{2.82}
$$

Note that in Eq. (2.82), the Eq. (2.55) is used.

In summary, Eq. (2.57) and Eq. (2.58) are equivalent. The update equation of the discrete-time Kalman filter is the same as the Bayesian estimation.

2.4 Linear Continuous-Time Kalman Filter

The continuous-time Kalman filter was developed by Kalman and Bucy after the discrete-time Kalman filter. It is not as widely used in practice as the discrete-time filter because of the prevalent implementation of the Kalman filter in real-time systems with digital computers. However, it provides valuable conceptual and theoretical perspectives on the filtering technique. The continuous-time Kalman filter can be readily derived as a limiting case of its discrete-time counterpart as the sampling time becomes very small. The detailed derivation is omitted in this book and it can be seen in other references, e.g. (Crassidis and Junkins 2012, Brown and Hwang 2012). The summary of the linear continuous-time Kalman filter is provided in Table 2.3.

The covariance Eq. (2.85) is called the continuous Riccati equation. To implement the continuous-time Kalman filter, first, given the initial conditions for the state estimate ^ and error covariance $\mathbf{P}_0$, the Kalman gain is calculated from Eq. (2.87). Next, the covariance Eq. (2.85) and the state estimate Eq. (2.86) are numerically integrated forward in time using the continuous measurement $\mathbf{y}(t)$ while in the meantime, updating the Kalman gain with the current

Table 2.3 Continuous-time Kalman filter.

Continuous-time linear dynamics and measurement model:

$$\dot{\mathbf{x}}(t) = \mathbf{A}(t)\mathbf{x}(t) + \mathbf{v}(t) \tag{2.83}$$

$$\mathbf{y}(t) = \mathbf{H}(t)\mathbf{x}(t) + \mathbf{n}(t) \tag{2.84}$$

where $\mathbf{x}(t)$ is the continuous random process and $\mathbf{y}(t)$ is the continuous measurement.

<u>Assumptions</u>: $\mathbf{v}(t)$ and $\mathbf{n}(t)$ are white Gaussian process and measurement noise, respectively.

$E\left[\mathbf{v}(t)\mathbf{v}^T(t)\right] = \mathbf{Q}(t)\delta(t-\tau)$, $E\left[\mathbf{n}(t)\mathbf{n}^T(t)\right] = \mathbf{R}(t)\delta(t-\tau)$. $\delta(t-\tau)$ is the Dirac delta function.

$\mathbf{v}(t)$ and $\mathbf{n}(t)$ are uncorrelated with $\mathbf{x}_0$ and with each other.

<u>Initialization</u>: $\hat{\mathbf{x}}(t_0) = \hat{\mathbf{x}}_0$ and $\mathbf{P}(t_0) = \mathbf{P}_0$

<u>Covariance Update</u>:

$$\dot{\mathbf{P}}(t) = \mathbf{A}(t)\mathbf{P}(t) + \mathbf{P}(t)\mathbf{A}^T(t) - \mathbf{P}(t)\mathbf{H}^T(t)\mathbf{R}^{-1}(t)\mathbf{H}(t)\mathbf{P}(t) + \mathbf{Q}(t) \tag{2.85}$$

<u>State Estimate</u>:

$$\dot{\hat{\mathbf{x}}}(t) = \hat{\mathbf{x}}(t) + \mathbf{K}(t)\left[\mathbf{y}(t) - \mathbf{H}(t)\hat{\mathbf{x}}(t)\right]. \tag{2.86}$$

where $$\mathbf{K}(t) = \mathbf{P}(t)\mathbf{H}^T(t)\mathbf{R}^{-1}(t) \tag{2.87}$$

covariance $\mathbf{P}(t)$. The integration continues until the final measurement time is reached. It can be proved that if $\mathbf{R}(t)$ is positive definite and $\mathbf{Q}(t)$ is at least positive semi-definite, the continuous-time Kalman filter is stable (Crassidis and Junkins 2012).

It is worth noting the difference between the discrete-time covariance matrices $\mathbf{Q}_k$ and $\mathbf{R}_k$, and continuous-time covariance matrices $\mathbf{Q}(t)$ and $\mathbf{R}(t)$. They play the same role but have different numerical values. The relationship among them is related to the sampling interval Δt:

$$\mathbf{Q}_k = \mathbf{Q}(t)\Delta t \tag{2.88}$$

$$\mathbf{R}_k = \frac{\mathbf{R}(t)}{\Delta t} \tag{2.89}$$

It may seem counterintuitive to have the discrete measurement covariance approach ∞ as $\Delta t \to 0$. However, this is offset by the sampling rate approaching infinity at the same time (Brown and Hwang 2012).

It is also noted that as $\Delta t \to 0$, $\mathbf{P}_{k|k-1} \to \mathbf{P}_{k-1|k-1}$ from the covariance prediction equation $\mathbf{P}_{k|k-1} = \mathbf{F}_{k-1}\mathbf{P}_{k-1|k-1}\mathbf{F}_{k-1}^T + \mathbf{Q}_{k-1}$. Therefore, there is no need to make a distinction between *a priori* and *a posterior* covariance matrices in the continuous filter.

References

Bar-Shalom, Y., X. Li and T. Kirubarajan. 2001. Estimation with Application to Tracking and Navigation: Theory, Algorithms and Software. John Wiley & Sons, Inc., New York.

Brown, R.G. and P.Y.C. Hwang. 2012. Introduction to Random Signals and Applied Kalman Filtering. 4th Edition. John Wiley & Sons, New York.

Crassidis, J. and J. Junkins. 2012. Optimal Estimation of Dynamic Systems. 2nd Edition, CRC Press, Boca Raton.

Kailath, T., A.H. Sayed and B. Hassibi. 2000. Linear Estimation. Prentice Hall, Upper Saddle River.

Mutambara, G.O. 1998. Decentralized Estimation and Control for Multisensor Systems. 1st Edition, CRC Press, Boca Raton.

Šimandl, M. 2006. Lecture Notes on State Estimation of Nonlinear Non-Gaussian Stochastic Systems. Retrieved from semanticscholar.org.

Conventional Nonlinear Filters

3

Most of the estimation problems in practice involve nonlinear models including nonlinear dynamics and nonlinear measurement functions of the states. The state estimation of such nonlinear systems becomes very difficult because many useful statistic properties as a result of linearity and Gaussianity will not be applicable in the filtering algorithm after nonlinear transformation. As illustrated in the Bayesian approach, Chapter 1.3, the estimation requires the construction of the conditional PDF because the posterior PDF embodies all available statistical information. This makes it a formidable task to find an optimal estimate analytically and in the meantime develop a filtering algorithm implementable in real-time applications. Therefore, approximate filters are necessary. The research in this respect has been conducted for decades. A variety of nonlinear filtering algorithms were derived and applied based on different approximation techniques.

In this chapter, some conventional nonlinear filters that have been well recognized and extensively applied are presented. The first nonlinear filter is a natural extension of the Kalman filter derived from the linearization of nonlinear dynamic and measurement equations, which is called the extended Kalman filter (EKF) and its variants. The EKF, in spite of not being optimum, has been successfully applied to many nonlinear systems over the past decades. It is based on the notion that the true state is sufficiently close to the estimated state. Thus, the error dynamics can be represented fairly accurately by the first-order Taylor series expansion. The second class of nonlinear filters is the PDF approximation based method including the point-mass filter and the particle filter, which use an analytically designed deterministic grid and a large number of random Monte Carlo samples to approximate PDF, respectively. Another sampling point based nonlinear filter, the ensemble Kalman filter, is reviewed as a simple extension of the classical Kalman filter to solve large-scale nonlinear estimation problems. Two continuous-time nonlinear filtering methods based on the Zakai equation and Fokker Planck equation are presented at the end.

3.1 Extended Kalman Filter

The EKF adapts the linear Kalman filter so that it can be applied to nonlinear problems. It is an analytical approximation based on the linearization of the nonlinear dynamics and the nonlinear measurement model. For common applications, the dynamic system can be presented by either the continuous-time system or discrete-time system. The measurement equation is usually described by the discrete-time equation. Although we mainly use the discrete-time system description to introduce the estimation methods, the continuous-discrete time system can also use the result via discretization.

We consider a class of nonlinear discrete-time dynamical systems described by:

$$\mathbf{x}_k = \boldsymbol{f}_{k-1}\left(\mathbf{x}_{k-1}\right) + \mathbf{v}_{k-1} \tag{3.1}$$

$$\mathbf{y}_k = \boldsymbol{h}_k\left(\mathbf{x}_k\right) + \mathbf{n}_k \tag{3.2}$$

where $\mathbf{x}_k \in \mathbb{R}^{n_x}; \mathbf{y}_k \in \mathbb{R}^m$; $\mathbf{v}_{k-1}$ and $\mathbf{n}_k$ are independent white Gaussian process noise and measurement noise with covariance $\mathbf{Q}_{k-1}$ and $\mathbf{R}_k$, respectively.

The EKF is based on the assumption that local linearization is a sufficient description of nonlinearity. The filtering algorithm follows the same prediction and update steps.
The predicted state and covariance can be approximated by (Gelb 1974)

$$\hat{\mathbf{x}}_{k|k-1} = \boldsymbol{f}_{k-1}\left(\hat{\mathbf{x}}_{k-1|k-1}\right) \tag{3.3}$$

$$\mathbf{P}_{k|k-1} = \mathbf{F}_{k-1}\mathbf{P}_{k-1}\mathbf{F}_{k-1}^T + \mathbf{Q}_{k-1} \tag{3.4}$$

where $\hat{\mathbf{x}}_{k-1|k-1}$ is the *a posteriori* estimate at the time $k-1$, $\mathbf{F}_{k-1}$ is the Jacobian matrix of the nonlinear function f evaluated at $\hat{\mathbf{x}}_{k-1|k-1}$.

$$\mathbf{F}_{k-1} \triangleq \left.\frac{\partial \boldsymbol{f}_{k-1}}{\partial \mathbf{x}}\right|_{\hat{\mathbf{x}}_{k-1|k-1}} \tag{3.5}$$

The updated state and covariance are given by (Gelb 1974)

$$\hat{\mathbf{x}}_{k|k} = \hat{\mathbf{x}}_{k|k-1} + \mathbf{K}_k\left(\mathbf{y}_k - \hat{\mathbf{y}}_k\right) \tag{3.6}$$

$$\mathbf{P}_{k|k} = \mathbf{P}_{k|k-1} - \mathbf{K}_k\mathbf{H}_k\mathbf{P}_{k|k-1} \tag{3.7}$$

where the predicted observation $\hat{\mathbf{y}}_k$ is given by

$$\hat{\mathbf{y}}_k = \boldsymbol{h}_k\left(\hat{\mathbf{x}}_{k|k-1}\right) \tag{3.8}$$

$\mathbf{H}_k$ is the Jacobian matrix of the nonlinear function $\boldsymbol{h}$ evaluated at $\hat{\mathbf{x}}_{k|k-1}$.

$$\mathbf{H}_k \triangleq \left.\frac{\partial \boldsymbol{h}_k}{\partial \mathbf{x}}\right|_{\hat{\mathbf{x}}_{k|k-1}} \tag{3.9}$$

The Kalman gain $\mathbf{K}_k$ is given by

$$\mathbf{K}_k = \mathbf{P}_{k|k-1}\mathbf{H}_k^T\left(\mathbf{H}_k\mathbf{P}_{k|k-1}\mathbf{H}_k^T + \mathbf{R}_k\right)^{-1} \tag{3.10}$$

There are some observations about the EKF:

(1) The Kalman gain $\mathbf{K}_k$ and the resultant error covariance matrix $\mathbf{P}_{k|k}$ are random variables because they are dependent on the estimate $\hat{\mathbf{x}}_{k|k-1}$. Consequently, $\mathbf{K}_k$ and $\mathbf{P}_{k|k}$ have to be calculated in real time while for the linear Kalman filter, they can be precomputed and stored before the measurements are obtained.

(2) The EKF may degrade or diverge if the initial estimate $\hat{\mathbf{x}}_{0|0}$ is poor because it is the reference about which the linearization takes place. If the error in $\hat{\mathbf{x}}_{0|0}$ is large, the first-order approximation based on linearization will be poor. This error can be propagated through the prediction and update steps and cause further performance degradation, especially in the presence of large nonlinearity.

(3) Since the covariance matrix $\mathbf{P}_{k|k}$ is only an approximation to the true covariance matrix, there is no guarantee that the actual estimate will be close to the truly optimal estimate. However, the EKF has been used in a large number of practical applications with great success. It is usually the first try for nonlinear estimation problems.

Several EKF variants have been proposed to improve the EKF performance. One of the methods is to apply local iterations to repeatedly calculate $\mathbf{P}_{k|k}$, $\mathbf{K}_k$, and $\hat{}_{k k}$, each time linearizing about the most recent estimate. This method is called the iterative EKF.

3.2 Iterated Extended Kalman Filter

Note that the update equations of the EKF are functions of the estimate $\hat{\mathbf{x}}_{k|k-1}$, i.e.,

$$\hat{\mathbf{x}}_{k|k} = \hat{\mathbf{x}}_{k|k-1} + \mathbf{P}_{k|k-1}\mathbf{H}_k^T\left(\hat{\mathbf{x}}_{k|k-1}\right)\left[\mathbf{H}_k\left(\hat{\mathbf{x}}_{k|k-1}\right)\mathbf{P}_{k|k-1}\mathbf{H}_k^T\left(\hat{\mathbf{x}}_{k|k-1}\right) + \mathbf{R}_k\right]^{-1}\left[\mathbf{y}_k - \boldsymbol{h}\left(\hat{\mathbf{x}}_{k|k-1}\right)\right] \tag{3.11}$$

$$\mathbf{P}_{k|k} = \mathbf{P}_{k|k-1} - \mathbf{P}_{k|k-1}\mathbf{H}_k^T\left(\hat{\mathbf{x}}_{k|k-1}\right)\left[\mathbf{H}_k\left(\hat{\mathbf{x}}_{k|k-1}\right)\mathbf{P}_{k|k-1}\mathbf{H}_k^T\left(\hat{\mathbf{x}}_{k|k-1}\right)+\mathbf{R}_k\right]^{-1}\mathbf{H}_k\left(\hat{\mathbf{x}}_{k|k-1}\right)\mathbf{P}_{k|k-1} \tag{3.12}$$

The reference about which the linearization is performed is $\hat{\mathbf{x}}_{k|k-1}$, which may not be accurate. Hence, the estimation can be improved by updating the linearization reference point iteratively with $\hat{\mathbf{x}}_{k|k}$. Specifically, the update equations are rewritten as

$$\hat{\mathbf{x}}_{k|k}^{i+1} = \hat{\mathbf{x}}_{k|k-1} + \mathbf{P}_{k|k-1}\mathbf{H}_k^T\left(\hat{\mathbf{x}}_{k|k}^i\right)\left[\mathbf{H}_k\left(\hat{\mathbf{x}}_{k|k}^i\right)\mathbf{P}_{k|k-1}\mathbf{H}_k^T\left(\hat{\mathbf{x}}_{k|k}^i\right)+\mathbf{R}_k\right]^{-1}\left[\mathbf{y}_k - \boldsymbol{h}\left(\hat{\mathbf{x}}_{k|k}^i\right)\right] \tag{3.13}$$

$$\mathbf{P}_{k|k}^{i+1} = \mathbf{P}_{k|k-1} - \mathbf{P}_{k|k-1}\mathbf{H}_k^T\left(\hat{\mathbf{x}}_{k|k}^i\right)\left[\mathbf{H}_k\left(\hat{\mathbf{x}}_{k|k}^i\right)\mathbf{P}_{k|k-1}\mathbf{H}_k^T\left(\hat{\mathbf{x}}_{k|k}^i\right)+\mathbf{R}_k\right]^{-1}\mathbf{H}_k\left(\hat{\mathbf{x}}_{k|k}^i\right)\mathbf{P}_{k|k-1} \tag{3.14}$$

where $\hat{\mathbf{x}}_{k|k}^i$ and $\mathbf{P}_{k|k}^i$ denote the mean and covariance update at the ith iteration, respectively. The iteration is stopped when it converges, i.e., $\left\|\hat{\mathbf{x}}_{k|k}^{i+1} - \hat{\mathbf{x}}_{k|k}^i\right\| < \varepsilon$ with ε being a given threshold.

Remark 3.1: The EKF is derived from retaining the first-order terms in the Taylor series expansion of the nonlinear dynamics and measurement function. Thus, it is possible to improve the accuracy of estimation by including higher-order terms such as the second-order EKF. However, it involves solutions with extra complexity. One has to consider the extra computation cost to use such filters.

Another class of nonlinear filtering methods involves approximation of the *a posteriori* PDF directly. It does not rely on the Gaussian assumption and can be applied to more general nonlinear estimation problems. In the following, two types of such filters are introduced. The first one is the point-mass filter based on deterministic grid points, and the second one is the particle filter based on the sequential Monte Carlo method.

3.3 Point-Mass Filter

The key idea of the point-mass filter (PMF) is to represent the PDF by a grid of points G_0, a set of volume masses for each point's neighborhood, M_0, and a set of probability density values at each point, $V_{0|0}$. Using such representations, the Bayesian estimation equations can be rewritten. The PMF can be briefly described as follows.

Initialization

The initial PDF $p_{0|0}(\mathbf{x})$ is described by a grid of points $G_0 = \{\mathbf{p}_0^{(i)}; \mathbf{p}_0^{(i)} \in \mathbb{R}^{n_x}, i = 1, \cdots, N_0\}$, a set of volume masses for each point's neighborhood is given by $M_0 = \{\Delta\mathbf{p}_0^{(i)}, i = 1, \cdots, N_0\}$, and a set of probability density values at each point is given by $V_{0|0} = \{v_{0|0}^{(i)}; v_{0|0}^{(i)} = p_{0|0}(\mathbf{p}_0^{(i)}), \mathbf{p}_0^{(i)} \in G_0, i = 1, \cdots, N_0\}$. N_0 is the initial number of grid points.

Prediction of grid

Each point of the grid is propagated by

$$\mathbf{p}_{k|k-1}^{(i)} = \boldsymbol{f}\left(\mathbf{p}_{k-1}^{(i)}\right) \tag{3.15}$$

Due to the nonlinear transformation, the grid structure of $\{\mathbf{p}_{k|k-1}^{(i)}; \mathbf{p}_{k|k-1}^{(i)} \in \mathbb{R}^{n_x}, i = 1, \cdots, N_{k-1}\}$ is not the same as G_{k-1}.

Grid Redefinition

The grid $\{\mathbf{p}_{k|k-1}^{(i)}; \mathbf{p}_{k|k-1}^{(i)} \in \mathbb{R}^{n_x}, i = 1, \cdots, N_{k-1}\}$ is refined to the grid $G_k = \{\mathbf{p}_k^{(j)}; \mathbf{p}_k^{(j)} \in \mathbb{R}^{n_x}, j = 1, \cdots, N_k\}$ to maintain the structure of the grid G_{k-1} and consider the effect of process noise on the spread of the PDF.

Prediction of the density value

Given $V_{k-1|k-1}$, the predicted density value for the grid point $\mathbf{p}_k^{(j)}$ is then rewritten as

$$v_{k|k-1}^{(j)} = \sum_{i=1}^{N_{k-1}} \Delta\mathbf{p}_{k-1}^{(i)} \cdot v_{k-1|k-1}^{(i)} \cdot p_v\left(\mathbf{p}_k^{(j)} - \mathbf{p}_{k|k-1}^{(i)}\right), \quad j = 1, \cdots, N_k \tag{3.16}$$

where N_k is the number of grid points at the time k. $p_v(\cdot)$ denotes the PDF of the process noise. Note that M_{k-1} can be determined when the grid of points G_{k-1} is known.

Update of the density value

The posterior PDF at the ith grid point $\mathbf{p}_k^{(j)}$ are given by

$$v_{k|k}^{(j)} = c_k^{-1} \cdot v_{k|k-1}^{(j)} \cdot p_n\left(\mathbf{y}_k - \boldsymbol{h}\left(\mathbf{p}_k^{(j)}\right)\right), \quad j = 1, \cdots, N_k \tag{3.17}$$

where $c_k = \sum_{j=1}^{N_k} \Delta\mathbf{p}_k^{(j)} \cdot v_{k|k-1}^{(j)} \cdot p_n\left(\mathbf{y}_k - \boldsymbol{h}\left(\mathbf{p}_k^{(j)}\right)\right)$. $p_n(\cdot)$ denotes the PDF of the measurement noise.

From the above equations, one can see that design of the grid $G_k = \left\{ \mathbf{p}_k^{(j)}; \mathbf{p}_k^{(j)} \in \mathbb{R}^{n_x}, j = 1, \cdots, N_k \right\}$ is critical for the PMF. The floating grid technique was proposed in (Šimandl et al. 2002, 2006), where a rectangular equally-spaced grid is used. In addition, the grid is shifted and rotated based on the predicted expected value and the predicted covariance matrix.

Remark 3.2: In order to achieve good accuracy and efficiency, the PMF needs the sophisticated design of the grid (Šimandl et al. 2006). The computational load is high compared to other methods, such as the particle filter.

3.4 Particle Filter

In the Bayesian estimation, the prediction and update of the estimate involve integrals of PDFs in both the Chapman-Kolmogorov equation and the Bayesian update rule. These integrals can be calculated by the Monte Carlo method. Assume that we are able to generate N_s independent and identically distributed random particles $\{\mathbf{x}_k^{(i)}; i = 1, \cdots, N_s\}$ according to the posterior PDF $p(\mathbf{x}_k|\mathbf{y}_{1:k})$. An approximation of this distribution is given by

$$p_{Ns}(\mathbf{x}_k \mid \mathbf{y}_{1:k}) = \frac{1}{N_s} \sum_{i=1}^{N_s} \delta\left(\mathbf{x}_k \mid \mathbf{x}_k^{(i)}\right) \tag{3.18}$$

Then, the integral $I(f) = \int f(\mathbf{x}_k) p(\mathbf{x}_k \mid \mathbf{y}_{1:k}) d\mathbf{x}_k$ can be approximated by

$$I_{N_s}(f) = \int f(\mathbf{x}_k) p_{N_s}(\mathbf{x}_k \mid \mathbf{y}_{1:k}) d\mathbf{x}_k = \frac{1}{N_s} \sum_{i=1}^{N_s} f\left(\mathbf{x}_k^{(i)}\right) \tag{3.19}$$

using the set of random particles $\{\mathbf{x}_k^{(i)}; i = 1, \cdots, N_s\}$. When the number of particles N_s goes to infinity, $I_{N_S}(f)$ will approach the true $I(f)$.

The merit of the particle filter is that it has a convergence rate independent of the dimension of the integrand by the central limit theorem. In contrast, any deterministic numerical integration method has a rate of convergence that decreases with the increase of dimension (Arulampalam et al. 2002).

Nevertheless, for a general non-Gaussian, multivariate or multi-modal PDF $p(\mathbf{x}_k \mid \mathbf{y}_{1:k})$, it is difficult to generate samples directly from the posterior $p(\mathbf{x}_k \mid \mathbf{y}_{1:k})$. Alternatively, we can utilize the concept of importance sampling, which is a method to compute expectations with respect to one distribution using random samples drawn from another. Specifically, we choose a proposal PDF $q(\mathbf{x}_k \mid \mathbf{y}_{1:k})$ that approximates the target posterior distribution $p(\mathbf{x}_k \mid \mathbf{y}_{1:k})$ as closely as possible. This proposal PDF $q(\mathbf{x}_k \mid \mathbf{y}_{1:k})$ is referred to as the importance sampling distribution, since it samples the target distribution $p(\mathbf{x}_k \mid \mathbf{y}_{1:k})$ non-uniformly to give "more importance" to some values of $p(\mathbf{x}_k \mid \mathbf{y}_{1:k})$ than others. $q(\mathbf{x}_k \mid \mathbf{y}_{1:k})$ should be easy to sample and its support covers that of $p(\mathbf{x}_k \mid \mathbf{y}_{1:k})$, or the samples drawn from $q(\mathbf{x}_k \mid \mathbf{y}_{1:k})$ overlap the same region (or more) corresponding to the samples of $p(\mathbf{x}_k \mid \mathbf{y}_{1:k})$. $p(\mathbf{x}_k \mid \mathbf{y}_{1:k})$ and $q(\mathbf{x}_k \mid \mathbf{y}_{1:k})$ have the same support if the following condition is satisfied.

$$p(\mathbf{x}_k \mid \mathbf{y}_{1:k}) > 0 \Rightarrow q(\mathbf{x}_k \mid \mathbf{y}_{1:k}) > 0 \quad \forall \mathbf{x}_k \in \mathbb{R}^{n_x} \tag{3.20}$$

This is a necessary condition for the importance sampling theory to hold (Ristic et al. 2004).

We can write $p(\mathbf{x}_k \mid \mathbf{y}_{1:k}) \propto q(\mathbf{x}_k \mid \mathbf{y}_{1:k})$, which means $p(\mathbf{x}_k \mid \mathbf{y}_{1:k})$ is proportional to $q(\mathbf{x}_k \mid \mathbf{y}_{1:k})$ at every $\mathbf{x}_k$. A scaling factor can be defined as

$$\tilde{w}(\mathbf{x}_k) \triangleq \frac{p(\mathbf{x}_k \mid \mathbf{y}_{1:k})}{q(\mathbf{x}_k \mid \mathbf{y}_{1:k})} \tag{3.21}$$

Hence,

$$E\left[\mathbf{f}(\mathbf{x}_k)\right]_{p(\mathbf{x}_k \mid \mathbf{y}_{1:k})} = \frac{\int \mathbf{f}(\mathbf{x}_k)\tilde{w}(\mathbf{x}_k)q(\mathbf{x}_k \mid \mathbf{y}_{1:k})d\mathbf{x}_k}{\int \tilde{w}(\mathbf{x}_k)q(\mathbf{x}_k \mid \mathbf{y}_{1:k})d\mathbf{x}_k} \tag{3.22}$$

where $E[\cdot]_p$ denotes the expected value with respect to the PDF $p(\mathbf{x}_k \mid \mathbf{y}_{1:k})$.

If N_s particles $\left\{\mathbf{x}_k^{(i)}, i = 1, \cdots, N_s\right\}$ are generated from $q(\mathbf{x}_k \mid \mathbf{y}_{1:k})$, Eq. (3.22) can be rewritten as

$$E\left[\mathbf{f}(\mathbf{x}_k)\right]_{p(\mathbf{x}_k \mid \mathbf{y}_{1:k})} \approx \frac{\frac{1}{N_s}\sum_{i=1}^{N_s} \mathbf{f}\left(\mathbf{x}_k^{(i)}\right)\tilde{w}\left(\mathbf{x}_k^{(i)}\right)}{\frac{1}{N_s}\sum_{i=1}^{N_s} \tilde{w}\left(\mathbf{x}_k^{(i)}\right)} = \sum_{i=1}^{N_s} \mathbf{f}\left(\mathbf{x}_k^{(i)}\right) w\left(\mathbf{x}_k^{(i)}\right) \tag{3.23}$$

where

$$w\left(\mathbf{x}_k^{(i)}\right) = \frac{\tilde{w}\left(\mathbf{x}_k^{(i)}\right)}{\sum_{i=1}^{N_s} \tilde{w}\left(\mathbf{x}_k^{(i)}\right)} \tag{3.24}$$

Hence, *a posterior* PDF can be approximated by a set of particles and weights for times up to $k - 1$.

$$p(\mathbf{x}_{k-1} \mid \mathbf{y}_{1:k-1}) \approx \sum_{i=1}^{N_s} w_{k-1}^{(i)} \delta\left(\mathbf{x}_{k-1} - \mathbf{x}_{k-1|k-1}^{(i)}\right) \tag{3.25}$$

with

$$w_{k-1}^{(i)} \propto \frac{p\left(\mathbf{x}_{k-1|k-1}^{(i)}\right)}{q\left(\mathbf{x}_{k-1|k-1}^{(i)}\right)} \tag{3.26}$$

The importance sampling method can be modified to calculate an approximation of the current PDF without modifying past simulated trajectories.

$$q(\mathbf{x}_{1:k} \mid \mathbf{y}_{1:k}) = q(\mathbf{x}_{1:k-1} \mid \mathbf{y}_{1:k-1}) q(\mathbf{x}_k \mid \mathbf{x}_{1:k-1}, \mathbf{y}_{1:k}) \tag{3.27}$$

Equation (3.27) can be rewritten as follows:

$$q(\mathbf{x}_{1:k} \mid \mathbf{y}_{1:k}) = q(\mathbf{x}_1) \prod_{i=2}^{k} q(\mathbf{x}_i \mid \mathbf{x}_{1:i-1}, \mathbf{y}_{1:i}) \tag{3.28}$$

When the measurement is available, $p(\mathbf{x}_k \mid \mathbf{y}_{1:k})$ is approximated by the new set of samples and weights. The weight update equation for each particle becomes

$$\tilde{w}_k^{(i)} \propto w_{k-1}^{(i)} \frac{p\left(\mathbf{y}_k \mid \mathbf{x}_{k|k-1}^{(i)}\right) p\left(\mathbf{x}_{k|k-1}^{(i)} \mid \mathbf{x}_{k-1|k-1}^{(i)}\right)}{q\left(\mathbf{x}_{k|k-1}^{(i)} \mid \mathbf{x}_{k-1|k-1}^{(i)}, \mathbf{y}_k\right)} \tag{3.29}$$

where $\mathbf{x}_{k|k-1}^{(i)}$ are given by

$$\mathbf{x}_{k|k-1}^{(i)} = \boldsymbol{f}\left(\mathbf{x}_{k-1|k-1}^{(i)}, \mathbf{v}_{k-1}^{(i)}\right) \tag{3.30}$$

with $\mathbf{v}_{k-1}^{(i)}$ being the random sample from the PDF of the process noise. Finally, the posterior PDF $p(\mathbf{x}_k \mid \mathbf{y}_{1:k})$ can be approximated by

$$p(\mathbf{x}_k \mid \mathbf{y}_{1:k}) \approx \sum_{i=1}^{N_s} w_k^{(i)} \delta\left(\mathbf{x}_k - \mathbf{x}_{k|k-1}^{(i)}\right) \tag{3.31}$$

Unfortunately, the unconditional variance of the importance weights increases over time (Arulampalam et al. 2002). In practice, after a few iterations of the algorithm, all but one of the normalized importance weights are very close to zero and large computational effort is devoted to updating trajectories whose contribution to the final estimate is almost zero. This results in the degeneracy problem. Therefore, the resampling method is necessary to eliminate trajectories or samples with small normalized importance weights and concentrates upon those with large weights. A suitable measure of the degeneracy of the algorithm is the effective sample size N_{eff} that is defined as

$$N_{eff} = \frac{1}{\sum_{i=1}^{N_s} \left(w_k^{(i)}\right)^2} \tag{3.32}$$

If the effective sample size is less than a given threshold N_{thr}, resampling is performed. Other resampling strategies can be applied as well (Arulampalam et al. 2002).

A generic particle filter is given in Table 3.1.

There are many sampling techniques. The optimal proposal distribution includes current information of the state estimation and observation, which can be written as

Table 3.1 The generic particle filter.

Initialization: For $i = 1,\cdots,N_s$, sample $\mathbf{x}_0^{(i)} \sim p(\mathbf{x}_0)$

Importance Sampling Step:

1. For $i = 1,\cdots,N_s$, sample $\mathbf{x}_{k|k-1}^{(i)}$ from $q\left(\mathbf{x}_k \mid \mathbf{x}_{k-1|k-1}^{(i)}, \mathbf{y}_k\right)$.
2. For $i = 1,\cdots,N_s$, evaluate the importance weights $\tilde{w}_k^{(i)} \propto w_{k-1}^{(i)} \dfrac{p\left(\mathbf{y}_k \mid \mathbf{x}_{k|k-1}^{(i)}\right) p\left(\mathbf{x}_{k|k-1}^{(i)} \mid \mathbf{x}_{k-1|k-1}^{(i)}\right)}{q\left(\mathbf{x}_{k|k-1}^{(i)} \mid \mathbf{x}_{k-1|k-1}^{(i)}, \mathbf{y}_k\right)}$.
3. Normalize the importance weights, $w_k^{(i)} = \tilde{w}_k^{(i)} \Big/ \sum_{j=1}^{N_s} \tilde{w}_k^{(i)}$.

Resampling step:

1. Obtain N_s replacement particles $\left(\left[\mathbf{x}_k^{(i)}\right]_{new}; i = 1,\cdots,N_s\right)$ according to the old set $\left(\left[\mathbf{x}_k^{(i)}\right]_{old}; i = 1,\cdots,N_s\right)$ and $w_k^{(i)}$;
2. Let $w_k^{(i)} = 1/N_s$.
3. Use the N_s replacement particles $\left(\left[\mathbf{x}_{0:k}^{(i)}\right]_{new}; i = 1,\cdots,N_s\right)$ and $w_k^{(i)}$ in the importance sampling step to propagate the particles.

$$q\left(\mathbf{x}_k \mid \mathbf{x}_{k-1}^{(i)}, \mathbf{y}_k\right) = p\left(\mathbf{x}_k \mid \mathbf{x}_{k-1}^{(i)}, \mathbf{y}_k\right) = \frac{p\left(\mathbf{y}_k \mid \mathbf{x}_k\right) p\left(\mathbf{x}_k \mid \mathbf{x}_{k-1}^{(i)}\right)}{p\left(\mathbf{y}_k \mid \mathbf{x}_{k-1}^{(i)}\right)} \tag{3.33}$$

Although Eq. (3.33) is optimal, it is hard to implement because $\left(\mathbf{y}_k \quad \mathbf{x}_k^{()}\right)$ is hard to obtain and requires further integration operations.

Another choice for the importance/proposal distribution is the transition prior, which is the most common sampling technique due to its simplicity and is given by

$$q\left(\mathbf{x}_k \mid \mathbf{x}_{k-1}^{(i)}, \mathbf{y}_k\right) = p\left(\mathbf{x}_k \mid \mathbf{x}_{k-1}^{(i)}\right) \tag{3.34}$$

This sampling strategy is termed sequential importance resampling (SIR) and the particle filter using SIR is the bootstrap particle filter.

Using the transition prior as the importance distribution can cause new problems because it is not conditioned on the measurement data as shown in Eq. (3.34). Ignoring the latest available information from the most recent measurement to propose new values for the states may lead to only a few particles having significant weights when their likelihood is calculated. The transition prior has a much broader distribution than the likelihood, implying that only a few particles will be given a large weight. Thus, the algorithm may degenerate rapidly and lead to poor performance especially when data outliers exist or measurement noise is small.

To resolve this issue, many different proposal distributions and resampling strategies have been proposed (Arulampalam et al. 2002). A typical method is to use the auxiliary sampling strategy. It introduces the importance density

$q\left(\mathbf{x}_k, i \mid \mathbf{y}_k\right)$, which samples the pair $\left\{\mathbf{x}_k^{(j)}, i^j\right\}_{j=1}^{M_s}$ where i^j denotes the index of the particle at $k-1$ and M_s is the number of particles. The importance distribution is defined by

$$q\left(\mathbf{x}_k, i \mid \mathbf{y}_k\right) \propto p\left(\mathbf{y}_k \mid \boldsymbol{\mu}_k^{(i)}\right) p\left(\mathbf{x}_k \mid \mathbf{x}_{k-1}^{(i)}\right) w_{k-1}^{(i)} \tag{3.35}$$

where $\boldsymbol{\mu}_k^{(i)}$ is the characterization of $\mathbf{x}_k$, for example $E\left[\mathbf{x}_k \mid \mathbf{x}_{k-1}^{(i)}\right]$. In addition, the posterior density $p\left(\mathbf{x}_k, i \mid \mathbf{y}_k\right)$ can be rewritten as

$$p\left(\mathbf{x}_k, i \mid \mathbf{y}_k\right) \propto p\left(\mathbf{y}_k \mid \mathbf{x}_k\right) p\left(\mathbf{x}_k \mid \mathbf{x}_{k-1}^{(i)}\right) w_{k-1}^{(i)} \tag{3.36}$$

Using the sample pair $\left\{\mathbf{x}_k^{(j)}, i^j\right\}_{j=1}^{M_s}$, the weights are then updated by (Gustafsson 2010, Arulampalam et al. 2002)

$$w_k^{(j)} \propto \frac{p\left(\mathbf{x}_k^{(j)}, i^j \mid \mathbf{y}_k\right)}{q\left(\mathbf{x}_k^{(j)}, i^j \mid \mathbf{y}_k\right)} \propto \frac{p\left(\mathbf{y}_k \mid \mathbf{x}_k^{(j)}\right) p\left(\mathbf{x}_k^{(j)} \mid \mathbf{x}_{k-1}^{(i^j)}\right) w_{k-1}^{(i)}}{p\left(\mathbf{y}_k \mid \boldsymbol{\mu}_k^{(i^j)}\right) p\left(\mathbf{x}_k^{(j)} \mid \mathbf{x}_{k-1}^{(i^j)}\right) w_{k-1}^{(i)}} = \frac{p\left(\mathbf{y}_k \mid \mathbf{x}_k^{(j)}\right)}{p\left(\mathbf{y}_k \mid \boldsymbol{\mu}_k^{(i^j)}\right)} \tag{3.37}$$

Note that the measurement information is used in Eq. (3.37), which is helpful to improve the accuracy of the filter. If the process noise is small, the auxiliary sampling strategy usually performs better than the SIR because $p\left(\mathbf{x}_k \mid \mathbf{x}_{k-1}^{(i)}\right)$ is well characterized by $\boldsymbol{\mu}_k^{(i)}$. However, if the process noise is large, a single point $\boldsymbol{\mu}_k^{(i)}$ does not characterize $p\left(\mathbf{x}_k \mid \mathbf{x}_{k-1}^{(i)}\right)$ well and the auxiliary sampling strategy may degrade. To increase the diversity of the samples, the Markov chain Monte Carlo (MCMC) step can be used (Doucet and Johansen 2011).

Remark 3.3: The PMF and the particle filter (PF) both use the grid to represent the uncertainty distribution. However, PMF uses deterministic grids while the PF uses random grids. In addition, the new grid is specially designed for PMF. For PF, instead, the new grid is obtained by resampling (Gustafsson 2010, Arulampalam et al. 2002). The PF, roughly speaking, is easier to implement compared to the PMF, and thus is more widely used in engineering applications.

3.5 Combined Particle Filter

Instead of using the sequential Monte Carlo strategies alone, classical filtering technique can be combined with the PF to improve the filtering accuracy. As we know, the classical Kalman filter is optimal for linear Gaussian systems. Hence, if the system can be separated into two parts, the nonlinear and linear systems, the PF can work with the classical Kalman filter. In addition, for the

PF, the typical Gaussian approximation filter, such as the extended Kalman filter, can be used to provide the proposal distribution in order to obtain samples. In this section, we introduced these two techniques.

3.5.1 Marginalized Particle Filter

The marginalized particle filter (MPF) utilizes the potential linear Gaussian structure of the system. The state vector for the system is partitioned by $\mathbf{x}_k = \left[\left(\mathbf{x}_k^n\right)^T, \left(\mathbf{x}_k^l\right)^T\right]^T$ in which $\mathbf{x}_k^l$ and $\mathbf{x}_k^n$ are the corresponding linear and nonlinear states, respectively. Similarly, the process noise is partitioned by $\mathbf{v}_k = \left[\left(\mathbf{v}_k^n\right)^T, \left(\mathbf{v}_k^l\right)^T\right]^T$. The MPF can be summarized in Table 3.2.

Table 3.2 Steps of marginalized particle filter.

System Model:

$$\mathbf{x}_k^n = f_{k-1}^n\left(\mathbf{x}_{k-1}^n\right) + \mathbf{F}_{k-1}^n\left(\mathbf{x}_{k-1}^n\right)\mathbf{x}_{k-1}^l + \mathbf{G}_{k-1}^n\left(\mathbf{x}_{k-1}^n\right)\mathbf{v}_{k-1}^n \tag{3.38}$$

$$\mathbf{x}_k^l = f_{k-1}^l\left(\mathbf{x}_{k-1}^n\right) + \mathbf{F}_{k-1}^l\left(\mathbf{x}_{k-1}^n\right)\mathbf{x}_{k-1}^l + \mathbf{G}_{k-1}^l\left(\mathbf{x}_{k-1}^n\right)\mathbf{v}_{k-1}^l \tag{3.39}$$

$$\mathbf{y}_k = \boldsymbol{h}_k\left(\mathbf{x}_k^n\right) + \mathbf{H}_k\left(\mathbf{x}_k^n\right)\mathbf{x}_k^l + \mathbf{n}_k \tag{3.40}$$

Particle Filter Prediction:

For the nonlinear part, the PF can be used. When $p\left(\mathbf{x}_{k-1}^l \mid \mathbf{x}_{k-1}^n, \mathbf{y}_{k-1}\right)$ is given, the nonlinear state can be predicted using Eq. (3.38)

$$\begin{aligned} p\left(\mathbf{x}_k^n \mid \mathbf{y}_{k-1}\right) &= p\left(\mathbf{x}_{k-1}^n \mid \mathbf{y}_{k-1}\right) p\left(\mathbf{x}_k^n \mid \mathbf{x}_{k-1}^n, \mathbf{y}_{k-1}\right) \\ &= p\left(\mathbf{x}_{k-1}^n \mid \mathbf{y}_{k-1}\right) \int p\left(\mathbf{x}_k^n \mid \mathbf{x}_{k-1}^l, \mathbf{x}_{k-1}^n, \mathbf{y}_{k-1}\right) p\left(\mathbf{x}_{k-1}^l \mid \mathbf{x}_{k-1}^n, \mathbf{y}_{k-1}\right) d\mathbf{x}_{k-1}^l \end{aligned} \tag{3.41}$$

Note that $p\left(\mathbf{x}_{k-1}^l \mid \mathbf{x}_{k-1}^n, \mathbf{y}_{k-1}\right)$ is usually described by a Gaussian distribution.

Kalman Filter Prediction:

For each $\mathbf{x}_{k-1}^{n,(i)}$, the PDF of the linear state $\mathbf{x}_k^l$ can be predicted from $p\left(\mathbf{x}_{k-1}^l \mid \mathbf{x}_{k-1}^n, \mathbf{y}_{k-1}\right)$ using the Kalman filter and Eq. (3.39). Note that the predicted PDF $p\left(\mathbf{x}_k^l \mid \mathbf{x}_{k-1}^n, \mathbf{y}_{k-1}\right)$ $N\left(\mathbf{x}_k^l; \hat{\mathbf{x}}_{k|k-1}^l\left(\mathbf{x}_{k-1}^{n,(i)}\right), \mathbf{P}_{k|k-1}^l\left(\mathbf{x}_{k-1}^{n,(i)}\right)\right)$.

Model Interaction:

For each given particle $\mathbf{x}_{k-1}^{n,(i)}$, $p\left(\mathbf{x}_k^l \mid \mathbf{x}_{k-1}^n, \mathbf{y}_{k-1}\right)$ is updated as $p\left(\mathbf{x}_k^l \mid \mathbf{x}_k^n, \mathbf{y}_{k-1}\right)$ using the Kalman filter update equation if we treat Eq. (3.38) as an extra measurement equation (Hendeby et al. 2007).

Particle Filter Update:

Given $p\left(\mathbf{x}_k^l \mid \mathbf{x}_k^n, \mathbf{y}_{k-1}\right)$, the $p\left(\mathbf{x}_k^n \mid \mathbf{y}_{k-1}\right)$ can be updated using the PF and Eq. (3.40). The updated state distribution is denoted as $p\left(\mathbf{x}_k^n \mid \mathbf{y}_k\right)$.

Kalman Filter update:

For each particle $\mathbf{x}_k^{n,(i)}$, $p\left(\mathbf{x}_k^l \mid \mathbf{x}_k^n, \mathbf{y}_{k-1}\right)$ is updated using Eq. (3.40) and the Kalman filter. The updated state distribution is denoted as $p\left(\mathbf{x}_k^l \mid \mathbf{x}_k^n, \mathbf{y}_k\right)$.

After one filtering cycle, the state distribution is denoted by

$$p\left(\mathbf{x}_k^l, \mathbf{x}_k^n \mid \mathbf{y}_k\right) = p\left(\mathbf{x}_k^l \mid \mathbf{x}_k^n, \mathbf{y}_k\right) p\left(\mathbf{x}_k^n \mid \mathbf{y}_k\right) \tag{3.42}$$

Specifically, the nonlinear state distribution is described by particles.

$$p\left(\mathbf{x}_k^n \mid \mathbf{y}_k\right) \approx \sum_{i=1}^{N} w_k^{(i)} \delta\left(\mathbf{x}_k^n - \mathbf{x}_k^{n,(i)}\right) \tag{3.43}$$

where N is the number of particles and $w_k^{(i)}$ is the weight of the particle. Given particles of the nonlinear state, the $p\left(\mathbf{x}_k^l \mid \mathbf{y}_k\right)$ is approximated by

$$p\left(\mathbf{x}_k^l \mid \mathbf{y}_k\right) \approx \sum_{i=1}^{N} w_k^{(i)} N\left(\mathbf{x}_k^l; \hat{\mathbf{x}}_k^l\left(\mathbf{x}_k^{n,(i)}\right), \mathbf{P}_k^l\left(\mathbf{x}_k^{n,(i)}\right)\right) \tag{3.44}$$

Note that $\hat{\mathbf{x}}_k^l\left(\mathbf{x}_k^{n,(i)}\right)$ and $\mathbf{P}_k^l\left(\mathbf{x}_k^{n,(i)}\right)$ are the state mean and covariance for the linear part given particle $\mathbf{x}_k^{n,(i)}$. For MPF, the Kalman filter and particle filter are processed in an interleaved way. Hence, the structure of the system and the order to use the Kalman filter and particle filter are important. More details can be found in the references (Gustafsson 2010, Schon et al. 2005, Nordlund and Gustafsson 2009, Hendeby et al. 2007).

3.5.2 Gaussian Filter Aided Particle Filter

In Section 3.4, it is mentioned that the optimal proposal distribution considers the latest observation information. Alternatively, for each particle, the EKF or other Gaussian approximation filters can be used to provide the proposal distribution (Van Der Merwe et al. 2001). For each particle, the proposal distribution can be chosen as

$$q\left(\mathbf{x}_k^{(i)} \mid \mathbf{x}_{k-1}^{(i)}, \mathbf{y}_k\right) = N\left(\mathbf{x}_k; \hat{\mathbf{x}}_{k|k}^{(i)}, \mathbf{P}_{k|k}^{(i)}\right) \quad i = 1, \cdots, N_s \tag{3.45}$$

For the ith particle, the EKF or unscented Kalman filter (UKF) is employed to compute the mean value $\hat{\mathbf{x}}_{k|k}^{(i)}$ and covariance value $\mathbf{P}_{k|k}^{(i)}$. Then, $\mathbf{x}_k^{(i)}$ is sampled from the distribution $N\left(\mathbf{x}_k; \hat{\mathbf{x}}_{k|k}^{(i)}, \mathbf{P}_{k|k}^{(i)}\right)$.
The weights are given by

$$\tilde{w}_k^{(i)} \propto w_{k-1}^{(i)} \frac{p\left(\mathbf{y}_k \mid \mathbf{x}_{k|k-1}^{(i)}\right) p\left(\mathbf{x}_{k|k-1}^{(i)} \mid \mathbf{x}_{k-1|k-1}^{(i)}\right)}{N\left(\mathbf{x}_k; \hat{\mathbf{x}}_{k|k}^{(i)}, \mathbf{P}_{k|k}^{(i)}\right)} \tag{3.46}$$

The weights are then normalized to $w_k^{(i)}$.

Using the final points $w_k^{(i)}$ and $\hat{\mathbf{x}}_k^{(i)}$, the state estimation can be obtained. Note that the resampling may be necessary to avoid the sample depletion problem.

3.6 Ensemble Kalman Filter

For high dimensional estimation problems, the EKF is hard to use due to the computational complexity. To overcome this problem, the ensemble Kalman filter (EnKF) can be used. The EnKF also includes two steps: the prediction and update. Conventionally, these two steps for the EnKF are termed as the forecast step and analysis step, respectively, in the literature. Initially, we use $\mathbf{x}_{0,a}^{(1:N_s)}$ to denote the independent and identically distributed (iid) samples from the distribution $p(\mathbf{x}_0 \mid \mathbf{y}_0)$.

Forecast step:

The samples are then propagated via the dynamic equation by

$$\mathbf{x}_{k,f}^{(1:N_s)} = \boldsymbol{f}\left(\mathbf{x}_{k-1,a}^{(1:N_s)}\right) + \mathbf{q}_{k-1}^{(1:N_s)} \tag{3.47}$$

where $\mathbf{q}_{k-1}^{(1:N_s)}$ are the iid samples from the process noise distribution $N(\mathbf{v}_{k-1};\mathbf{0},\mathbf{Q}_{k-1})$.

Analysis step:

The samples are updated using the measurement value by

$$\mathbf{x}_{k,a}^{(1:N_s)} = \mathbf{x}_{k,f}^{(1:N_s)} + \overline{\mathbf{K}}_k\left(\mathbf{y}_k - \mathbf{r}_k^{(1:N_s)} - \boldsymbol{h}\left(\mathbf{x}_{k,f}^{(1:N_s)}\right)\right) \tag{3.48}$$

where $\mathbf{r}_k^{(1:N_s)}$ is the iid samples from the measurement noise distribution $N(\mathbf{n}_k;\mathbf{0},\mathbf{R}_k)$. The ensemble-estimated Kalman gain is given by

$$\overline{\mathbf{K}}_k = \overline{\mathbf{P}}_{k,f}\mathbf{H}_k^T\left(\mathbf{H}_k\overline{\mathbf{P}}_{k,f}\mathbf{H}_k^T + \mathbf{R}_k\right)^{-1} \tag{3.49}$$

where $\overline{\mathbf{P}}_{k,f}$ is the ensemble-estimate prediction covariance. $\mathbf{H}_k$ is the Jacobian matrix given by Eq. (3.9).

Equation (3.49) can be rewritten by (Raanes 2015)

$$\overline{\mathbf{K}}_k = \mathbf{A}_k\mathbf{Y}_k^T\left(\mathbf{Y}_k\mathbf{Y}_k^T + (N_s - 1)\mathbf{R}_k\right)^{-1} \tag{3.50}$$

where

$$\mathbf{A}_k = \mathbf{E}_{k,f}\left(\mathbf{I}_{N_s} - \mathbf{1}\mathbf{1}^T / N_s\right) \tag{3.51}$$

$$\mathbf{E}_{k,f} = \left[\mathbf{x}_{k,f}^1, \mathbf{x}_{k,f}^2, \cdots, \mathbf{x}_{k,f}^{N_s}\right] \tag{3.52}$$

$$\mathbf{Y}_k = \boldsymbol{h}\left(\mathbf{E}_{k,f}\right)\left(\mathbf{I}_{N_s} - \mathbf{1}\mathbf{1}^T / N_s\right) \tag{3.53}$$

$\mathbf{I}_{N_s}$ is the identity matrix with N_s diagonal elements. **1** is the vector of ones of length N_s.

***Remark 3.4*:** By using Eq. (3.50), the ensemble-estimate prediction covariance $\overline{\mathbf{P}}_{k,f}$ and the Jacobian matrix $\mathbf{H}_k$ are not explicitly computed. Hence, the computational complexity is reduced. If necessary, the covariance can be obtained from $\mathbf{x}_{k,a}^{(1:N_s)}$.

Note that the EnKF is often used in large-scale estimation systems, such as the problems with the state dimension on the order of 109 (Raanes 2015), where the original Kalman filter is hard to use. The size of the ensemble N_s is often much smaller than the system dimension n_x. Roughly speaking, the more samples are used, the more accurate estimation can be obtained. Experiments show that 10 to 100 samples are sufficient for the large geoscientific model (Raanes 2015). In addition, unlike the EKF, computation of the Jacobian matrix of the dynamic system or the measurement equation is not necessary. Moreover, the EnKF is highly parallelizable, which can be very efficient in implementation.

3.7 Zakai Filter and Fokker Planck Equation

In this section, two continuous-time nonlinear filtering techniques are presented. We consider the continuous-time dynamic system given by (Bao et al. 2014)

$$d\mathbf{x} = \boldsymbol{f}(t,\mathbf{x})dt + \boldsymbol{g}(t,\mathbf{x})d\mathbf{v}(t) \tag{3.54}$$

$$d\mathbf{y} = \boldsymbol{h}(t,\mathbf{x})dt + d\mathbf{n}(t) \tag{3.55}$$

where $\mathbf{v}(t)$ represents the zero mean Brownian motion process with the correlation function $\mathbf{Q}\delta(t_1 - t_2)$.

The Zakai filtering method solves the Zakai equation to represent the PDF of the nonlinear filtering solution. Many methods have been proposed to solve the Zakai equation. A typical numerical approximation method, the splitting-up approximation scheme, was used in (Bensoussan et al. 1992, Bao et al. 2014). The Zakai equation is split into the second order PDEs. By solving the second order PDEs, the posterior PDF can be obtained. The numerical approximation to the Zakai equation often uses the grid-based method to represent the PDF. For high dimensional nonlinear filtering problems, the number of points used to represent the PDF is large. To alleviate the increase of points with the increase of dimension, the hierarchical sparse grid method is used in (Bao et al. 2014). The Zakai equation is given by (Bao et al. 2014)

$$dp(t,\mathbf{x}) = L^T p(t,\mathbf{x})dt + \boldsymbol{h}^T(t,\mathbf{x})p(t,\mathbf{x})d\mathbf{y} \tag{3.56}$$

where

$$Lp(t,\mathbf{x}\mid\mathbf{y})=\frac{1}{2}\sum_{i,j=1}^{n}\left(\boldsymbol{g}\boldsymbol{g}^{T}\right)_{i,j}\frac{\partial^2}{\partial x_i\partial x_j}p(t,\mathbf{x}\mid\mathbf{y})+\sum_{i=1}^{n}f_i\frac{\partial}{\partial x_i}p(t,\mathbf{x}\mid\mathbf{y}) \quad (3.57)$$

Note that the superscript 'T' denotes the transpose operation.

As an alternative description of the nonlinear filter, the Fokker-Planck equation (FPE) describes the propagation/prediction phase of the PDF (Kalender and Schottl 2013). However, solving FPE is tedious. In (Beard et al. 1999), the FPE was solved via the Galerkin projection. The measurement update is based on the Bayesian rule. The solver of FPE in (Beard et al. 1999), however, is hard to use for high-dimensional nonlinear filtering problems. In (Kumar and Chakravorty 2012), the FPE was solved by the meshless particle partition of the unity finite element technique in near-real time. The measurement update uses the Bayesian rule with the Markov-Chain Monte Carlo (MCMC) sampling technique.

Note that the Chapman-Kolmogorov equation (CKE) for the discrete-time system is reduced to FPE for the continuous-time dynamic system. The FPE is a parabolic partial differential equation, which is given by (Kumar and Chakravorty 2012)

$$\frac{\partial}{\partial t}p(t,\mathbf{x}\mid\mathbf{y})=L_{FP}\,p(t,\mathbf{x}\mid\mathbf{y}) \quad (3.58)$$

where

$$L_{FP}=-\sum_{i=1}^{n}\frac{\partial}{\partial x_i}D_i^{(1)}(\cdot,\cdot)+\sum_{i,j=1}^{n}\frac{\partial^2}{\partial x_i\partial x_j}D_{i,j}^{(2)}(\cdot,\cdot) \quad (3.59)$$

$$D^{(1)}(t,\mathbf{x})=\boldsymbol{f}(t,\mathbf{x})+\frac{1}{2}\frac{\partial \boldsymbol{g}(t,\mathbf{x})}{\partial \mathbf{x}}\mathbf{Q}\boldsymbol{g}(t,\mathbf{x}) \quad (3.60)$$

$$D^{(2)}(t,\mathbf{x})=\boldsymbol{g}(t,\mathbf{x})\mathbf{Q}\boldsymbol{g}(t,\mathbf{x}) \quad (3.61)$$

L_{FP} is the Fokker-Planck operator. $D^{(1)}$ is the drift coefficient vector and $D^{(2)}$ is the diffusion coefficient matrix.

3.8 Summary

The Kalman filter and EKF are the most widely used for real applications. For the general nonlinear estimation problem, the point-mass filter and the particle filter are reviewed. The point-mass filter requires the sophisticated design of the grid while the particle filter utilizes Monte Carlo sampling and the sampling method is of utmost importance. Both filters aim to represent the posterior PDF and are computationally intensive, especially for high-dimensional estimation problems. One advantage of such filters is that both filters work for nonlinear non-Gaussian estimation problems. The particle filter has the sample depletion problem. The effective number of samples becomes very small with the increase of time. Hence, many samples become useless and contribute trivial information to the estimation results. Roughly speaking, the particle filter works well when the noise is high. The low measurement noise may lead to the narrow likelihood function, which makes many particles useless. Besides using different proposal distributions and integrating the particle filter with other nonlinear filters, the resampling scheme is revised in (Reich 2013) to alleviate this problem. Ensemble Kalman filter is introduced since it is very effective to solve large-scale estimation problems. Zakai equation and Fokker Planck equation based filters provide rigorous frameworks to solve continuous-time estimation problems. However, these equations are usually hard to solve. In the next chapter, we will focus on the Gaussian nonlinear filters that can yield many accurate and efficient approximate solutions to the nonlinear estimation problems.

References

Arulampalam, M.S., S. Maskell, N. Gordon and T. Clapp. 2002. A tutorial on particle filters for online nonlinear/non-Gaussian Bayesian tracking. IEEE Transactions on Signal Processing 50: 174–188.

Bao, F., Y. Cao, C. Webster and G. Zhang. 2014. A hybrid sparse-grid approach for nonlinear filtering problems based on adaptive-domain of the Zakai equation approximations. SIAM/ASA Journal on Uncertainty Quantification 2: 784–804.

Beard, R., J. Kenney, J. Gunther, J. Lawton and W. Stirling. 1999. Nonlinear projection filter based on Galerkin approximation. Journal of Guidance, Control, and Dynamics 22: 258–266.

Bensoussan, A., R. Glowinski and A. Raşcanu. 1992. Approximation of some stochastic differential equations by the splitting up method. Applied Mathematics and Optimization 25: 81–106.

Doucet, A., and A.M. Johansen. 2011. A tutorial on particle filtering and smoothing: fifteen years later. pp. 656–704. *In*: Crisan, D. and B. Rozovskii (eds.). The Oxford Handbook of Nonlinear Filtering. Oxford University Press, Oxford.

Gelb, A. 1974. Applied Optimal Estimation. MIT Press, Cambridge.

Gustafsson, F. 2010. Particle filter theory and practice with positioning applications. IEEE Aerospace and Electronic Systems Magazine 25: 53–82.

Hendeby, G., R. Karlsson and F. Gustafsson. 2007. A new formulation of the Rao-Blackwellized particle filter. IEEE/SP 14th Workshop on Statistical Signal Processing. USA, 26–29.

Kalender, C. and A. Schottl. 2013. Sparse grid-based nonlinear filtering. IEEE Transactions on Aerospace and Electronic Systems 49: 2386–2396.

Kumar, M. and S. Chakravorty. 2012. Nonlinear filter based on the Fokker-Planck equation. Journal of Guidance, Control, and Dynamics 35: 68–79.

Nordlund, P.J. and F. Gustafsson. 2009. Marginalized particle filter for accurate and reliable terrain-aided navigation. IEEE Transactions on Aerospace and Electronic Systems 45: 1385–1399.

Raanes, P.N. 2015. Improvements to ensemble methods for data assimilation in the geosciences. M.S. Thesis, University of Oxford, Oxford, England.

Reich, S. 2013. A nonparametric ensemble transform method for Bayesian inference. SIAM Journal on Scientific Computing 35: A2013–A2024.

Schon, T., F. Gustafsson and P.J. Nordlund. 2005. Marginalized particle filters for mixed linear/nonlinear state-space models. IEEE Transactions on Signal Processing 53: 2279–2289.

Simandl, M., J. Kralovec and T. Soderstrom. 2002. Anticipative grid design in point-mass approach to nonlinear state estimation. IEEE Transactions on Automatic Control 47: 699–702.

Šimandl, M., J. Královec and T. Söderström. 2006. Advanced point-mass method for nonlinear state estimation. Automatica 42: 1133–1145.

Van Der Merwe, R., A. Doucet, N.D. Freitas and E.A. Wan. 2001. The unscented particle filter. Proceedings of the 13th International Conference on Neural Information Processing Systems. USA, 563–569.

4 Grid-based Gaussian Nonlinear Estimation

In this chapter, we present another class of approximate Bayesian estimation algorithms under the Gaussian assumption. It can significantly improve the accuracy of the linearization based nonlinear filters such as EKF schemes. Besides, it is simpler to implement than the sequential Monte Carlo methods such as the particle filters. This class of nonlinear Gaussian filters is a statistical transformation approach based on the perception that "it is easier to approximate a probability distribution, than to approximate an arbitrary nonlinear function of transformation" (Julier and Uhlmann 2004). The EKF and its variants are based on linearizing nonlinear functions of the state and measurements to provide estimates of the underlying statistics, while the statistical transformation approach is based on a set of selected sample points that capture statistical properties of the underlying distribution. This set of sample points is nonlinearly transformed or propagated to a new space. The statistics of the new samples are then calculated to provide the required estimates. Nevertheless, in terms of using sample points to approximate the statistics, this method is fundamentally different from the particle filter, in which random samples are drawn from the prior distribution and updated through the likelihood function to produce a sample from the posterior distribution. Here, the samples are not drawn at random, but according to certain deterministic numerical rules. It still keeps the recursive Bayesian filtering framework with prediction and update steps but provides easy-to-implement filtering algorithms.

Under the Gaussian assumption, the filtering problem reduces to the estimation of the mean and covariance of the state vector, which involves evaluation of the Gaussian-weighted integrals. Various numerical integration rules can be applied to generate deterministic sampling points to approximate the Gaussian-weighted integrals. The remarkable merit of these Gaussian approximation filters is that the estimation accuracy can be readily analyzed, predicted, and controlled because there is a suite of numerical rules of different accuracy degrees to select for practical estimation problems according to the filtering accuracy requirement and computational constraints.

4.1 General Gaussian Approximation Nonlinear Filter

We consider a class of nonlinear discrete-time dynamical systems described by:

$$\mathbf{x}_k = \boldsymbol{f}(\mathbf{x}_{k-1}) + \mathbf{v}_{k-1} \tag{4.1}$$

$$\mathbf{y}_k = \boldsymbol{h}(\mathbf{x}_k) + \mathbf{n}_k \tag{4.2}$$

where $\mathbf{x}_k \in \mathbb{R}^n; \mathbf{y}_k \in \mathbb{R}^m$; $\mathbf{v}_{k-1}$ and $\mathbf{n}_k$ are independent white Gaussian process noise and measurement noise with covariance $\mathbf{Q}_{k-1}$ and $\mathbf{R}_k$, respectively.

Recall that the optimal filtering problem in the sense of minimizing MSE is to find the conditional mean $\hat{\mathbf{x}}_{k|k}$

$$\hat{\mathbf{x}}_{k|k} = E[\mathbf{x}_k \mid \mathbf{y}_{1:k}] = \int \mathbf{x}_k p(\mathbf{x}_k \mid \mathbf{y}_{1:k}) d\mathbf{x}_k \tag{4.3}$$

The corresponding covariance $\mathbf{P}_{k|k}$ is needed to evaluate the accuracy of the state estimation.

$$\mathbf{P}_{k|k} = \int (\mathbf{x}_k - \hat{\mathbf{x}}_k)(\mathbf{x}_k - \hat{\mathbf{x}}_k)^T p(\mathbf{x}_k \mid \mathbf{y}_{1:k}) d\mathbf{x}_k \tag{4.4}$$

Under the Gaussian assumption, only the mean and covariance are required to be predicted and updated.

The state and covariance predictions are respectively given by

$$\begin{aligned}
\hat{\mathbf{x}}_{k|k-1} &= E(\mathbf{x}_{k|k-1}) \\
&= \int \mathbf{x}_{k|k-1} p(\mathbf{x}_{k|k-1}) d\mathbf{x}_{k|k-1} \\
&= \int \left(\int \mathbf{x}_{k|k-1} N(\mathbf{x}_{k|k-1}; \boldsymbol{f}(\mathbf{x}_{k-1|k-1}), \mathbf{Q}_{k-1}) d\mathbf{x}_{k|k-1} \right) \\
&\quad N(\mathbf{x}_{k-1|k-1}; \hat{\mathbf{x}}_{k-1|k-1}, \mathbf{P}_{k-1|k-1}) d\mathbf{x}_{k-1|k-1} \\
&= \int \boldsymbol{f}(\mathbf{x}_{k-1|k-1}) N(\mathbf{x}_{k-1|k-1}; \hat{\mathbf{x}}_{k-1|k-1}, \mathbf{P}_{k-1|k-1}) d\mathbf{x}_{k-1|k-1}
\end{aligned} \tag{4.5}$$

$$\begin{aligned}\mathbf{P}_{k|k-1} &= E\left[\left(\mathbf{x}_{k|k-1}-\hat{\mathbf{x}}_{k|k-1}\right)\left(\mathbf{x}_{k|k-1}-\hat{\mathbf{x}}_{k|k-1}\right)^T\right]\\ &= \int\left(\mathbf{x}_{k|k-1}-\hat{\mathbf{x}}_{k|k-1}\right)\left(\mathbf{x}_{k|k-1}-\hat{\mathbf{x}}_{k|k-1}\right)^T p\left(\mathbf{x}_{k|k-1}\right)d\mathbf{x}_{k|k-1}\\ &= \int\left(\int\left(\mathbf{x}_{k-1|k-1}-\hat{\mathbf{x}}_{k|k-1}\right)\left(\mathbf{x}_{k-1|k-1}-\hat{\mathbf{x}}_{k|k-1}\right)^T\right.\\ &\quad \left. N\left(\mathbf{x}_{k|k-1};\boldsymbol{f}\left(\mathbf{x}_{k-1|k-1}\right),\mathbf{Q}_{k-1}\right)d\mathbf{x}_{k|k-1}\right)p\left(\mathbf{x}_{k-1|k-1}\right)d\mathbf{x}_{k-1|k-1}\\ &= \int\left(\boldsymbol{f}\left(\mathbf{x}_{k-1|k-1}\right)-\hat{\mathbf{x}}_{k|k-1}\right)\left(\boldsymbol{f}\left(\mathbf{x}_{k-1|k-1}\right)-\hat{\mathbf{x}}_{k|k-1}\right)^T\\ &\quad N\left(\mathbf{x}_{k-1|k-1};\hat{\mathbf{x}}_{k-1|k-1},\mathbf{P}_{k-1|k-1}\right)d\mathbf{x}_{k-1|k-1}+\mathbf{Q}_{k-1}\end{aligned} \tag{4.6}$$

Next, similar to the Kalman filter, the update equations of the Gaussian approximation filter are given by (Ito and Xiong 2000).

$$\hat{\mathbf{x}}_{k|k} = \hat{\mathbf{x}}_{k|k-1} + \mathbf{K}_k\left(\mathbf{y}_k - \hat{\mathbf{y}}_k\right) \tag{4.7}$$

$$\mathbf{P}_{k|k} = \mathbf{P}_{k|k-1} - \mathbf{K}_k\mathbf{P}_{xy,k}^T \tag{4.8}$$

where

$$\mathbf{K}_k = \mathbf{P}_{xy,k}\left(\mathbf{R}_k + \mathbf{P}_{yy,k}\right)^{-1} \tag{4.9}$$

$$\begin{aligned}\hat{\mathbf{y}}_k &= E\left[\mathbf{y}_k\right]\\ &= \int \boldsymbol{h}\left(\mathbf{x}_{k|k-1}\right)N\left(\mathbf{x}_{k|k-1};\hat{\mathbf{x}}_{k|k-1},\mathbf{P}_{k|k-1}\right)d\mathbf{x}_{k|k-1}\end{aligned} \tag{4.10}$$

$$\begin{aligned}\mathbf{P}_{yy,k} &= E\left[\left(\mathbf{y}_k-\hat{\mathbf{y}}_k\right)\left(\mathbf{y}_k-\hat{\mathbf{y}}_k\right)^T\right]\\ &= \int\left(\boldsymbol{h}\left(\mathbf{x}_{k|k-1}\right)-\hat{\mathbf{y}}_k\right)\left(\boldsymbol{h}\left(\mathbf{x}_{k|k-1}\right)-\hat{\mathbf{y}}_k\right)^T N\left(\mathbf{x}_{k|k-1};\hat{\mathbf{x}}_{k|k-1},\mathbf{P}_{k|k-1}\right)d\mathbf{x}_{k|k-1}\end{aligned} \tag{4.11}$$

$$\begin{aligned}\mathbf{P}_{xy,k} &= E\left[\left(\mathbf{x}_{k|k-1}-\hat{\mathbf{x}}_{k|k-1}\right)\left(\mathbf{y}_k-\hat{\mathbf{y}}_k\right)^T\right]\\ &= \int\left(\mathbf{x}_{k|k-1}-\hat{\mathbf{x}}_{k|k-1}\right)\left(\boldsymbol{h}\left(\mathbf{x}_{k|k-1}\right)-\hat{\mathbf{y}}_k\right)^T N\left(\mathbf{x}_{k|k-1};\hat{\mathbf{x}}_{k|k-1},\mathbf{P}_{k|k-1}\right)d\mathbf{x}_{k|k-1}\end{aligned} \tag{4.12}$$

It can be seen that the Gaussian weighted integrals need to be calculated in Eqs. (4.5), (4.6) and (4.10)–(4.12) in order to obtain the estimate under the Gaussian assumption.

Generally, analytical solutions to these Gaussian weighted integrals are intractable. There are many different numerical techniques to compute them, though, including the Gauss-Hermite Quadrature (GHQ) rule (Ito and Xiong 2000), Unscented Transformation (UT) (Julier and Uhlmann 2004) and Spherical-Radial Cubature Rule (Arasaratnam and Haykin 2009). The main idea of such rules is to approximate the integrals by summation of weighted functional evaluations on selected quadrature points.

Considering the following integral,

$$I(g) = \int_{\mathbb{R}^n} g(\mathbf{x}) N(\mathbf{x}; \mathbf{0}, \mathbf{I}) d\mathbf{x}. \tag{4.13}$$

This is an integral of a general nonlinear function $g(\mathbf{x})$ with respect to a normal distribution with zero mean and unity covariance. It can be approximated by a quadrature rule

$$I(g) \approx \sum_{i=1}^{N_p} W_i g(\gamma_i), \tag{4.14}$$

where γ_i and W_i are the quadrature points and weights, respectively. N_p is the number of points.

For an integral with respect to a more general Gaussian distribution, we can convert it to the standard integral (4.13) by an affine transformation

$$\int_{\mathbb{R}^n} g(\mathbf{x}) N(\mathbf{x}; \hat{\mathbf{x}}, \mathbf{P}) d\mathbf{x} = \int_{\mathbb{R}^n} g(\mathbf{z}) N(\mathbf{z}; \mathbf{0}, \mathbf{I}) d\mathbf{z} \tag{4.15}$$

where $\mathbf{z} = \mathbf{S}\mathbf{x} + \hat{\mathbf{x}}$ and $\mathbf{P} = \mathbf{S}\mathbf{S}^T$ (see Chapter 1.2). Then, the quadrature approximation (4.14) can be applied.

With the quadrature approximation, the prediction step and the update step of the nonlinear Gaussian filter can be rewritten as follows, given $\hat{\mathbf{x}}_{k-1|k-1}$ and $\mathbf{P}_{k-1|k-1}$.

Prediction:

$$\hat{\mathbf{x}}_{k|k-1} = \sum_{i=1}^{N_p} W_i f(\xi_{k-1,i}) \tag{4.16}$$

$$\mathbf{P}_{k|k-1} = \sum_{i=1}^{N_p} W_i \left[f(\xi_{k-1,i}) - \hat{\mathbf{x}}_{k|k-1} \right] \left[f(\xi_{k-1,i}) - \hat{\mathbf{x}}_{k|k-1} \right]^T + \mathbf{Q}_{k-1} \tag{4.17}$$

where N_p is the total number of quadrature points; $\xi_{k-1,i}$ is the transformed quadrature point obtained by

$$\xi_{k-1,i} = \mathbf{S}_{k-1} \gamma_i + \hat{\mathbf{x}}_{k-1|k-1}; \quad \mathbf{P}_{k-1|k-1} = \mathbf{S}_{k-1} \mathbf{S}_{k-1}^T \tag{4.18}$$

Update:

$$\hat{\mathbf{x}}_{k|k} = \hat{\mathbf{x}}_{k|k-1} + \mathbf{K}_k (\mathbf{y}_k - \hat{\mathbf{y}}_k) \tag{4.19}$$

$$\mathbf{P}_{k|k} = \mathbf{P}_{k|k-1} - \mathbf{K}_k \mathbf{P}_{xy}^T \tag{4.20}$$

where

$$\hat{\mathbf{y}}_k = \sum_{i=1}^{N_p} W_i \boldsymbol{h}\left(\tilde{\xi}_{k,i}\right) \tag{4.21}$$

$$\mathbf{P}_{xy} = \sum_{i=1}^{N_p} W_i \left[\tilde{\xi}_{k,i} - \hat{\mathbf{x}}_{k|k-1}\right]\left[\boldsymbol{h}\left(\tilde{\xi}_{k,i}\right) - \hat{\mathbf{y}}_k\right]^T \tag{4.22}$$

$$\mathbf{P}_{yy} = \sum_{i=1}^{N_p} W_i \left(\boldsymbol{h}\left(\tilde{\xi}_{k,i}\right) - \hat{\mathbf{y}}_k\right)\left(\boldsymbol{h}\left(\tilde{\xi}_{k,i}\right) - \hat{\mathbf{y}}_k\right)^T \tag{4.23}$$

$\tilde{\xi}_{k,i}$ is the transformed quadrature point using the predicted state and covariance

$$\tilde{\xi}_{k,i} = \tilde{\mathbf{S}}_k \boldsymbol{\gamma}_i + \hat{\mathbf{x}}_{k|k-1};\ \mathbf{P}_{k|k-1} = \tilde{\mathbf{S}}_k \tilde{\mathbf{S}}_k^T \tag{4.24}$$

Using different numerical rules to compute $\boldsymbol{\gamma}_i$ and W_i in the above algorithm leads to different nonlinear Gaussian approximation filters.

Remark 4.1: The covariance decomposition in (4.18) and (4.24) can be done by Cholesky decomposition or singular value decomposition (SVD).

4.2 General Gaussian Approximation Nonlinear Smoother

There are two alternative forms of the optimal Bayesian smoothing equation: two-filter smoother and forward-backward smoother (Särkkä 2008). The forward-backward smoother is introduced here and the form is described as follows.

$$p(\mathbf{x}_k \mid \mathbf{y}_{1:T}) = p(\mathbf{x}_k \mid \mathbf{y}_{1:k}) \int \frac{p(\mathbf{x}_{k+1} \mid \mathbf{x}_k)\, p(\mathbf{x}_{k+1} \mid \mathbf{y}_{1:T})}{p(\mathbf{x}_{k+1} \mid \mathbf{y}_{1:k})} d\mathbf{x}_{k+1} \tag{4.25}$$

where $p(\mathbf{x}_k \mid \mathbf{y}_{1:T})$ and $p(\mathbf{x}_k \mid \mathbf{y}_{1:k})$ are the smoothed PDF and the filtering PDF at the time k, respectively; $p(\mathbf{x}_{k+1} \mid \mathbf{y}_{1:k})$ is the predicted PDF at the time k. The smoothing state can be recursively obtained backward from the last time $k = T$.

Due to the difficulty of integration, Eq. (4.25) is hard to implement in practice. Hence, the following approximation is often used (Särkkä 2008).

When $\mathbf{x}_{k+1|k+1}$ and $\mathbf{y}_{1:k}$ are given, the conditional distribution of $\mathbf{x}_{k|k}$ is given by

$$p(\mathbf{x}_k \mid \mathbf{x}_{k+1}, \mathbf{y}_{1:k}) = \frac{p(\mathbf{x}_k, \mathbf{x}_{k+1} \mid \mathbf{y}_{1:k})}{p(\mathbf{x}_{k+1} \mid \mathbf{y}_{1:k})} \tag{4.26}$$

where $p(\mathbf{x}_{k+1} \mid \mathbf{y}_{1:k})$ can be obtained by

$$p(\mathbf{x}_{k+1} \mid \mathbf{y}_{1:k}) = \int p(\mathbf{x}_{k+1} \mid \mathbf{x}_k)\, p(\mathbf{x}_k \mid \mathbf{y}_{1:k}) d\mathbf{x}_k \tag{4.27}$$

Note, $p(\mathbf{x}_k \mid \mathbf{x}_{k+1}, \mathbf{y}_{1:k}) = p(\mathbf{x}_k \mid \mathbf{x}_{k+1}, \mathbf{y}_{1:T})$, due to the Markov property of the state-space model (Särkkä 2008).

Hence,

$$p(\mathbf{x}_k, \mathbf{x}_{k+1} \mid \mathbf{y}_{1:T}) = p(\mathbf{x}_k \mid \mathbf{x}_{k+1}, \mathbf{y}_{1:T}) p(\mathbf{x}_{k+1} \mid \mathbf{y}_{1:T}) \tag{4.28}$$

By marginalizing the joint distribution $p(\mathbf{x}_k, \mathbf{x}_{k+1} \mid \mathbf{y}_{1:T})$ over $\mathbf{x}_{k+1}$, $p(\mathbf{x}_k \mid \mathbf{y}_{1:T})$ can be obtained as

$$p(\mathbf{x}_k \mid \mathbf{y}_{1:T}) = \int p(\mathbf{x}_k \mid \mathbf{x}_{k+1}, \mathbf{y}_{1:T}) p(\mathbf{x}_{k+1} \mid \mathbf{y}_{1:T}) d\mathbf{x}_{k+1} \tag{4.29}$$

We assume $p(\mathbf{x}_k \mid \mathbf{y}_{1:T})$ is a Gaussian distribution $N(\mathbf{x}_k; \hat{\mathbf{x}}_k^s, \mathbf{P}_k^s)$. After some algebra, by Eqs. (4.26)–(4.29), the $\hat{\mathbf{x}}_k^s$ and $\mathbf{P}_k^s$ can be obtained as follows (Särkkä 2008).

$$\hat{\mathbf{x}}_k^s = \hat{\mathbf{x}}_{k|k} + \mathbf{D}_k \left(\hat{\mathbf{x}}_{k+1}^s - \hat{\mathbf{x}}_{k+1|k} \right) \tag{4.30}$$

$$\mathbf{P}_k^s = \mathbf{P}_k + \mathbf{D}_k \left(\mathbf{P}_{k+1}^s - \mathbf{P}_{k+1|k} \right) \mathbf{D}_k^T \tag{4.31}$$

where

$$\mathbf{D}_k = \mathbf{C}_{k+1} \mathbf{P}_{k+1|k}^{-1} \tag{4.32}$$

$$\mathbf{C}_{k+1} = \sum_{i=1}^{N_p} W_i \left(\xi_{k,i} - \hat{\mathbf{x}}_{k|k} \right) \left(f(\xi_{k,i}) - \hat{\mathbf{x}}_{k+1|k} \right)^T \tag{4.33}$$

Note that the recursive smoother starts from the last time step T, i.e., $\hat{\mathbf{x}}_T^s = \hat{\mathbf{x}}_{T|T}$ and $\mathbf{P}_T^s = \mathbf{P}_{T|T}$.

As can be seen, the Gaussian weighted integration is important for both Bayesian filter and smoother. As mentioned in Section 4.1, many numerical rules, such as the UT (Julier and Uhlmann 2004), the GHQ rule (Ito and Xiong 2000), the sparse-grid quadrature (SGQ) rule (Jia et al. 2012b), and the spherical-radial cubature rule (Arasaratnam and Haykin 2009, Jia et al. 2013) can be applied to approximate these integrals. In the following sections, each of these numerical rules is introduced.

4.3 Unscented Transformation

The unscented transformation (UT) in the unscented Kalman filter (UKF) is frequently used as an alternative method to predict the uncertainty after the nonlinear mapping. It uses the so-called sigma points to calculate the mean and covariance of the Gaussian distribution. The derivation of the UT has been given in many references (Julier 2003, Julier and Uhlmann 2004, Julier and Uhlmann 2002).

Using the UT, the quadrature points and weights are given by Eqs. (4.34) and (4.35), respectively.

$$\boldsymbol{\gamma}_i = \begin{cases} \mathbf{0} & i = 1 \\ \sqrt{n+\kappa}\,\mathbf{e}_{i-1} & i = 2, \cdots n+1 \\ -\sqrt{n+\kappa}\,\mathbf{e}_{i-n-1} & i = n+2, \cdots, 2n+1 \end{cases} \tag{4.34}$$

$$W_i = \begin{cases} \dfrac{\kappa}{n+\kappa} & i=1 \\ \dfrac{1}{2(n+\kappa)} & i=2,\cdots,2n+1 \end{cases} \tag{4.35}$$

where $\mathbf{e}_{i-1}$ is the unit vector in $\mathbb{R}^n$ with the $(i-1)$th element being 1 and κ is a tunable parameter of which the suggested optimal value for Gaussian distributions is $\kappa = 3 - n$. When the points (4.34) and weights (4.35) are used in the nonlinear Gaussian filtering algorithm (4.16)–(4.24), it leads to the UKF. Note that the number of points using the UT increases linearly with the increase of the dimension n. The UKF using the UT can achieve the same performance as the Kalman filter for the linear Gaussian system. Different forms of the UT are also available (Julier 2003, Julier and Uhlmann 2004).

4.4 Gauss-Hermite Quadrature

The Gauss-Hermite Quadrature (GHQ) is a numerical rule to approximate the Gaussian integral. For one dimensional (univariate) GHQ, different degrees of accuracy can be achieved by using a different number of quadrature points. For convenience, we use m_q to denote the number of univariate points, γ and w to denote the point and weight, respectively. The univariate GHQ point and weight can be calculated as follows.

If $m_q = 1$, then $\gamma_1 = 0$ and $w_1 = 1$. If $m_q > 1$, first construct a symmetric tri-diagonal matrix J with zero diagonal elements and $J_{i,i+1} = J_{i+1,i} = \sqrt{i/2}, 1 \le i \le m_q - 1$. Then the ith quadrature point γ_i is calculated by $\gamma_i = \sqrt{2}\varepsilon_i$, where ε_i is the ith eigenvalue of J. The corresponding weight w_i is calculated by $w_i = (\mathbf{v}_i)_1^2$ where $(\mathbf{v}_i)_1$ is the first element of the ith normalized eigenvector of J. The univariate GHQ rule with m_q quadrature points is exact up to the $(2m_q - 1)$th order of polynomials (Ito and Xiong 2000). The multivariate GHQ rule extends the above univariate GHQ rule by the direct tensor product (Ito and Xiong 2000). For example, the two dimensional GHQ points $\boldsymbol{\gamma}$ and weights W constructed from 3 univariate GHQ points $\gamma_1, \gamma_2, \gamma_3$ with corresponding weights w_1, w_2, w_3 by the tensor product rule contain $3^2 = 9$ points

$$\boldsymbol{\gamma} = \begin{Bmatrix} (\gamma_1,\gamma_1) & (\gamma_1,\gamma_2) & (\gamma_1,\gamma_3) \\ (\gamma_2,\gamma_1) & (\gamma_2,\gamma_2) & (\gamma_2,\gamma_3) \\ (\gamma_3,\gamma_1) & (\gamma_3,\gamma_2) & (\gamma_3,\gamma_3) \end{Bmatrix} \tag{4.36}$$

and 9 corresponding weights

$$W = \begin{Bmatrix} w_1 \cdot w_1 & w_1 \cdot w_2 & w_1 \cdot w_3 \\ w_2 \cdot w_1 & w_2 \cdot w_2 & w_2 \cdot w_3 \\ w_3 \cdot w_1 & w_3 \cdot w_2 & w_3 \cdot w_3 \end{Bmatrix} \tag{4.37}$$

where $\gamma_1 = -\sqrt{3}$, $\gamma_2 = 0$, and $\gamma_3 = \sqrt{3}$; $w_1 = 1/6$, $w_2 = 2/3$, and $w_3 = 2/3$. The multivariate GHQ rule is exact for all polynomials of the form $x_1^{i_1} x_2^{i_1} \cdots x_n^{i_n}$ with $1 \le i_j \le 2m_q - 1$ (Ito and Xiong 2000). However, the total number of points $N_p = (m_q)^n$ increases exponentially with the dimension n, which is the curse of dimensionality problem.

The GHQ is, in general, more accurate than other quadrature rules. It can be utilized as the starting point or benchmark to optimize other quadrature rules.

4.5 Sparse-Grid Quadrature

The sparse-grid quadrature (SGQ) is based on the Smolyak's rule for multivariate extension of the univariate quadrature rule and integration operators. The Smolyak rule is given by (Nobile et al. 2008b, Heiss and Winschel 2008)

$$\int_{\mathbb{R}^n} f(\mathbf{x}) N(\mathbf{x}; \mathbf{0}, \mathbf{I}_n) d\mathbf{x} \approx I_{n,L}(f) \triangleq \sum_{\Xi \in \Upsilon_{n,L}} \left(\Delta^{i_1} \otimes \cdots \otimes \Delta^{i_n}\right)(f) \tag{4.38}$$

where $I_{n,L}(f)$ is the multi-dimensional integral of the function f with respect to a standard normal distribution with the accuracy level L, $L \in \mathbb{N}$; $\mathbb{N}$ is the set of natural numbers. By accuracy level L, it means that $I_{n,L}(f)$ is exact for all polynomials of the form $x_1^{i_1} x_2^{i_2} \cdots x_n^{i_n}$ with $\sum_{j=1}^{n} i_j \le 2L - 1$ (Heiss and Winschel 2008). The operator '$\otimes$' denotes the tensor product and $\Delta^{i_j} = I_{i_j} - I_{i_j - 1}$, $\forall i_j \in \mathbb{N}, (j = 1, \cdots, n)$, where $I_0 = 0$ and

$$I_{i_j} \triangleq \sum_{\gamma_{s_j} \in \mathbf{X}_{i_j}} f\left(\gamma_{s_j}\right) w_{s_j} \approx \int_{\mathbb{R}} f(x) N(x; 0, 1) dx \quad (s_j \in \mathbb{N}) \tag{4.39}$$

is the univariate quadrature rule with the accuracy level $i_j \in \Xi$, $\Xi \triangleq (i_1, \cdots, i_n)$ is the accuracy level sequence; X_{i_j} is the univariate quadrature point set with the accuracy level i_j; γ_{s_j} and w_{s_j} are the univariate quadrature point and weight, respectively. By the accuracy level i_j for the univariate quadrature rule, it means that I_{i_j} is exact to at least the $(2i_j - 1)$th order of all univariate polynomials. The minimum number of quadrature points for I_{i_j} is i_j. The choice here is that the univariate quadrature rule I_1 uses only one point with the value of 0 and weight of 1 (Jia et al. 2012b).

The accuracy level set $\Upsilon_{n,L}$ in Eq. (4.38) is defined by (Nobile et al. 2008b)

$$\Upsilon_{n,L} \triangleq \left\{ \Xi \subset \mathbb{N}^n ; i_j \geq 1, \sum_{j=1}^{n} \left(i_j - 1 \right) \leq \left(L - 1 \right) \right\} \tag{4.40}$$

where $\mathbb{N}^n$ denotes a sequence of natural numbers with n-elements.

In order to use the univariate quadrature rule, Eq. (4.38) can be rewritten as (Nobile et al. 2008b)

$$I_{n,L}(f) = \sum_{\Xi \in \mathbf{Y}_{n,L}} (-1)^{L-1+n-|\Xi|} \binom{n-1}{L-1+n-|\Xi|} \left(I_{i_1} \otimes \cdots I_{i_n} \right)(f) \tag{4.41}$$

where $|\Xi|$ denotes the summation of the elements in Ξ;

$$\mathbf{Y}_{n,L} \triangleq \left\{ \Xi \subset \mathbb{N}^n : i_j \geq 1, L - n \leq \sum_{j=1}^{n} \left(i_j - 1 \right) \leq \left(L - 1 \right) \right\} \tag{4.42}$$

and $$\left(I_{i_1} \otimes \cdots I_{i_n} \right)(f) \triangleq \sum_{\gamma_{s_1} \in X_{i_1}} \cdots \sum_{\gamma_{s_n} \in X_{i_n}} w_{s_1} \cdots w_{s_n} f\left(\gamma_{s_1}, \cdots, \gamma_{s_n} \right) \tag{4.43}$$

where X_{i_j} is the univariate point set of I_{i_j}; w_{s_j} is the corresponding weight of the point $\gamma_{s_j} \in X_{i_j}$.

Equation (4.41) is equivalent to the following equation that is used in Refs. (Jia et al. 2012b, Heiss and Winschel 2008)

$$I_{n,L}(f) = \sum_{q=L-n}^{L-1} (-1)^{L-1-q} \binom{n-1}{L-1-q} \sum_{\Xi \in \mathrm{N}_q^n} \left(I_{i_1} \otimes \cdots I_{i_n} \right)(f) \tag{4.44}$$

where q is an auxiliary parameter to prescribe the range of the univariate accuracy level i_j.
N_q^n in Eq. (4.44) is defined as (Heiss and Winschel 2008)

$$\mathrm{N}_q^n \triangleq \begin{cases} \left\{ \Xi \subset \mathbb{N}^n : \sum_{j=1}^{n} i_j = n + q \right\} & \text{for } q \geq 0 \\ \varnothing & \text{for } q < 0 \end{cases} \tag{4.45}$$

The corresponding sparse grid point set defined by $\mathbf{X}_{n,L}$ is given by

$$\mathbf{X}_{n,L} = \bigcup_{\Xi \in \mathbf{Y}_{n,L}} \left(X_{i_1} \otimes \cdots \otimes X_{i_n} \right) \tag{4.46}$$

The weight W_i for each multivariate quadrature point γ_i in $\mathbf{X}_{n,L}$ is the sum of the weights on the point over all combinations of $X_{i_1} \otimes \cdots \otimes X_{i_n}$ containing that point (Jia et al. 2012b, Heiss and Winschel 2008). Moreover, for one specific combination Ξ, the weight on the point γ_i is calculated by $W_i = (-1)^{L-1+n-|\Xi|}\binom{n-1}{L-1+n-|\Xi|}\left(w_{s_1} \times \cdots \times w_{s_j} \times \cdots \times w_{s_n}\right)$.

Note that the number of SGQ points increases polynomially (Heiss and Winschel 2008) with the dimension, which alleviates the curse of dimensionality of the GHQ. Moreover, the accuracy of the SGQ is guaranteed by the following theorem.

Theorem 4.1 (Heiss and Winschel 2008): Assume that the sequence of univariate quadrature rules $I = \left\{I_{i_j} : i_j \in \mathrm{N}\right\}$ is defined such that I_{i_j} is exact for all univariate polynomials of the order up to $2i_j - 1$. The Smolyak rule $I_{n,L}$ in Eq. (4.44) using I as the univariate basis sequence is exact for n-variate polynomials of the total order up to $2L - 1$.

Remark 4.2: Theorem 4.1 shows that the quadrature accuracy using the SGQ increases with the accuracy level L. The level-2 SGQ ($L = 2$) using 1 and 3 univariate GHQ points for I_1 and I_2 is identical to the UT using the suggested optimal parameter $\kappa = 3 - n$, which will be shown in Theorem 4.2.

The univariate quadrature can be obtained by the GHQ rule (Ito and Xiong 2000) or the moment matching method.

For the n-dimensional SGQ with the accuracy level L to be exact for all multivariate polynomials with the total order of $2L$-1, the univariate quadrature rule I_{i_j} ($i_j \in \Xi$ and $1 \le i_j \le L$) needs to be exact for all univariate polynomials of the order up to $2i_j - 1$, which can be satisfied using the following moment matching method.

The general moment matching formula used here (one-dimensional Gaussian type integral) is

$$M_j = \int_{-\infty}^{\infty} x^j N(x;0,1)\,dx = \sum_{i=1}^{N_u} \bar{w}_i \bar{p}_i^j \tag{4.47}$$

where M_j is the jth order moment; N_u is the number of univariate quadrature points; $\bar{p}_i$ and $\bar{w}_i$ are the univariate quadrature points and weights, respectively. We require that the quadrature rule be exact for all univariate polynomials of the order up to $N_u - 1$. Hence, $\bar{p}_i$ and $\bar{w}_i$ should satisfy the following equation.

$$\begin{bmatrix} 1 & 1 & \cdots & 1 \\ \bar{p}_1 & \bar{p}_2 & \cdots & \bar{p}_{N_u} \\ \vdots & \vdots & & \vdots \\ \bar{p}_1^{N_u-1} & \bar{p}_2^{N_u-1} & \cdots & \bar{p}_{N_u}^{N_u-1} \end{bmatrix} \begin{bmatrix} \bar{w}_1 \\ \bar{w}_2 \\ \vdots \\ \bar{w}_{N_u} \end{bmatrix} = \begin{bmatrix} M_0 \\ M_1 \\ \vdots \\ M_{N_u-1} \end{bmatrix} \tag{4.48}$$

Note a normalization constraint exists since $\sum_{i=1}^{N_u} \bar{w}_i = M_0 = 1$.

The values of $\bar{p}_i$ and $\bar{w}_i\left(i=1\cdots N_u\right)$ can be obtained by solving the nonlinear equations (4.48). But the solution is not unique in general. Note that when we choose distinct points $\bar{p}_i\left(i=1\cdots N_u\right)$, the coefficient matrix is the Vandermonde's matrix whose determinant is nonzero and the weights $\bar{w}_i\left(i=1\cdots N_u\right)$ can be uniquely determined. The univariate GHQ rule is obtained when $\bar{p}_i$ are chosen to be the roots of Hermite polynomials (Arasaratnam et al. 2007). Because the N_u-point univariate GHQ is exact for polynomials of the order up to $\left(2N_u-1\right)$, the GHQ points can match up to the $(2N_u-1)$th moments.

In the following, we use $N_u=2l-1$ symmetric points about the origin for the univariate quadrature point set with the accuracy level l. The locations of the points are treated as free parameters; the weights are uniquely determined from the point locations. Other point selection methods are also possible, which will be discussed afterward. Because symmetric points are used, we use different notations from those in Eq. (4.48) to represent the points ($\tilde{p}_i$) and weights ($\tilde{w}_i$). We also assume that the weights for $\tilde{p}_i$ and $-\tilde{p}_i$ are both $\tilde{w}_i$. Using this point-selection strategy, Eqs. (4.47) and (4.48) can be rewritten as Eqs. (4.49) and (4.50), respectively.

$$M_j=\begin{cases}\tilde{w}_1+\sum\limits_{i=2}^{(N_u-1)/2+1}2\tilde{w}_i & j=0\\ 0 & j \text{ is odd}\\ \tilde{w}_1 0^j+\sum\limits_{i=2}^{(N_u-1)/2+1}2\tilde{w}_i\tilde{p}_i^j & j \text{ is even}\end{cases} \tag{4.49}$$

Because of $N_u=2l-1$, we have

$$\begin{bmatrix}1 & 1 & \cdots & 1\\ 0^2 & \tilde{p}_2^2 & \cdots & \tilde{p}_l^2\\ \vdots & \vdots & & \vdots\\ 0^{2l-2} & \tilde{p}_2^{2l-2} & \cdots & \tilde{p}_l^{2l-2}\end{bmatrix}\begin{bmatrix}\tilde{w}_1\\ 2\tilde{w}_2\\ \vdots\\ 2\tilde{w}_l\end{bmatrix}=\begin{bmatrix}M_0\\ M_2\\ \vdots\\ M_{2l-2}\end{bmatrix} \tag{4.50}$$

In Eq. (4.50), only the even moments are considered because the symmetry of the points implies that the odd moments are matched automatically. If we choose $2l-1$ symmetric and distinct points, the coefficient matrix of Eq. (4.50) is a Vandermonde's matrix. The weights $\tilde{w}_i\left(i=1,\cdots,l\right)$ can be uniquely determined since the inverse of the Vandermonde's matrix has an analytical form (Macon and Spitzbart 1958).

In the following, we use the SGQ with $L=3$ as an example to illustrate the moment matching method. According to the Smolyak's rule Eq. (4.44), or Eq. (4.45), we need the univariate quadrature rules at level-1 (I_1), level-2 (I_2), and level-3 (I_3). For the univariate quadrature rule I_1, the point set is chosen to be {0} with the corresponding weight of 1. For the univariate quadrature rule I_2, we choose the symmetric univariate quadrature point set as $\{-\hat{p}_1, 0, \hat{p}_1\}$ with the corresponding weight sequence $(\hat{w}_2, \hat{w}_1, \hat{w}_2)$. For the univariate quadrature rule I_3, we choose the symmetric univariate quadrature point set as $\{-\hat{p}_3, -\hat{p}_2, 0, \hat{p}_2, \hat{p}_3\}$ with the corresponding weight sequence $(\hat{w}_5, \hat{w}_4, \hat{w}_3, \hat{w}_4, \hat{w}_5)$. To satisfy the condition of Theorem 4.1, these univariate point sets for level-2 and level-3 should match univariate polynomials up to the 3rd order and the 5th order respectively.

Note that the above set of notations for the univariate quadrature points and weights is different from the notations used in the general univariate moment matching Eqs. (4.47)–(4.50) because the SGQ involves different levels of univariate quadrature points and weights and is difficult to adopt a uniform set of notations for all levels. From Eq. (4.50), for level-2, $\tilde{p}_2 = \hat{p}_1$, $\tilde{w}_1 = \hat{w}_1$, and $\tilde{w}_2 = \hat{w}_2$, the following equations should be satisfied.

$$\begin{cases} \hat{w}_1 + 2\hat{w}_2 = M_0 = 1 \\ 2\hat{w}_2\hat{p}_1^2 = M_2 = 1 \end{cases} \tag{4.51}$$

Solving these two equations for $\hat{w}_1$ and $\hat{w}_2$ leads to

$$\begin{cases} \hat{w}_1 = 1 - 1/\hat{p}_1^2 \\ \hat{w}_2 = 1/(2\hat{p}_1^2) \end{cases} \tag{4.52}$$

If the point set $\{-\hat{p}_1, 0, \hat{p}_1\}$ is the set of the univariate GHQ points, $\hat{p}_1 = \sqrt{3}$, $\hat{w}_1 = \dfrac{2}{3}$, $\hat{w}_2 = \dfrac{1}{6}$.

Similarly, for level-3, $\tilde{p}_2 = \hat{p}_2$, $\tilde{p}_3 = \hat{p}_3$, $\tilde{w}_1 = \hat{w}_3$, $\tilde{w}_2 = \hat{w}_4$ and $\tilde{w}_3 = \hat{w}_5$, the following equations should be satisfied

$$\begin{cases} \hat{w}_3 + 2\hat{w}_4 + 2\hat{w}_5 = M_0 = 1 \\ 2\hat{w}_4\hat{p}_2^2 + 2\hat{w}_5\hat{p}_3^2 = M_2 = 1 \\ 2\hat{w}_4\hat{p}_2^4 + 2\hat{w}_5\hat{p}_3^4 = M_4 = 3 \end{cases} \tag{4.53}$$

Solving these equations (if $\hat{p}_3 \neq \hat{p}_2$) for $\hat{w}_3$, $\hat{w}_4$, $\hat{w}_5$ yields

$$\begin{cases} \hat{w}_3 = 1 - 2\hat{w}_4 - 2\hat{w}_5 \\ \hat{w}_4 = (3 - \hat{p}_3^2)/[2\hat{p}_2^2(\hat{p}_2^2 - \hat{p}_3^2)] \\ \hat{w}_5 = (3 - \hat{p}_2^2)/[2\hat{p}_3^2(\hat{p}_3^2 - \hat{p}_2^2)] \end{cases} \tag{4.54}$$

If $\hat{p}_3 = \hat{p}_2$, for Eq. (4.53) to remain valid, we must have $\hat{p}_3 = \hat{p}_2 = \sqrt{3}$. Then, $\hat{w}_3 = \dfrac{2}{3}$, $\hat{w}_4 + \hat{w}_5 = \dfrac{1}{6}$.

Note that the points $\hat{p}_1$, $\hat{p}_2$ and $\hat{p}_3$ are tunable parameters. When the locations of these points are given, the weights can be determined from Eqs. (4.52) and (4.54).

Remark 4.3: The moment matching method has been used to analyze (Julier and Uhlmann 2004) or extend the UT (Tenne and Singh 2003). The main difference of the SGQ method from the methods used in those references is that only univariate moment matching is needed here and then the resultant univariate point sets and weights are extended to the *n*-dimensional point set and weights using the sparse-grid method. Thus, the computations of the points and weights are greatly simplified.

Remark 4.4: Other univariate point-selection methods can be used if the condition of Theorem 4.1 is satisfied. For example, we can use $4l-1$ symmetric points for the univariate quadrature point set with the accuracy level l, which can match the $(4l-2)$th order moments. Increasing the number of univariate quadrature points increases the accuracy level of the univariate integral approximation. But it may not be sufficient to increase the accuracy level of the *n*-dimensional SGQ. It is worth noting that the univariate $\boldsymbol{I}_l$ with the accuracy level l requires the minimum number of l univariate quadrature points. In fact, one can choose l symmetric univariate GHQ points and weights as the univariate quadrature points and weights to satisfy the level-l requirement. Therefore, the univariate GHQ is a subset of the univariate quadrature given by the moment matching method. The sparse Gauss-Hermite Quadrature (Jia et al. 2011) for multi-dimensional problems is a subset of the SGQ since they are based on the same sparse-grid method.

The SGQ points and weights can be generated by the algorithm given in Table 4.1 (Jia et al. 2012b).

Next, we show an example to construct a two-dimensional sparse-grid with an accuracy level 3. Let's assume the univariate quadrature point is obtained via the GHQ rule. To construct the sparse grid with the accuracy level 3, three different quadrature point sequences and weight sequences are required. Hence, we can choose 1-point, 3-point, 5-point GHQs for level-1, level-2, level-3 univariate quadrature rule, respectively. They are

$$(0), (1)$$

$$\left(-\sqrt{3},0,\sqrt{3}\right),\ (1/6,2/3,1/6)$$

$$(-2.8570,-1.3556,0,1.3556,2.8570),(0.0113,0.2221,0.5332,0.2221,0.0113).$$

Based on Eq. (4.44), there are two different accuracy combination sequences, N_1^2 and N_2^2. For N_1^2, there are two possible accuracy sequences $\{(2,1),(1,2)\}$. For N_2^2, there are three possible accuracy sequences $\{(3,1),(2,2),(1,3)\}$. Note

Table 4.1 Generate SGQ points and weights.

$[\chi, W] = \text{SGQ}[n, L]$

(χ: SGQ point set with the element of χ_i; W: weight sequence with the element of W_i)

Obtain univariate quadrature points $\hat{p}_s \in \hat{p}, (s \in \mathbb{N})$ and weights $\hat{w}_s \in \hat{w}$ for the univariate quadrature rule I_l, $1 \le l \le L$, where $\hat{p}$ and $\hat{w}$ are the univariate SGQ point set and weight set, respectively.

FOR $q = L-n : L-1$

Determine N_q^n

FOR each element $\Xi = (i_1, \cdots, i_n)$ in N_q^n, form

$X_{i_1} \otimes X_{i_2} \cdots \otimes X_{i_n}$

FOR each point $\left[\hat{p}_{s_1}, \cdots, \hat{p}_{s_j}, \cdots, \hat{p}_{s_n}\right]^T$ $(\hat{p}_{s_j} \in \hat{p})$ in

$X_{i_1} \otimes \cdots X_{i_j} \cdots \otimes X_{i_n}$

IF the point is new, add it to χ, assign a new index i to this point and calculate the weight of χ_i as

$$W_i = (-1)^{L-1-q} \mathrm{C}_{n-1}^{L-1-q} \prod_{j=1}^{n} \hat{w}_{s_j} \quad (\hat{w}_{s_j} \in \hat{w}) \tag{4.55}$$

ELSE (the point already exists)

Update the old weight by

$$W_i = W_i + (-1)^{L-1-q} \mathrm{C}_{n-1}^{L-1-q} \prod_{j=1}^{n} \hat{w}_{s_j} \tag{4.56}$$

END IF

END FOR

END FOR

END FOR

that each sequence represents the way to construct the points and weights. For example, for the accuracy sequence $(2,1)$, it corresponds to the tensor product rule $I_2 \otimes I_1$. The points for $I_2 \otimes I_1$ are $\left[-\sqrt{3}, 0\right]^T$, $[0,0]^T$, $\left[\sqrt{3}, 0\right]^T$ with the weights 1/6, 2/3, and 1/6, respectively. Other tensor product rules can be constructed similarly. The final integration rule is given by

$$I_{2,3} = (-1)\mathbf{C}_1^1 (I_2 \otimes I_1 + I_1 \otimes I_2) + (-1)\mathbf{C}_1^0 (I_3 \otimes I_1 + I_2 \otimes I_2 + I_1 \otimes I_3) \tag{4.57}$$

The final grid (17 points) and weights are shown in Fig. 4.1. The symbol '*' denotes the location of the points on the x-y plane. The symbol 'o' denotes the weights corresponding to the points.

Remark 4.5: It is not necessary to use the same univariate quadrature for different dimensions, which further increases the flexibility of the quadrature construction.

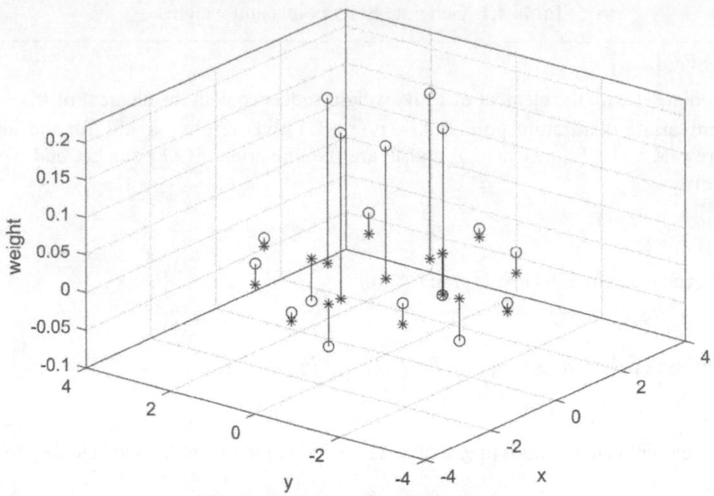

Fig. 4.1 Points and weights for $I_{2,3}$.

The level-2 SGQ using the univariate quadrature point set $\{-\hat{p}_1, 0, \hat{p}_1\}$ with the weight $(\hat{w}_2, \hat{w}_1, \hat{w}_2)$ for I_2 is given by (Jia et al. 2012b)

$$I_{n,2} = (1 - n + n\hat{w}_1)\boldsymbol{f}(\mathbf{0}) + w_2 \sum_{j=1}^{n} \left[\boldsymbol{f}(\hat{p}_1 \mathbf{e}_j) + \boldsymbol{f}(-\hat{p}_1 \mathbf{e}_j) \right] \tag{4.58}$$

Note that (4.58) is identical to the UT if $\hat{p}_1 = \sqrt{n+\kappa}$, $\hat{w}_1 = 1 - 1/(n+\kappa)$, and $\hat{w}_2 = 1/\left[2(n+\kappa)\right]$ (Julier and Uhlmann 2004).

Theorem 4.2: The points and weights generated by the UT are identical to the points and weights generated by the SGQ rule with level-2 accuracy if one point and three symmetric points are used for the level-1 and level-2 univariate quadrature point set respectively.

Proof: For the level-1 univariate quadrature rule I_1, the point set is {0} with the corresponding weight of 1. For the level-2 univariate quadrature I_2, the univariate quadrature point set is $\{-\hat{p}_1, 0, \hat{p}_1\}$ with the corresponding weight sequence $(\hat{w}_2, \hat{w}_1, \hat{w}_2)$.

Since the accuracy level of the SGQ is $L = 2$, the value of q in N_q^n can be 0 or 1. When $q = 0$, $\mathrm{N}_0^n = \left\{ \underbrace{(1,1,\cdots,1,1)}_{n\ \text{elements}} \right\}$, the n-dimensional SGQ point corresponding to N_0^n will be $[0,0,\cdots 0,0]^T$ with the corresponding weight of $(-1)^{2-1-0} \times \mathrm{C}_{n-1}^{2-1-0} \times \underbrace{\left(\overbrace{1\times 1\times \cdots \times 1\times 1}^{n-1\ \text{elements}} \right)}_{\text{product of weights}} = -(n-1)$ (using Eq. (4.55)).

When $q = 1$, the set of accuracy level sequences becomes

$$\mathrm{N}_1^n = \underbrace{\left\{ \overbrace{(2,1,\cdots 1,1)}^{n}, \overbrace{(1,2,\cdots 1,1)}^{n}, \cdots \overbrace{(1,\cdots 1,2,1,\cdots 1)}^{n}, \cdots \overbrace{(1,1,\cdots 1,2)}^{n} \right\}}_{n \text{ elements}}.$$

Corresponding to the sequence $(2,1,\cdots 1,1)$, there are three n-dimensional SGQ points $[-\hat{p}_1,0,\cdots 0,0]^T$, $[0,0,\cdots 0,0]^T$, and $[\hat{p}_1,0,\cdots 0,0]^T$. The weights corresponding to $[-\hat{p}_1,0,\cdots 0,0]^T$ and $[\hat{p}_1,0,\cdots 0,0]^T$ are the same, i.e., $W_2 = (-1)^{2-1-1} \times \mathrm{C}_{n-1}^{2-1-1} \times \underbrace{\left(\overbrace{1 \times 1 \times \cdots \times 1}^{n-1 \text{ elements}} \times \hat{w}_2 \right)}_{\text{product of weights}} = \hat{w}_2$ (using Eq. (4.55)). Moreover, for this sequence, the SGQ can match the polynomials of the form $\left(ax_1^3 + bx_1^2 + cx_1 + d\right) x_2^{k_1} x_3^{k_2} \cdots x_n^{k_n}$, where a,b,c,d are real numbers and $k_1,\cdots,k_n$ can be 0 or 1, because I_2 is exact for all univariate polynomials of the order up to $2 \times 2 - 1 = 3$ and I_1 is exact for all univariate polynomials of the order up to $2 \times 1 - 1 = 1$ according to Theorem 4.1. Similarly, for other sequences in N_1^n like $(1,\cdots,1,2,1,\cdots 1)$, there are three n-dimensional SGQ points $[0,\cdots,0,-\hat{p}_1,0,\cdots 0]^T$, $[0,\cdots,0,0,0,\cdots 0]^T$, and $[0,\cdots,0,\hat{p}_1,0,\cdots 0]^T$. The weights corresponding to $[0,\cdots,0,-\hat{p}_1,0,\cdots 0]^T$ and $[0,\cdots,0,\hat{p}_1,0,\cdots 0]^T$ are the same, i.e., $W_2 = \hat{w}_2$ (using Eq. (4.55)). There are n such combinations of accuracy level sequences in N_1^n with the similar calculation of n-dimensional SGQ points and weights. Note that since the point $[0,\cdots,0,0,0,\cdots 0]^T$ is a repeated point appearing once in N_0^n and n times in N_1^n, Eq. (4.56) is used to calculate its weight

$$\begin{aligned} W_1 &= (-1)^{2-1-0} \times \mathrm{C}_{n-1}^{2-1-0} \times \underbrace{\left(\overbrace{1 \times 1 \times \cdots \times 1 \times 1}^{n \text{ elements}} \right)}_{\text{product of weights}} + (-1)^{2-1-1} \times \mathrm{C}_{n-1}^{2-1-1} \\ &\times \left\{ \overbrace{ \underbrace{\left(\hat{w}_1 \times \overbrace{1 \times 1 \times \cdots \times 1}^{n-1 \text{ elements}} \right)}_{\text{product of weights}} + \underbrace{\left(1 \times \hat{w}_1 \times \overbrace{1 \times 1 \times \cdots \times 1}^{n-2 \text{ elements}} \right)}_{\text{product of weights}} + \cdots + \underbrace{\left(\overbrace{1 \times 1 \times \cdots \times 1}^{n-1 \text{ elements}} \times \hat{w}_1 \right)}_{\text{product of weights}} }^{n \text{ elements}} \right\} \\ &= -(n-1) + n \cdot \hat{w}_1 \end{aligned}$$

To summarize it, the points and weights of the SGQ with level-2 accuracy are

$$\boldsymbol{\gamma}_i = \begin{cases} [0,0,\cdots 0,0]^T; & i=1 \\ \hat{p}_1 \mathbf{e}_{i-1}; & 2 \le i \le n+1 \\ -\hat{p}_1 \mathbf{e}_{i-n-1}; & n+2 \le i \le 2n+1 \end{cases} \tag{4.59}$$

and
$$W_i = \begin{cases} -(n-1)+n\cdot\hat{w}_1 & i=1 \\ \hat{w}_2 & i=2\cdots 2n+1 \end{cases} \tag{4.60}$$

respectively. If we choose $\hat{p}_1 = \sqrt{n+\kappa}$, by Eq. (4.52), $\hat{w}_1 = 1 - \frac{1}{(n+\kappa)}$ and $\hat{w}_2 = \frac{1}{2(n+\kappa)}$. Hence

$$W_i = \begin{cases} -(n-1)+n\cdot\left(1-\frac{1}{(n+\kappa)}\right) = \frac{\kappa}{n+\kappa} & i=1 \\ \frac{1}{2(n+\kappa)} & i=2\cdots 2n+1 \end{cases} \tag{4.61}$$

Comparing Eqs. (4.59) and (4.61) with the sigma points and weights of the UT in Eqs. (4.34) and (4.35), they are identical. ■

The significance of Theorem 4.2 is that the UKF based on UT is a subset of the level-2 SGQ filter.

The fifth-degree SGQ using the univariate quadrature point set $\{-\hat{p}_2, -\hat{p}_1, 0, \hat{p}_1, \hat{p}_2\}$ with the weight $(\hat{w}_5, \hat{w}_4, \hat{w}_3, \hat{w}_4, \hat{w}_5)$ for I_3 is given by (Jia et al. 2012b)

$$I_{n,3}(\boldsymbol{f}) = A_1 \boldsymbol{f}(\mathbf{0}) + B_1 \sum_{j=1}^{2n} \boldsymbol{f}(\mathbf{P}_{j,1}) + C_1 \sum_{j=1}^{2n} \boldsymbol{f}(\mathbf{P}_{j,2}) + D_1 \sum_{j=1}^{4C_n^2} \boldsymbol{f}(\mathbf{P}_{j,3}) \tag{4.62}$$

where $A_1 = 0.5(n-1)(n-2+n\hat{w}_1^2) - n(n-1)\hat{w}_1 + n\hat{w}_3$;

$B_1 = (n-1)\hat{w}_2(\hat{w}_1 - 1) + \hat{w}_4$; $C_1 = \hat{w}_5$; $D_1 = \hat{w}_2^2$;

$$\mathbf{P}_{j,1} = \begin{cases} \hat{p}_1 \mathbf{e}_j, & j=1,\cdots,n \\ -\hat{p}_1 \mathbf{e}_{j-n}, & j=n+1,\cdots,2n \end{cases} \tag{4.63a}$$

$$\mathbf{P}_{j,2} = \begin{cases} \hat{p}_2 \mathbf{e}_j, & j=1,\cdots,n \\ -\hat{p}_2 \mathbf{e}_{j-n}, & j=n+1,\cdots,2n \end{cases} \tag{4.63b}$$

$$\mathbf{P}_{j,3} = \begin{cases} \hat{p}_1(\mathbf{e}_i + \mathbf{e}_k) & i,k = 1,\cdots n; i < k \\ \hat{p}_1(\mathbf{e}_i - \mathbf{e}_k) & i,k = 1,\cdots n; i < k \\ \hat{p}_1(-\mathbf{e}_i + \mathbf{e}_k) & i,k = 1,\cdots n; i < k \\ \hat{p}_1(-\mathbf{e}_i - \mathbf{e}_k) & i,k = 1,\cdots n; i < k \end{cases} \tag{4.63c}$$

Note that $\hat{p}_1$ and $\hat{p}_2$ are tunable parameters (Jia et al. 2012b).
An example of (4.62) is given by (Jia et al. 2012b)

$$I_{n,3}(f) = C_0 f(\mathbf{0}) + C_1 \sum_{j=1}^{2n} f(\mathbf{P}_{j,4}) + C_2 \sum_{j=1}^{2n(n-1)} f(\mathbf{P}_{j,5}) \tag{4.64}$$

where $C_0 = n^2/18 - 7n/18 + 1$, $C_1 = -n/18 + 2/9$, $C_2 = 1/36$ and

$$\mathbf{P}_{j,4} = \begin{cases} \sqrt{3}\mathbf{e}_j, & j = 1,\cdots,n \\ -\sqrt{3}\mathbf{e}_{j-n}, & j = n+1,\cdots,2n \end{cases} \tag{4.65a}$$

$$\mathbf{P}_{j,5} = \begin{cases} \sqrt{3}(\mathbf{e}_i + \mathbf{e}_k) & i,k = 1,\cdots n; i < k \\ \sqrt{3}(\mathbf{e}_i - \mathbf{e}_k) & i,k = 1,\cdots n; i < k \\ \sqrt{3}(-\mathbf{e}_i + \mathbf{e}_k) & i,k = 1,\cdots n; i < k \\ \sqrt{3}(-\mathbf{e}_i - \mathbf{e}_k) & i,k = 1,\cdots n; i < k \end{cases} \tag{4.65b}$$

Note that (4.62) is identical to (4.64) when $\hat{p}_1 = \hat{p}_2 = \sqrt{3}$ is used.

4.6 Anisotropic Sparse-Grid Quadrature and Accuracy Analysis

The anisotropic sparse-grid quadrature (ASGQ) can be used to further improve the computation efficiency of the SGQ. The SGQ is isotropic in the sense that all dimensions are assumed to be equally important and it uses an isotropic sparse-grid, which may result in more points than necessary. There are many practical problems in which different dimensions are not equally important. Motivated by this observation, the ASGQ provides a mechanism for distributing more quadrature points or weights on more important dimensions and allows for better trade-offs between the computational efficiency and the estimation accuracy. The number of ASGQ points can be flexibly controlled by a tunable importance vector and is, in general, less than that of the SGQ.

4.6.1 Anisotropic Sparse-Grid Quadrature

The ASGQ is given by (Nobile et al. 2008a, Jia et al. 2012a)

$$I_{n,L^{\boldsymbol{\alpha}}}^{\boldsymbol{\alpha}}(f)=\sum_{\Xi^{\boldsymbol{\alpha}}\in\Upsilon_{n,L^{\boldsymbol{\alpha}}}^{\boldsymbol{\alpha}}}\left(\Delta^{i_1}\otimes\cdots\otimes\Delta^{i_n}\right)(f) \tag{4.66}$$

with the accuracy level set

$$\Upsilon_{n,L^{\boldsymbol{\alpha}}}^{\boldsymbol{\alpha}}\triangleq\left\{\Xi^{\boldsymbol{\alpha}}\subset\mathbb{N}^n:i_j\geq1,\sum_{j=1}^{n}\left(i_j-1\right)\alpha_j\leq\left(L^{\boldsymbol{\alpha}}-1\right)\underline{\alpha}\right\},\quad \underline{\alpha}\triangleq\min(\boldsymbol{\alpha}). \tag{4.67}$$

In Eq. (4.67), $\Xi^{\boldsymbol{\alpha}}\triangleq\left(i_1,\cdots,i_n\right)$ and the superscript '$\boldsymbol{\alpha}$' denotes the anisotropic sparse-grid associated with the importance vector $\boldsymbol{\alpha}=\left[\alpha_1,\cdots,\alpha_j,\cdots,\alpha_n\right]$ $\left(\alpha_j>0\right)$, in which each element of $\boldsymbol{\alpha}$ represents the relative importance of the corresponding state variable and min($\boldsymbol{\alpha}$) denotes the minimum element in $\boldsymbol{\alpha}$. Without loss of generality, $\underline{\alpha}$ is set to 1. The accuracy level of the ASGQ is denoted by $L^{\boldsymbol{\alpha}}\in\mathbb{N}$. Equation (4.66) can be rewritten as (Nobile et al. 2008a)

$$I_{n,L^{\boldsymbol{\alpha}}}^{\boldsymbol{\alpha}}(f)=\sum_{\Xi^{\boldsymbol{\alpha}}\in\mathbf{Y}_{n,L^{\boldsymbol{\alpha}}}^{\boldsymbol{\alpha}}}c^{\boldsymbol{\alpha}}\left(\Xi^{\boldsymbol{\alpha}}\right)\left(I_{i_1}\otimes\cdots\otimes I_{i_n}\right)(f) \tag{4.68}$$

where

$$c^{\boldsymbol{\alpha}}\left(\Xi^{\boldsymbol{\alpha}}\right)\triangleq\sum_{\Psi\in\{0,1\}^n,\,\Xi^{\boldsymbol{\alpha}}+\Psi\in\Upsilon_{n,L^{\boldsymbol{\alpha}}}^{\boldsymbol{\alpha}}}(-1)^{|\Psi|} \tag{4.69}$$

$$\mathbf{Y}_{n,L^{\boldsymbol{\alpha}}}^{\boldsymbol{\alpha}}\triangleq\Upsilon_{n,L^{\boldsymbol{\alpha}}}^{\boldsymbol{\alpha}}\setminus\Upsilon_{n,L^{\boldsymbol{\alpha}}-\frac{|\boldsymbol{\alpha}|}{\underline{\alpha}}}^{\boldsymbol{\alpha}} \tag{4.70}$$

Note that $\{0,1\}^n$ denotes the set of all n-dimensional sequences with each dimension's value being 0 or 1. $|\Psi|$ and $|\boldsymbol{\alpha}|$ denote the summation of the elements in Ψ and $\boldsymbol{\alpha}$ respectively; '\' denotes the subtraction operation of two sets. Equations (4.68)–(4.70) are valid whether $\boldsymbol{\alpha}$ is an integer vector or not, but for the convenience of formulation, $\boldsymbol{\alpha}$ is assumed to be an integer vector such that $\frac{|\boldsymbol{\alpha}|}{\underline{\alpha}}$ and $L^{\boldsymbol{\alpha}}-\frac{|\boldsymbol{\alpha}|}{\underline{\alpha}}$ are integers.

By Eqs. (4.67) and (4.70),

$$\mathbf{Y}_{n,L^{\boldsymbol{\alpha}}}^{\boldsymbol{\alpha}}=\Upsilon_{n,L^{\boldsymbol{\alpha}}}^{\boldsymbol{\alpha}}\setminus\Upsilon_{n,L^{\boldsymbol{\alpha}}-\frac{|\boldsymbol{\alpha}|}{\underline{\alpha}}}^{\boldsymbol{\alpha}}=\left\{\Xi^{\boldsymbol{\alpha}}\subset\mathbb{N}^n;i_j\geq1,\left(L^{\boldsymbol{\alpha}}-\frac{|\boldsymbol{\alpha}|}{\underline{\alpha}}-1\right)\underline{\alpha}<\sum_{j=1}^{n}\left(i_j-1\right)\alpha_j\leq\left(L^{\boldsymbol{\alpha}}-1\right)\underline{\alpha}\right\} \tag{4.71}$$

From Eq. (4.68), the corresponding ASGQ point set defined by $\mathbf{X}_{n,L^{\boldsymbol{\alpha}}}^{\boldsymbol{\alpha}}$ is

$$\mathbf{X}_{n,L^{\boldsymbol{\alpha}}}^{\boldsymbol{\alpha}}=\bigcup_{\Xi^{\boldsymbol{\alpha}}\in\mathbf{Y}_{n,L^{\boldsymbol{\alpha}}}^{\boldsymbol{\alpha}}}\left(X_{i_1}\otimes\cdots\otimes X_{i_n}\right) \tag{4.72}$$

The weight W_i for each point $\boldsymbol{\gamma}_i$ in $\mathbf{X}_{n,L^{\boldsymbol{\alpha}}}^{\boldsymbol{\alpha}}$ is the sum of the weights on the point over all combinations of $X_{i_1}\otimes\cdots\otimes X_{i_n}$ containing the point. Moreover, for one specific combination, the weight on the point γ_i is calculated by $W_i = c^{\boldsymbol{\alpha}}\left(\Xi^{\boldsymbol{\alpha}}\right)\left(w_{s_1} \times \cdots \times w_{s_j} \times \cdots \times w_{s_n}\right)$.

Remark 4.6: Given the accuracy level-L^{α} and dimension n, the final ASGQ point set is determined by $\mathbf{Y}^{\boldsymbol{\alpha}}_{n,L^{\alpha}}$. When the importance vector $\boldsymbol{\alpha}$ is also chosen, the upper bound of the inequality in Eq. (4.71) is fixed. This implies that for a larger α_j, the allowable range of i_j is smaller. In other words, if $\alpha_i < \alpha_j$, by Eq. (4.71), the ASGQ can use a univariate quadrature rule $I_{\max(i_i)}$ with a higher accuracy level for dimension i than the univariate quadrature rule $I_{\max(i_j)}$ for dimension j. Note that $I_{\max(i_i)}$ uses more points than $I_{\max(i_j)}$ in this case. Hence, a smaller element in $\boldsymbol{\alpha}$ indicates that the corresponding dimension is more important and uses more quadrature points.

Remark 4.7: By Eqs. (4.71) and (4.72), the effect of $\boldsymbol{\alpha}$ on the ASGQ point set is determined by the ratio of $\alpha_j/\underline{\alpha}$ or $\underline{\alpha}/\alpha_j$ rather than the absolute value of α_j.

To better illustrate the ASGQ rule and compare it with the SGQ, $n = 2$, $L^{\alpha} = 3$ are used as an example to show how to construct the ASGQ point set. Assume $\boldsymbol{\alpha} = [2,1]$. Then,

$$\Upsilon^{\boldsymbol{\alpha}}_{2,3} = \left\{\Xi^{\boldsymbol{\alpha}} \subset \mathbb{N}^2 : i_j \geq 1, 2(i_1 - 1) + (i_2 - 1) \leq 2\right\} = \left\{\Xi^{\boldsymbol{\alpha}} \subset \mathbb{N}^2 : i_j \geq 1, i_1 + 0.5 i_2 \leq 2.5\right\},$$

It can be rewritten as $\Upsilon^{\boldsymbol{\alpha}}_{2,3} = \{(1,1),(1,2),(2,1),(1,3)\}$.
By Eq. (4.71),

$$\mathbf{Y}^{\boldsymbol{\alpha}}_{2,3} = \Upsilon^{\boldsymbol{\alpha}}_{2,3} \setminus \Upsilon^{\boldsymbol{\alpha}}_{2,3-\frac{|\boldsymbol{\alpha}|}{\underline{\alpha}}}, \text{ and } \Upsilon^{\boldsymbol{\alpha}}_{2,3-\frac{|\boldsymbol{\alpha}|}{\underline{\alpha}}} = \Upsilon^{\boldsymbol{\alpha}}_{2,0} = \left\{\Xi^{\boldsymbol{\alpha}} \subset \mathbb{N}^2 ; i_j \geq 1, i_1 + 0.5 i_2 \leq 1\right\} = \varnothing.$$

Thus, $\mathbf{Y}^{\boldsymbol{\alpha}}_{2,3} = \left\{\Xi^{\boldsymbol{\alpha}} \subset \mathbb{N}^2 : i_j \geq 1, 1 < i_1 + 0.5 i_2 \leq 2.5\right\} = \{(1,1),(1,2),(2,1),(1,3)\}$

By Eq. (4.69), $c^{\boldsymbol{\alpha}}((1,1)) = -1$, $c^{\boldsymbol{\alpha}}((1,2)) = 0$, $c^{\boldsymbol{\alpha}}((2,1)) = 1$ and $c^{\boldsymbol{\alpha}}((1,3)) = 1$. Then, $I^{\boldsymbol{\alpha}}_{2,3} = -(I_1 \otimes I_1) + (I_2 \otimes I_1) + (I_1 \otimes I_3)$.

The final ASGQ point set is shown in Fig. 4.2. Note that Fig. 4.2 verifies Remark 4.6. That is, the value '2' in $\boldsymbol{\alpha} = [2,1]$ makes i_1 take a smaller range of values than i_2, and thus generates fewer points in the vertical dimension than the points that the value '1' in $\boldsymbol{\alpha}$ generates in the horizontal dimension.

Compared with the conventional SGQ, ASGQ does not contain the tensor product $X_1 \otimes X_2$, owing to the fact that $c^{\boldsymbol{\alpha}}((1,2)) = 0$, and also does not contain $X_1 \otimes X_2$ and $X_3 \otimes X_1$ owing to the fact that $\{(2,2),(3,1)\} \not\subset \mathbf{Y}^{\boldsymbol{\alpha}}_{2,3} = \{(1,1),(1,2),(2,1),(1,3)\}$.

Note that the actual accuracy of the level-L^{α} ASGQ may be different from that of the level-L SGQ when $L = L^{\alpha}$. For example, when $L^{\alpha} = L = 3$ and $n = 2$,

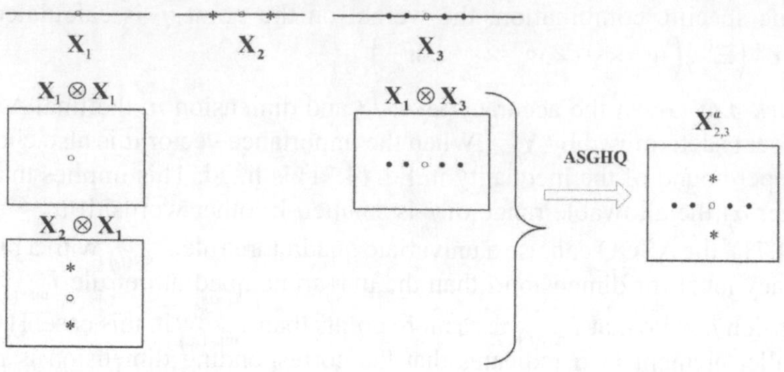

Fig. 4.2 ASGQ with the accuracy level-3 and $\boldsymbol{\alpha} = [2,1]$ for 2-dimensional problems.

by Theorem 4.1, SGQ can calculate the polynomial $x_1^2 x_2^2$ exactly, whereas the ASGQ cannot, as seen from the above expression of $I^{\boldsymbol{\alpha}}_{2,3}$.

4.6.2 Analysis of Accuracy of the Anisotropic Sparse-Grid Quadrature

The conventional SGQ can be viewed as a special case of the ASGQ. Note that in Eq. (4.71) and Eq. (4.42), when all elements in $\boldsymbol{\alpha}$ are the same, $L-n-1<\sum_{j=1}^{n}(i_j-1)$ is the same as $L-n\le\sum_{j=1}^{n}(i_j-1)$.

Remark 4.8: If $\left(\dfrac{\alpha_j}{\underline{\alpha}}\right) > L^{\boldsymbol{\alpha}}-1$, by Eq. (4.71), i_j must be 1, which implies that there will be only one point in dimension j, a case of degenerated accuracy. To prevent this case from happening, a constraint should be satisfied for the ratio, i.e., $1\le\left(\dfrac{\alpha_j}{\underline{\alpha}}\right)\le L^{\boldsymbol{\alpha}}-1$. Under this constraint, the level-2 ($L^{\boldsymbol{\alpha}} = 2$) ASGQ is the same as the level-2 ($L = 2$) SGQ since $\dfrac{\alpha_j}{\underline{\alpha}}=1$, and every α_j is the same. Note that if $L^{\boldsymbol{\alpha}} = L$ and all elements in $\boldsymbol{\alpha}$ are the same, by Eq. ((4.71)), the maximum value of i_j is $L^{\boldsymbol{\alpha}}$ in each dimension, which becomes SGQ. In most cases, $\alpha_j/\underline{\alpha}>1$, by Eq. (4.71), $i_j<L^{\boldsymbol{\alpha}}$, which means ASGQ is less accurate than SGQ in dimension j in these cases.

The relationship between the ASGQ and the SGQ can be revealed by the following proposition and theorem.

Proposition 4.1: If $L^{\alpha}=L$ and $L-n\le 0$, the ASGQ point set is a subset of the SGQ point set.

Proof: By Eq. (4.71), $\mathbf{Y}^{\boldsymbol{\alpha}}_{n,L^{\alpha}}$ can be rewritten as

$$\mathbf{Y}^{\boldsymbol{\alpha}}_{n,L^{\alpha}} = \left\{ \Xi^{\boldsymbol{\alpha}} \subset \mathbb{N}^n; i_j \geq 1, \left(L^{\boldsymbol{\alpha}} - 1\right) - \frac{|\boldsymbol{\alpha}|}{\underline{\alpha}} < \sum_{j=1}^{n} \left(i_j - 1\right) \frac{\alpha_j}{\underline{\alpha}} \leq \left(L^{\boldsymbol{\alpha}} - 1\right) \right\}.$$

If $L - n \leq 0$, $\left(L^{\boldsymbol{\alpha}} - 1\right) - \frac{|\boldsymbol{\alpha}|}{\underline{\alpha}} \leq \left(L^{\boldsymbol{\alpha}} - 1\right) - n = L^{\boldsymbol{\alpha}} - n - 1 < 0$.

Since $\sum_{j=1}^{n} \left(i_j - 1\right) \frac{\alpha_j}{\underline{\alpha}} \geq 0$, $\mathbf{Y}^{\boldsymbol{\alpha}}_{n,L^{\alpha}} = \left\{ \Xi^{\boldsymbol{\alpha}} \subset \mathbb{N}^n; i_j \geq 1, 0 \leq \sum_{j=1}^{n} \left(i_j - 1\right) \frac{\alpha_j}{\underline{\alpha}} \leq \left(L^{\boldsymbol{\alpha}} - 1\right) \right\}$.

Recall that $\mathbf{Y}_{n,L} = \left\{ \Xi \subset \mathbb{N}^n; i_j \geq 1, L - n \leq \sum_{j=1}^{n} \left(i_j - 1\right) \leq \left(L - 1\right) \right\}$ for the SGQ. Since $L - n \leq 0$ and $\sum_{j=1}^{n} \left(i_j - 1\right) \geq 0$, $\mathbf{Y}_{n,L} = \left\{ \Xi \subset \mathbb{N}^n; i_j \geq 1, 0 \leq \sum_{j=1}^{n} \left(i_j - 1\right) \leq \left(L - 1\right) \right\}$. Comparing the above $\mathbf{Y}^{\boldsymbol{\alpha}}_{n,L^{\alpha}}$ with $\mathbf{Y}_{n,L}$ shows $\mathbf{Y}^{\boldsymbol{\alpha}}_{n,L^{\alpha}} \subset \mathbf{Y}_{n,L}$ and thus $\mathbf{X}^{\boldsymbol{\alpha}}_{n,L^{\alpha}} \subset \mathbf{X}_{n,L}$ because $\frac{\alpha_j}{\underline{\alpha}} \geq 1$. ■

Remark 4.9: The condition $L - n \leq 0$ is needed for Proposition 4.1. For example, if $L^{\alpha} = L = 6, n = 2$ and $\boldsymbol{\alpha} = [1,2]$, which violates the condition, it can be verified that some points in ASGQ are not included in the SGQ point set.

It is worth noting that for typical multidimensional problems, the accuracy level-L is usually less than the dimension n. So the condition in Proposition 4.1 is satisfied in most cases.

The ASGQ accuracy depends on the accuracy level-$L^{\boldsymbol{\alpha}}$ and the parameter $\boldsymbol{\alpha}$, whereas the SGQ accuracy only depends on the accuracy level-L. The accuracy of the ASGQ is guaranteed by the following theorem.

Theorem 4.3: If $\boldsymbol{\alpha} = \left[\underbrace{1, \cdots, 1}_{n_1}, \underbrace{\alpha_m, \cdots, \alpha_m}_{n_2} \right]$, $(n_1 + n_2 = n)$ and $2 \leq \alpha_m + 1 \leq L^{\boldsymbol{\alpha}}$, ASGQ is either more accurate than or as accurate as the level-2 SGQ.

Proof: For this $\boldsymbol{\alpha}$, $\Upsilon^{\boldsymbol{\alpha}}_{n_1+n_2,L^{\boldsymbol{\alpha}}} = \left\{ \Xi^{\boldsymbol{\alpha}} \subset \mathbb{N}^{n_1+n_2}; i_j \geq 1, \sum_{j=1}^{n_1} \left(i_j - 1\right) + \alpha_m \sum_{j=n_1+1}^{n_1+n_2} \left(i_j - 1\right) \leq \left(L^{\boldsymbol{\alpha}} - 1\right) \right\}$. Since $i_j \geq 1$, $0 \leq \sum_{j=n_1+1}^{n_1+n_2} \left(i_j - 1\right) \leq \frac{1}{\alpha_m} \left(\left(L^{\boldsymbol{\alpha}} - 1\right) - \sum_{j=1}^{n_1} \left(i_j - 1\right) \right)$. Hence, each $i_j \left(j = n_1 + 1, \cdots, n_1 + n_2\right)$ can be $1, \cdots, \left\lfloor \frac{L^{\boldsymbol{\alpha}} - 1}{\alpha_m} + 1 \right\rfloor$, where '$\lfloor \ \rfloor$' denotes the floor operation, which returns an integer less than or equal to $\frac{L^{\boldsymbol{\alpha}} - 1}{\alpha_m} + 1$. Note that $\alpha_m + 1 \le L^{\boldsymbol{\alpha}}$ or $\alpha_m \le L^{\boldsymbol{\alpha}} - 1$ implies $\frac{L^{\boldsymbol{\alpha}} - 1}{\alpha_m} + 1 \ge 2$. There are three possible cases with this $\boldsymbol{\alpha}$:

Case (1): All $i_j\left(j = 1, \cdots, n_1\right)$ are 1 and $\sum_{j=n_1+1}^{n_1+n_2}\left(i_j - 1\right) \le \frac{L^{\boldsymbol{\alpha}} - 1}{\alpha_m}$. Define this set of $\Xi^{\boldsymbol{\alpha}}$ as

$$\Upsilon_{s1} = \left\{\Xi^{\boldsymbol{\alpha}} \subset \Upsilon_{n_1+n_2, L^{\boldsymbol{\alpha}}}; i_j \ge 1, \sum_{j=1}^{n_1}\left(i_j - 1\right) = 0, \quad 0 \le \sum_{j=n_1+1}^{n_1+n_2}\left(i_j - 1\right) \le \frac{L^{\boldsymbol{\alpha}} - 1}{\alpha_m}\right\}. \quad (4.73)$$

By Eq. (4.40), the range of $i_j\left(j = n_1 + 1, \cdots, n_1 + n_2\right)$ in Υ_{s1} is identical to that of

$$\Upsilon_{n_2, \left\lfloor \frac{L^{\boldsymbol{\alpha}} - 1}{\alpha_m} + 1 \right\rfloor} = \left\{\Xi \subset \mathbb{N}^{n_2}; i_j \ge 1, \sum_{j=n_1+1}^{n_1+n_2}\left(i_j - 1\right) \le \left\lfloor \frac{L^{\boldsymbol{\alpha}} - 1}{\alpha_m} \right\rfloor\right\} \quad (4.74)$$

for the SGQ.

Case (2): All $i_j\left(j = n_1 + 1, \cdots, n_1 + n_2\right)$ are 1 and $\sum_{j=1}^{n_1}\left(i_j - 1\right) \le L^{\boldsymbol{\alpha}} - 1$. Define this set of $\Xi^{\boldsymbol{\alpha}}$ as

$$\Upsilon_{s2} = \left\{\Xi^{\boldsymbol{\alpha}} \subset \Upsilon^{\boldsymbol{\alpha}}_{n_1+n_2, L^{\boldsymbol{\alpha}}}; i_j \ge 1, \quad 0 \le \sum_{j=1}^{n_1}\left(i_j - 1\right) \le L^{\boldsymbol{\alpha}} - 1, \quad \sum_{j=n_1+1}^{n_1+n_2}\left(i_j - 1\right) = 0\right\}. \quad (4.75)$$

Similarly, by Eq. (4.40), the range of $i_j\left(j = 1, \cdots, n_1\right)$ in Υ_{s2} is identical to that of

$$\Upsilon_{n_1, L^{\boldsymbol{\alpha}}} = \left\{\Xi \subset \mathbb{N}^{n_1}; i_j \ge 1, \sum_{j=1}^{n_1}\left(i_j - 1\right) \le L^{\boldsymbol{\alpha}} - 1\right\} \quad (4.76)$$

for the SGQ.

Note that Case (1) and Case (2) contain a duplicate case when all $i_j = 1, \left(j = 1, \cdots, n_1 + n_2\right)$, i.e.,

$$\Upsilon_{s3} \triangleq \Upsilon_{s1} \cap \Upsilon_{s2} = \left\{\Xi^{\boldsymbol{\alpha}} \subset \Upsilon^{\boldsymbol{\alpha}}_{n_1+n_2, L^{\boldsymbol{\alpha}}}; i_j \ge 1, \sum_{j=1}^{n_1+n_2}\left(i_j - 1\right) = 0\right\} \quad (4.77)$$

Case (3): $\sum_{j=1}^{n_1}\left(i_j - 1\right) \ge 1$ and $1 \le \sum_{j=n_1+1}^{n_1+n_2}\left(i_j - 1\right) \le \left\lfloor \frac{L^{\boldsymbol{\alpha}} - 2}{\alpha_m} \right\rfloor$. Define this set of $\Xi^{\boldsymbol{\alpha}}$ as

$$\Upsilon_{s4} = \left\{\Xi^{\boldsymbol{\alpha}} \subset \Upsilon^{\boldsymbol{\alpha}}_{n_1+n_2, L^{\boldsymbol{\alpha}}}; i_j \ge 1, \; 1 \le \sum_{j=1}^{n_1}\left(i_j - 1\right) \le L^{\boldsymbol{\alpha}} - \alpha_m - 1, \quad 1 \le \sum_{j=n_1+1}^{n_1+n_2}\left(i_j - 1\right) \le \left\lfloor \frac{L^{\boldsymbol{\alpha}} - 2}{\alpha_m} \right\rfloor\right\} \quad (4.78)$$

Note that $\Upsilon_{s1} \cap \Upsilon_{s4} = \varnothing$ and $\Upsilon_{s2} \cap \Upsilon_{s4} = \varnothing$.
In summary, $\Upsilon^{\alpha}_{n_1+n_2,L^{\alpha}} = \Upsilon_{s1} \cup \Upsilon_{s2} \setminus \Upsilon_{s3} \cup \Upsilon_{s4}$. Hence,

$$
\begin{aligned}
I^{\alpha}_{n_1+n_2,L^{\alpha}} &= \sum_{\Xi^{\alpha}\in\Upsilon^{\alpha}_{n_1+n_2,L^{\alpha}}} \left(\Delta^{i_1} \otimes \cdots \otimes \Delta^{i_{n_1}} \otimes \Delta^{i_{n_1+1}} \otimes \cdots \otimes \Delta^{i_{n_1+n_2}}\right) \\
&= \sum_{\Xi^{\alpha}\in(\Upsilon_{s1}\cup\Upsilon_{s2}\setminus\Upsilon_{s3}\cup\Upsilon_{s4})} \left(\Delta^{i_1} \otimes \cdots \otimes \Delta^{i_{n_1}} \otimes \Delta^{i_{n_1+1}} \otimes \cdots \otimes \Delta^{i_{n_1+n_2}}\right) \\
&= \left(\underbrace{\Delta^1 \otimes \cdots \otimes \Delta^1}_{n_1 \text{ elements}} \otimes \sum_{\Xi\in\Upsilon_{n_2,\left\lfloor \frac{L^{\alpha}-1}{\alpha_m}+1\right\rfloor}} \left(\Delta^{i_{n_1+1}} \otimes \cdots \otimes \Delta^{i_{n_1+n_2}}\right)\right) \\
&\quad + \left(\sum_{\Xi\in\Upsilon_{n_1,L^{\alpha}}} \left(\Delta^{i_1} \otimes \cdots \otimes \Delta^{i_{n_1}}\right) \otimes \underbrace{\Delta^1 \cdots \otimes \Delta^1}_{n_2 \text{ elements}}\right) \\
&\quad - \underbrace{\Delta^1 \otimes \Delta^1 \cdots \otimes \Delta^1}_{n_1+n_2} + \sum_{\Xi^{\alpha}\in\Upsilon_{s4}} \left(\Delta^{i_1} \otimes \cdots \otimes \Delta^{i_{n_1}} \otimes \Delta^{i_{n_1+1}} \otimes \cdots \otimes \Delta^{i_{n_1+n_2}}\right) \\
&= \left(\underbrace{I_1 \otimes \cdots \otimes I_1}_{n_1 \text{ elements}} \otimes \sum_{\Xi\in\Upsilon_{n_2,\left\lfloor \frac{L^{\alpha}-1}{\alpha_m}+1\right\rfloor}} \left(\Delta^{i_{n_1+1}} \otimes \cdots \otimes \Delta^{i_{n_1+n_2}}\right)\right) \\
&\quad + \left(\sum_{\Xi\in\Upsilon_{n_1,L^{\alpha}}} \left(\Delta^{i_1} \otimes \cdots \otimes \Delta^{i_{n_1}}\right) \otimes \underbrace{I_1 \cdots \otimes I_1}_{n_2 \text{ elements}}\right) \\
&\quad - \underbrace{I_1 \otimes I_1 \cdots \otimes I_1}_{n_1+n_2} + \sum_{\Xi^{\alpha}\in\Upsilon_{s4}} \left(\Delta^{i_1} \otimes \cdots \otimes \Delta^{i_{n_1}} \otimes \Delta^{i_{n_1+1}} \otimes \cdots \otimes \Delta^{i_{n_1+n_2}}\right) \\
&\quad I_{n_1,1} \otimes I_{n_2,\left\lfloor \frac{L^{\alpha}-1}{\alpha_m}+1\right\rfloor} + I_{n_1,L^{\alpha}} \otimes I_{n_2,1} - I_{n_1+n_2,1} + \\
&\quad \sum_{\Xi^{\alpha}\in\Upsilon_{s4}} \left(\Delta^{i_1} \otimes \cdots \otimes \Delta^{i_{n_1}} \otimes \Delta^{i_{n_1+1}} \otimes \cdots \otimes \Delta\right.
\end{aligned}
\tag{4.79}
$$

From the 3rd equality to the 4th equality in Eq. (4.79), the fact that $\Delta^1 = I_1 - I_0 = I_1$ is used. Note that $\underbrace{I_1 \otimes \cdots \otimes I_1}_{n_1 \text{ elements}} \otimes \sum_{\Xi\in\Upsilon_{n_2,\left\lfloor \frac{L^{\alpha}-1}{\alpha_m}+1\right\rfloor}} \left(\Delta^{i_{n_1+1}} \otimes \cdots \otimes \Delta^{i_{n_1+n_2}}\right)$

and $\sum_{\Xi\in\Upsilon_{n_1,L^\alpha}}\left(\Delta^{i_1}\otimes\cdots\otimes\Delta^{i_{n_1}}\right)\otimes\underbrace{I_1\cdots\otimes I_1}_{n_2\text{ elements}}$ correspond to Case (1) and Case (2) respectively. $\underbrace{I_1\otimes I_1\cdots\otimes I_1}_{n_1+n_2}$ is the duplicate part of Case (1) and Case (2). The last term $\sum_{\Xi^\alpha\in\Upsilon_{s4}}\left(\Delta^{i_1}\otimes\cdots\otimes\Delta^{i_{n_1}}\otimes\Delta^{i_{n_1+1}}\otimes\cdots\otimes\Delta^{i_{n_1+n_2}}\right)$ corresponds to Case (3) and has no duplicate part with Case (1) and Case (2). It is worth noting that when $L^\alpha < 2+\alpha_{m'}$, this last term vanishes and Υ_{s4} is empty, as seen from Eq. (4.78).

Now, compare Eq. (4.79) with the level-2 SGQ. The level-2 SGQ is

$$I_{n_1+n_2,2}=\sum_{\Xi\in\Upsilon_{n_1+n_2,2}}\left(\Delta^{i_1}\otimes\cdots\otimes\Delta^{i_{n_1}}\otimes\Delta^{i_{n_1+1}}\otimes\cdots\otimes\Delta^{i_{n_1+n_2}}\right)\text{ and }\sum_{j=1}^{n_1+n_2}\left(i_j-1\right)\le 1.$$

If all $i_j\left(j=1,\cdots,n_1\right)$ are 1, then $\sum_{j=n_1+1}^{n_1+n_2}\left(i_j-1\right)\le 1$; if $\sum_{j=1}^{n_1}\left(i_j-1\right)\le 1$, all $i_j\left(j=n_1+1,\cdots,n_1+n_2\right)$ must be 1. Note that both cases contain the duplicate case when all $i_j=1\left(j=1,\cdots,n_1+n_2\right)$. As with Eq. (4.79), these two cases lead to

$$\begin{aligned}I_{n_1+n_2,2}&=\sum_{\Xi\in\Upsilon_{n,2}}\left(\Delta^{i_1}\otimes\cdots\otimes\Delta^{i_{n_1}}\otimes\Delta^{i_{n_1+1}}\otimes\cdots\otimes\Delta^{i_{n_1+n_2}}\right)\\&=\left(\underbrace{I_1\otimes\cdots\otimes I_1}_{n_1\text{ elements}}\otimes\sum_{\Xi\in\Upsilon_{n_2,2}}\left(\Delta^{i_{n_1+1}}\otimes\cdots\otimes\Delta^{i_{n_1+n_2}}\right)\right)\\&\quad+\left(\sum_{\Xi\in\Upsilon_{n_1,2}}\left(\Delta^{i_1}\otimes\cdots\otimes\Delta^{i_{n_1}}\right)\otimes\underbrace{I_1\cdots\otimes I_1}_{n_2\text{ elements}}\right)\\&\quad-\underbrace{I_1\otimes I_1\cdots\otimes I_1}_{n_1+n_2}\\&=I_{n_1,1}\otimes I_{n_2,2}+I_{n_1,2}\otimes I_{n_2,1}-I_{n_1+n_2,1}\end{aligned}\tag{4.80}$$

Remark 4.10: By Theorem 4.1, $I_{n_1+n_2,2}$ is exact for all polynomials $x_1^{a_1}\cdots x_j^{a_j}\cdots x_{n_1+n_2}^{a_{n_1+n_2}}$ with $0\le\sum_{j=1}^{n_1+n_2}a_j\le 3$. Note, however, that the condition of Theorem 4.1 is sufficient but not necessary. In fact, from Eq. (4.80), it can be seen that $I_{n_1+n_2,2}$ is exact for all polynomials $x_1^{a_1}\cdots x_j^{a_j}\cdots x_{n_1+n_2}^{a_{n_1+n_2}}$ with $0\le\sum_{j=1}^{n_1}a_j\le 1, 0\le\sum_{j=n_1+1}^{n_1+n_2}a_j\le 3$ because of the exactness of $I_{n_1,1}\otimes I_{n_2,2}$ for these polynomials. $I_{n_1+n_2,2}$ is also exact for all polynomials $x_1^{a_1}\cdots x_j^{a_j}\cdots x_{n_1+n_2}^{a_{n_1+n_2}}$ with $0\le\sum_{j=1}^{n_1}a_j\le 3, 0\le\sum_{j=n_1+1}^{n_1+n_2}a_j\le 1$, because of the exactness of $I_{n_1,2}\otimes I_{n_2,1}$ for these polynomials. The above discussions also apply to the UT, which is exact up to the 3rd order polynomial, but may be exact for some higher order polynomials. On the other hand, the exactness of $I_{n_1+n_2,2}$ for the above polynomials cannot be further generalized to all the polynomials with $0\le\sum_{j=1}^{n_1+n_2}a_j\le 4$. For example, $I_{n_1+n_2,2}$ is not exact for all polynomials $x_1^{a_1}\cdots x_j^{a_j}\cdots x_{n_1+n_2}^{a_{n_1+n_2}}$ with $0\le\sum_{j=1}^{n_1}a_j\le 2, 0\le\sum_{j=n_1+1}^{n_1+n_2}a_j\le 2$. Note that the UT has the same accuracy as the level-2 SGQ (Jia et al. 2011) because all polynomials for which the UT is exact can be exactly calculated by the level-2 SGQ.

Comparing Eq. (4.79) with Eq. (4.80), one can see that when $L^{\boldsymbol{\alpha}} > 2$ and $\left\lfloor \frac{L^{\boldsymbol{\alpha}}-1}{\alpha_m}+1 \right\rfloor \ge 2$, there exist higher-order polynomials that can be exactly calculated by the first three terms in Eq. (4.79), but cannot by $I_{n_1+n_2,2}$ (level-2 SGQ). When $L^{\boldsymbol{\alpha}} = 2$, $\alpha_m = 1$, the first three terms in Eq. (4.79) are as accurate as $I_{n_1+n_2,2}$ (level-2 SGQ) is.

Now, let us consider the contribution of $\sum_{\Xi^{\boldsymbol{\alpha}}\in\Upsilon_{s4}}\left(\Delta^{i_1}\otimes\cdots\otimes\Delta^{i_{n_1}}\otimes\Delta^{i_{n_1+1}}\otimes\cdots\otimes\Delta^{i_{n_1+n_2}}\right)$, the last term of Eq. (4.79), to $I^{\boldsymbol{\alpha}}_{n_1+n_2,L^{\boldsymbol{\alpha}}}$. This term is generated from Case (3) where $1\le\sum_{j=1}^{n_1}(i_j-1)\le L^{\boldsymbol{\alpha}}-\alpha_m-1$, and $1\le\sum_{j=n_1+1}^{n_1+n_2}(i_j-1)\le\left\lfloor\frac{L^{\boldsymbol{\alpha}}-2}{\alpha_m}\right\rfloor$. When both inequalities are satisfied, there exist at least one $i_m\ge 2\,(1\le m\le n_1)$ and one $i_k\ge 2\,(n_1+1\le k\le n_1+n_2)$.

For the polynomials that can be exactly calculated by $I_{n_1,s}\otimes I_{n_2,1}$ $(s\in\mathbb{N})$ using Eq. (4.79), the contribution of the last term in Eq. (4.79) is 0, i.e.

$$\begin{aligned}
&\left(\Delta^{i_1}\otimes\cdots\otimes\Delta^{i_m}\otimes\cdots\otimes\Delta^{i_{n_1}}\right)\otimes\left(\Delta^{i_{n_1+1}}\otimes\cdots\otimes\Delta^{i_k}\otimes\cdots\otimes\Delta^{i_{n_1+n_2}}\right)\\
&=\left(\Delta^{i_1}\otimes\cdots\otimes\Delta^{i_m}\otimes\cdots\otimes\Delta^{i_{n_1}}\right)\otimes\left(\Delta^{i_{n_1+1}}\otimes\cdots\otimes 0\otimes\cdots\otimes\Delta^{i_{n_1+n_2}}\right)\\
&=0
\end{aligned}\tag{4.81}$$

That is because a polynomial f that can be exactly calculated by I_1 can be exactly calculated by $I_j (j \geq 1)$ as well, i.e., $I_j(f) = I_1(f)(j \geq 1)$ and thus $\Delta^{i_k}(f) = I_{i_k}(f) - I_{i_k - 1}(f) = 0\,(i_k \geq 2)$.

Likewise, for the polynomials that can be exactly calculated by $I_{n_1,1} \otimes I_{n_2,s}$ $(s \in \mathbb{N})$ using Eq. (4.79), the contribution of the last term in Eq. (4.79) is also 0, i.e.,

$$\begin{aligned}
&\left(\Delta^{i_1} \otimes \cdots \otimes \Delta^{i_m} \otimes \cdots \otimes \Delta^{i_{n_1}}\right) \otimes \left(\Delta^{i_{n_1+1}} \otimes \cdots \otimes \Delta^{i_k} \otimes \cdots \otimes \Delta^{i_{n_1+n_2}}\right) \\
&= \left(\Delta^{i_1} \otimes \cdots \otimes 0 \otimes \cdots \otimes \Delta^{i_{n_1}}\right) \otimes \left(\Delta^{i_{n_1+1}} \otimes \cdots \otimes \Delta^{i_k} \otimes \cdots \otimes \Delta^{i_{n_1+n_2}}\right) \\
&= 0
\end{aligned} \tag{4.82}$$

Hence, any polynomial that can be exactly calculated by $I_{n_1,2} \otimes I_{n_2,1}$, $I_{n_1,1} \otimes I_{n_2,2}$, or $I_{n_1+n_2,2}$ can be exactly calculated by Eq. (4.79). As discussed in Remark 4.10, the types of polynomials that can be exactly calculated by $I_{n_1+n_2,2}$ are all covered by the types of polynomials that can be exactly calculated by $I_{n_1,2} \otimes I_{n_2,1}$ or $I_{n_1,1} \otimes I_{n_2,2}$.

To summarize, when $2 \leq \alpha_m + 1 \leq L^{\alpha}$, because (1) the first three terms in Eq. (4.79) are more accurate than or as accurate as the level-2 SGQ $(I_{n_1+n_2,2})$ and (2) the last term in Eq. (4.79) is 0 for polynomials that can be exactly calculated by $I_{n_1,s} \otimes I_{n_2,1}$ or $I_{n_1,1} \otimes I_{n_2,s}$ $(s \in \mathbb{N})$, which include the first three terms in Eq. (4.79), ASGQ is more accurate than or as accurate as the level-2 SGQ. ■

Note that for sufficiently large L^{α}, there exist polynomials that can be exactly calculated by the last term in Eq. (4.79) but cannot be exactly calculated by either the first three terms of Eq. (4.79) or the level-2 SGQ $(I_{n_1+n_2,2})$. Examples of these polynomials include $x_m^2 x_k^2$ $(1 \leq m \leq n_1,\ n_1 + 1 \leq k \leq n_1 + n_2)$. Also recall that when $L^{\alpha} = 2$ and $\alpha_m = 1$, the ASGQ is identical to the level-2 SGQ as discussed in Remark 4.8.

Remark 4.11: As discussed in Remark 4.10, the UT has the same accuracy as the level-2 SGQ. Thus, the ASGQ is more accurate than the UT when $2 < \alpha_m + 1 \leq L^{\alpha}$.

When $L^{\alpha} = 3$ and $\boldsymbol{\alpha} = \left[\underbrace{1,\cdots,1}_{n_1},\underbrace{2,\cdots,2}_{n_2}\right]$, which may be used for many applications, the following equation can be obtained from Eq. (4.79).

$$I^{\boldsymbol{\alpha}}_{n_1+n_2,3} = \left(I_{n_1,1} \otimes I_{n_2,2}\right) + \left(I_{n_1,3} \otimes I_{n_2,1}\right) - I_{n_1+n_2,1} \tag{4.83}$$

The illustration of ASGQ in Fig. 4.2 can be viewed as a special case $(n_1 = 1,\ n_2 = 1,\ L^{\alpha} = 3)$ of Eq. (4.83).

Note that, by Eq. (4.83), $I^{\boldsymbol{\alpha}}_{n_1+n_2,3}$ is exact for all the polynomials of the form $x_1^{a_1}\cdots x_j^{a_j}\cdots x_{n_1+n_2}^{a_{n_1+n_2}}$ with $0\le\sum_{j=1}^{n_1}a_j\le 1, 0\le\sum_{j=n_1+1}^{n_1+n_2}a_j\le 3$ or $0\le\sum_{j=1}^{n_1}a_j\le 5, 0\le\sum_{j=n_1+1}^{n_1+n_2}a_j\le 1$. In contrast, the UT is exact for all polynomials with $0\le\sum_{j=1}^{n_1+n_2}a_j\le 3$. Therefore, ASGQ is more accurate than UT when $\boldsymbol{\alpha}=\left[\underbrace{1,\cdots,1}_{n_1},\underbrace{2,\cdots,2}_{n_2}\right]$ and $L^{\alpha}=3$, which satisfies the condition of Theorem 4.3.

4.7 Spherical-Radial Cubature

Equation (4.13) can be rewritten as

$$\int_{\mathbb{R}^n}\mathbf{g}(\mathbf{x})N(\mathbf{x};\mathbf{0},\mathbf{I})d\mathbf{x}=\frac{1}{\pi^{n/2}}\int_{\mathbb{R}^n}\mathbf{g}\left(\sqrt{2}\mathbf{x}\right)\exp\left(-\mathbf{x}^T\mathbf{x}\right)d\mathbf{x} \tag{4.84}$$

The spherical-radial cubature rule is a numerical integration rule to compute the following integral:

$$I(\mathbf{g})=\int_{\mathbb{R}^n}\mathbf{g}(\mathbf{x})\exp\left(-\mathbf{x}^T\mathbf{x}\right)d\mathbf{x} \tag{4.85}$$

Let $\mathbf{x}=r\mathbf{s}$ with $\mathbf{s}^T\mathbf{s}=1$ and $r=\sqrt{\mathbf{x}^T\mathbf{x}}$. Equation (4.85) can be transformed in the spherical-radial coordinate system (Arasaratnam and Haykin 2009, Jia et al. 2013)

$$I(\mathbf{g})=\int_0^{\infty}\int_{U_n}\mathbf{g}(r\mathbf{s})r^{n-1}\exp\left(-r^2\right)d\sigma(\mathbf{s})dr \tag{4.86}$$

where $\mathbf{s}=\left[s_1,s_2,\cdots,s_n\right]^T$, $U_n=\left\{\mathbf{s}\in\mathbb{R}^n: s_1^2+s_2^2+\cdots+s_n^2=1\right\}$, and $\sigma(.)$ is the area element on U_n (Arasaratnam and Haykin 2009).

Similar to the accuracy level of the SGQ rule, the approximation accuracy of the spherical-radial cubature rule to calculate the integral of (4.85) or (4.86) can be defined.

Definition 4.1. A numerical integration rule $\int_{\mathbb{R}^n}\mathbf{g}(\mathbf{x})w_g(\mathbf{x})d\mathbf{x}\approx\sum W_i\mathbf{g}(\boldsymbol{\gamma}_i)$ is a *d*th-degree rule if it is exact for $\mathbf{g}(\mathbf{x})$ whose components are linear combinations of monomials $x_1^{\alpha_1}x_2^{\alpha_2}\cdots x_n^{\alpha_n}$ with the total degree up to d ($\alpha_1,\alpha_2,\cdots,\alpha_n$ are nonnegative integers and $0\le\alpha_1+\alpha_2+\cdots+\alpha_n\le d$).

Note that a rule with accuracy level L is a $(2L-1)$th-degree rule.

Equation (4.86) contains two types of integrals: the radial integral $\int_0^{\infty}g_r(r)r^{n-1}\exp\left(-r^2\right)dr$ with the weighting function $w_g(r)=r^{n-1}\exp\left(-r^2\right)$

and the spherical integral $\int_{U_n} g_s(\mathbf{s})d\sigma(\mathbf{s})$ with the weighting function $w_g(\mathbf{s}) = 1$. The spherical-radial cubature rule is an approximation to Eq. (4.86) by

$$I(g) = \int_0^\infty r^{n-1}\exp(-r^2)\int_{U_n} g(r\mathbf{s})d\sigma(\mathbf{s})dr \approx \int_0^\infty r^{n-1}\exp(-r^2)\sum_{j=1}^{N_s} w_{s,j} g(r\mathbf{s}_j)dr$$

$$\approx \sum_{i=1}^{N_r}\sum_{j=1}^{N_s} w_{r,i}w_{s,j} g(r_i\mathbf{s}_j) \qquad (4.87)$$

where r_i and $w_{r,i}$ are the points and weights for calculating the radial integral; $\mathbf{s}_j$ and $w_{s,j}$ are the points and weights for calculating the spherical integral. N_r and N_s are the number of points for the radial rule and the spherical rule respectively.

Note that the spherical-radial cubature rule in Eq. (4.87) is a *d*th degree rule if the radial rule and the spherical rule are both *d*th degree rules (Arasaratnam and Haykin 2009, Jia et al. 2013).

Remark 4.12: A spherical-radial cubature rule is said to be fully symmetric if the points generated from the rule are fully symmetric. If a spherical rule is fully symmetric, the corresponding spherical-radial cubature rule in Eq. (4.87) is fully symmetric as well. A fully symmetric spherical-radial cubature rule is exact for any odd function $g(\mathbf{x})$ in Eq. (4.85) automatically. If a fully symmetric spherical rule is used in the spherical-radial cubature rule, the radial rule only needs to be exact for even-degree polynomials in r because the spherical rule vanishes when $g_s(\mathbf{s})$ are polynomials of odd degrees.

The arbitrary degree radial rule can be obtained by the generalized Gauss-Laguerre quadrature (Jia et al. 2013) or by the moment matching method (Jia et al. 2013).

The moment matching method is to find the points r_i and weights $w_{r,i}$ to satisfy the moment equation of the form

$$\sum_{i=1}^{N_r} w_{r,i} g_r(r_i) = \int_0^\infty g_r(r) r^{n-1}\exp(-r^2)dr \qquad (4.88)$$

where $g_r(r) = r^l$ is a monomial in r, with l an even integer. Note that for $g_r(r) = r^l$, the right-hand side of (4.88) reduces to $\frac{1}{2}\Gamma\left(\frac{n+l}{2}\right)$ where $\Gamma(z)$ is the gamma function defined by $\Gamma(z) = \int_0^\infty \exp(-\lambda)\lambda^{z-1}d\lambda$. Only even-degree monomials need to be matched. To obtain a $(2L-1)$th-degree radial rule for the $(2L-1)$th-degree spherical-radial cubature rule, Eq. (4.88) needs to be exact for $l = 0, 2, \ldots, 2(L-1)$, which contains L equations.

As an example, the third-degree and fifth-degree radial rules with the minimum number of points are given (Jia et al. 2013). For the third-degree radial rule, one radial point ($N_r = 1$) can be used and needs to satisfy the following equations:

$$\begin{cases} w_{r,1} r_1^0 = \frac{1}{2}\Gamma\left(\frac{1}{2}n\right) \\ w_{r,1} r_1^2 = \frac{1}{2}\Gamma\left(\frac{1}{2}n+1\right) = \frac{n}{4}\Gamma\left(\frac{1}{2}n\right) \end{cases} \tag{4.89}$$

where the last equality follows the identity $\Gamma(z+1) = z\Gamma(z)$.
Solving (4.89) gives the point and weight of the third-degree radial rule,

$$r_1 = \sqrt{\frac{n}{2}} \tag{4.90a}$$

$$w_{r,1} = \frac{\Gamma(n/2)}{2} \tag{4.90b}$$

For the fifth-degree radial rule ($N_r = 2$), the points and weights satisfy the following three equations:

$$\begin{cases} w_{r,1} r_1^0 + w_{r,2} r_2^0 = \frac{1}{2}\Gamma\left(\frac{1}{2}n\right) \\ w_{r,1} r_1^2 + w_{r,2} r_2^2 = \frac{1}{2}\Gamma\left(\frac{1}{2}n+1\right) = \frac{n}{4}\Gamma\left(\frac{1}{2}n\right) \\ w_{r,1} r_1^4 + w_{r,2} r_2^4 = \frac{1}{2}\Gamma\left(\frac{1}{2}n+2\right) = \frac{1}{2}\left(\frac{1}{2}n\right)\left(\frac{1}{2}n+1\right)\Gamma\left(\frac{1}{2}n\right) \end{cases} \tag{4.91}$$

Since there are three equations and four variables in Eq. (4.91), there is one free variable. We can choose r_1 as the free variable and set it to 0 to get the minimum number of points. Solving these three equations gives the points and weights for the fifth-degree radial rule,

$$\begin{cases} r_1 = 0 \\ r_2 = \sqrt{n/2+1} \end{cases} \tag{4.92a}$$

and

$$\begin{cases} w_{r,1} = \frac{1}{2}\Gamma\left(\frac{1}{2}n\right) - \frac{n\Gamma\left(\frac{1}{2}n\right)}{2(n+2)} = \frac{1}{(n+2)}\Gamma\left(\frac{1}{2}n\right) \\ w_{r,2} = \frac{\frac{n}{4}\Gamma\left(\frac{1}{2}n\right)}{\left(\frac{n}{2}+1\right)} = \frac{n}{2(n+2)}\Gamma\left(\frac{1}{2}n\right) \end{cases} \tag{4.92b}$$

The spherical rule of arbitrary degrees of accuracy can be given by the following Genz rule (Genz 2003).

Theorem 4.4 (Genz 2003): For the spherical integral

$$I_{U_n}(\boldsymbol{g}_s) \triangleq \int_{U_n} \boldsymbol{g}_s(\mathbf{s})\,d\sigma(\mathbf{s}), \tag{4.93}$$

$$I_{U_n,L}(\boldsymbol{g}_s) = \sum_{|\Xi|=n+L-1} w_{\Xi}\boldsymbol{G}\{\mathbf{u}_{\Xi}\}\ (L \geq 2) \tag{4.94}$$

is a $(2L-1)$th-degree rule, where I_{U_n} denotes a spherical integral and $I_{U_n,L}$ denotes the $(2L-1)$th-degree spherical rule used to approximate the integral. w_{Ξ} and $\boldsymbol{G}\{\mathbf{u}_{\Xi}\}$ are defined as

$$w_{\Xi} \triangleq I_{U_n}\left(\prod_{j=1}^{n}\prod_{m=0}^{i_j-2}\frac{s_j^2-u_m^2}{u_{i_j-1}^2-u_m^2}\right) \tag{4.95}$$

$$\boldsymbol{G}\{\ _{\Xi}\} \triangleq 2^{(\ )}\sum \boldsymbol{g}_s\left(v_1u_{i_1-1}, v_2u_{i_2-1}, \cdots, v_nu_{i\ -1}\right) \tag{4.96}$$

where the subscripts i_j in Eqs. (4.95) and (4.96) are natural numbers with $\Xi = (i_1, i_2, \cdots, i_n)$ and $|\Xi| = i_1 + i_2 + \cdots + i_n$; the superscript $c(\mathbf{u}_{\Xi})$ in Eq. (4.96) is the number of nonzero entries in $\mathbf{u}_{\Xi} = \left(u_{i_1-1}, u_{i_2-1}, \cdots, u_{i_n-1}\right)$; u_{i_j-1} are chosen to be $u_{i_j-1} = \sqrt{(i_j-1)/(L-1)}$ $(i_j = 1, \cdots, L)$; the points of the spherical rule $I_{U_n,L}$ are $\left[v_1u_{i_1-1}, v_2u_{i_2-1}, \cdots, v_nu_{i_n-1}\right]^T$ with $v_i = \pm 1$ and the weights on the points are $2^{-c(\mathbf{u}_{\Xi})}w_{\Xi}$. ■

The 1st-degree spherical rule is trivial and thus $L \geq 2$ is assumed in the Theorem 4.4.

Note that the subscript L in (4.94) represents the accuracy level that is the same as the concept of accuracy level L for the SGQ in (4.38).

Using Genz's rule in Theorem 4.4, the specific third-degree (accuracy level $L = 2$) spherical rule can be derived as (Jia et al. 2013)

$$I_{U_n,2}(\boldsymbol{g}_s) = \frac{A_n}{2n}\sum_{j=1}^{n}\left[\boldsymbol{g}_s(\mathbf{e}_j) + \boldsymbol{g}_s(-\mathbf{e}_j)\right] \tag{4.97}$$

where $A_n = 2\Gamma^n(1/2)/\Gamma(n/2) = 2\sqrt{\pi^n}/\Gamma(n/2)$ is the surface area of the unit sphere.

Similarly, the fifth-degree (accuracy level L = 3) spherical rule can be given by (Jia et al. 2013)

$$I_{U_n,3}(g_s)=\frac{(4-n)A_n}{2n(n+2)}\sum_{j=1}^{n}\left[g_s(\mathbf{e}_j)+g_s(-\mathbf{e}_j)\right]+\frac{A_n}{n(n+2)}\sum_{j=1}^{n(n-1)/2}\left[g_s(\mathbf{s}_j^+)+g_s(-\mathbf{s}_j^+)+g_s(\mathbf{s}_j^-)+g_s(-\mathbf{s}_j^-)\right] \tag{4.98}$$

where
$$\begin{cases}\mathbf{s}_j^+=\sqrt{\dfrac{1}{2}}(\mathbf{e}_k+\mathbf{e}_l) & k<l,\ k,l=1,2,\cdots,n\\ \mathbf{s}_j^-=\sqrt{\dfrac{1}{2}}(\mathbf{e}_k-\mathbf{e}_l) & k<l,\ k,l=1,2,\cdots,n\end{cases} \tag{4.99}$$

The spherical rule used in Theorem 4.4 is fully symmetric. Thus, the radial rule only needs to be exact for even-degree polynomials in r.

The spherical-radial cubature rule is a combination of the spherical rule and radial rule as shown in Eq. (4.87). The procedures to generate the cubature points and weights for approximating the integral Eq. (4.84) can be summarized in Table 4.2.

As an example, the third-degree and the fifth-degree spherical-radial cubature rules using Eqs. (4.90), (4.92), (4.97), (4.98), and the algorithm shown in Table 4.2, are given by Eqs. (4.100) and (4.101), respectively (Jia et al. 2013).

$$\int_{\mathbb{R}^n}\mathbf{g}(\mathbf{x})N(\mathbf{x};\mathbf{0},\mathbf{I})d\mathbf{x}\approx\frac{1}{2n}\sum_{j=1}^{n}\left[\mathbf{g}\left(\sqrt{n}\mathbf{e}_j\right)+\mathbf{g}\left(-\sqrt{n}\mathbf{e}_j\right)\right] \tag{4.100}$$

Table 4.2 Generate cubature points and weights.

$[\chi,W]=\text{CubatureRule}[n,L]$

(χ: cubature point set of which the kth element is χ_k; W: weight sequence of which the kth element is W_k)

(1) Obtain the spherical points and weights

Determine N_{L-1}^n

  FOR each element $\Xi=[i_1,i_2,\cdots,i_n]$ in N_{L-1}^n, form

$$\mathbf{s}_j=\left[v_1u_{i_1-1},v_2u_{i_2-1},\cdots,v_nu_{i_n-1}\right]^T$$

  Calculate the weight $w_{s,j}=2^{-c(\mathbf{u}_\Xi)}w_\Xi$.

  END

(2) Obtain the radial points r_i and weights $w_{r,i}$ by solving the moment matching equations.

Set k=1.

FOR ith point of the radial rule

  FOR jth point of the spherical rule

  $\chi_k=\sqrt{2}\,r_i\mathbf{s}_j$ and $W_k=w_{r,i}\,w_{s,j}/\pi^{n/2}$

  $k\leftarrow k+1$

  END

END

$$\begin{aligned}\int_{\mathbb{R}^n} \mathbf{g}(\mathbf{x}) N(\mathbf{x};\mathbf{0},\mathbf{I}) d\mathbf{x} &\approx \frac{2}{n+2}\mathbf{g}(\mathbf{0}) \\ &+ \frac{4-n}{2(n+2)^2}\sum_{j=1}^{n}\left[\mathbf{g}\left(\sqrt{n+2}\cdot\mathbf{e}_j\right)+\mathbf{g}\left(-\sqrt{n+2}\cdot\mathbf{e}_j\right)\right] \\ &+ \frac{1}{(n+2)^2}\sum_{j=1}^{n(n-1)/2}\left[\mathbf{g}\left(\sqrt{n+2}\cdot\mathbf{s}_j^{+}\right)+\mathbf{g}\left(-\sqrt{n+2}\cdot\mathbf{s}_j^{+}\right)\right] \\ &+ \frac{1}{(n+2)^2}\sum_{j=1}^{n(n-1)/2}\left[\mathbf{g}\left(\sqrt{n+2}\cdot\mathbf{s}_j^{-}\right)+\mathbf{g}\left(-\sqrt{n+2}\cdot\mathbf{s}_j^{-}\right)\right]\end{aligned} \tag{4.101}$$

Note that Eq. (4.100) is equivalent to the spherical-radial cubature rule of (Arasaratnam and Haykin 2009). The nonlinear Gaussian approximation filter using the spherical-radial cubature rule is called the cubature Kalman filter (CKF). The CKF in (Arasaratnam and Haykin 2009) is a third-degree CKF and the one given in (Jia et al. 2013) is a class of higher-degree (> 3) CKF.

4.8 The Relations Among Unscented Transformation, Sparse-Grid Quadrature, and Cubature Rule

In the following discussion, we assume $L \leq n+1$. This constraint is generally true for most applications.

4.8.1 From the Spherical-Radial Cubature Rule to the Unscented Transformation

As pointed out in (Jia et al. 2013), the third-degree spherical-radial cubature rule based on the single-point Gauss-Laguerre radial rule is identical to the UT with the parameter $\kappa = 0$. This coincidence indicates that this specific third-degree spherical-radial cubature rule can be regarded as a special case of the UT. As a matter of fact, there exists a more profound relation underlying the cubature rule and the UT the other way round. In the following, it will be shown that the UT can be constructed from the more general third-degree spherical-radial cubature rule using the Genz's spherical rule and a moment matching-based two-point radial rule.

***Theorem 4.5*:** The unscented transformation can be constructed from the third-degree spherical-radial cubature rule.

Proof: The third-degree spherical rule can be obtained by Eq. (4.97) using Genz's method (Genz 2003) and is repeated here

$$I_{U_n,2}(\mathbf{g}_s) = \frac{A_n}{2n}\sum_{j=1}^{n}\left[\mathbf{g}_s(\mathbf{e}_j)+\mathbf{g}_s(-\mathbf{e}_j)\right] \tag{4.102}$$

If two points r_1 and r_2 are used for the third-degree radial rule, by the moment-matching method, the following equations should be satisfied.

$$\begin{cases} w_{r,1}r_1^0 + w_{r,2}r_2^0 = \Gamma(n/2)/2 \\ w_{r,1}r_1^2 + w_{r,2}r_2^2 = n\Gamma(n/2)/4 \end{cases} \tag{4.103}$$

Note that the right-hand sides of the equations are the exact first and third moments of the radial rule and that the equations are the first two equations of Eq. (4.91). Since there are two equations with four unknown variables, one can choose r_1 and r_2 as

$$\begin{cases} r_1 = 0 \\ r_2 = \sqrt{(n+\kappa)/2} \end{cases} \tag{4.104}$$

Then the weights $w_{r,2}$ and $w_{r,1}$ can be solved from Eq. (4.92) as

$$\begin{cases} w_{r,1} = \kappa/(2(n+\kappa)) \cdot \Gamma(n/2) \\ w_{r,2} = n/(2(n+\kappa)) \cdot \Gamma(n/2) \end{cases} \tag{4.105}$$

Hence, the third-degree cubature rule can be obtained by the algorithm shown in Table 4.2 as follows.

$$\begin{aligned}
&\int_{\mathbb{R}^n} \mathbf{g}(\mathbf{x}) N(\mathbf{x};\mathbf{0},\mathbf{I}) d\mathbf{x} = \frac{1}{\pi^{n/2}} \int_{\mathbb{R}^n} \mathbf{g}\left(\sqrt{2}\mathbf{x}\right) \exp\left(-\mathbf{x}^T\mathbf{x}\right) d\mathbf{x} \\
&\approx \frac{1}{\pi^{n/2}} \sum_{i=1}^{N_r} \sum_{j=1}^{N_s} w_{r,i} w_{s,j} \mathbf{g}\left(\sqrt{2} r_i \mathbf{s}_j\right) \\
&= \frac{1}{\pi^{n/2}} \sum_{j=1}^{n} \frac{\Gamma(n/2)}{2} \frac{\kappa}{n+\kappa} \frac{A_n}{2n} \left[\mathbf{g}\left(\sqrt{2}\cdot 0 \cdot \mathbf{e}_j\right) + \mathbf{g}\left(-\sqrt{2}\cdot 0\cdot \mathbf{e}_j\right) \right] \\
&+ \frac{1}{\pi^{n/2}} \sum_{j=1}^{n} \frac{\Gamma(n/2)}{2} \cdot \frac{n}{n+\kappa} \cdot \frac{A_n}{2n} \left[\mathbf{g}\left(\sqrt{2}\cdot\sqrt{\frac{n+\kappa}{2}}\cdot \mathbf{e}_j\right) + \mathbf{g}\left(-\sqrt{2}\cdot\sqrt{\frac{n+\kappa}{2}}\cdot \mathbf{e}_j\right) \right] \\
&= \frac{\kappa}{n+\kappa} \mathbf{g}(\mathbf{0}) + \frac{1}{2(n+\kappa)} \sum_{j=1}^{n} \left(\mathbf{g}\left(\sqrt{n+\kappa}\cdot\mathbf{e}_j\right) + \mathbf{g}\left(-\sqrt{n+\kappa}\cdot \mathbf{e}_j\right) \right)
\end{aligned} \tag{4.106}$$

Equation (4.106) is the same as the unscented transformation. ■

Remark 4.13: All weights of the third-degree spherical rule are positive. To guarantee the weights of the spherical-radial cubature rule are positive, $w_{r,1} > 0$ and $w_{r,2} > 0$ should be satisfied. By Eq. (4.105), we have $w_{r,2} = n\Gamma(n/2)/(4r_2^2) > 0$ and $w_{r,1} = \Gamma(n/2)/(4r_2^2)\cdot(2r_2^2 - n)$. Hence, if $r_2 > \sqrt{n/2}$, all weights are positive.

4.8.2 The Connection between the Quadrature Rule and the Spherical Rule

In this section, we will show that the dth-degree spherical rule can be obtained from the general dth-degree quadrature rule by projecting the quadrature points onto the surface of the unit hypersphere.

Denote $\mathbf{x}^d$ and $\mathbf{s}^d$ as polynomials with total degree d. By Eq. (4.86),

$$\begin{aligned} I\left(\mathbf{x}^d\right) &= \int_{\mathbb{R}^n} x_1^{\alpha_1} x_2^{\alpha_2} \cdots x_n^{\alpha_n} \exp\left(-\mathbf{x}^T\mathbf{x}\right) d\mathbf{x} \\ &= \int_0^\infty r^{\alpha_1+\alpha_2+\cdots+\alpha_n} r^{n-1} \exp\left(-r^2\right) dr \int_{U_n} s_1^{\alpha_1} s_2^{\alpha_2} \cdots s_n^{\alpha_n} d\sigma(\mathbf{s}) \\ &= \int_0^\infty r^{n-1+d} \exp\left(-r^2\right) dr \int_{U_n} \mathbf{s}^d d\sigma(\mathbf{s}) \end{aligned} \tag{4.107}$$

where $d = \alpha_1 + \cdots + \alpha_n$.

Thus,
$$\int_{U_n} \mathbf{s}^d d\sigma(\mathbf{s}) = I\left(\mathbf{x}^d\right) \Big/ \int_0^\infty r^{n-1+d} \exp\left(-r^2\right) dr \tag{4.108}$$

Equation (4.108) implies that if a quadrature rule $\sum_{j=1}^{N_g} \bar{W}_j \boldsymbol{g}(\bar{\boldsymbol{\gamma}}_j)$ is exact for $I(\mathbf{g}(\mathbf{x}))$ with $\mathbf{g}(\mathbf{x}) = \mathbf{x}^d$ where $\bar{\boldsymbol{\gamma}}_j$ and $\bar{W}_j$ are the quadrature points and weights, respectively, and N_g is the number of quadrature points, it can be used to calculate the spherical integral $\int_{U_n} g_s(\mathbf{s}) d\sigma(\mathbf{s})$ exactly, where $g_s(\mathbf{s}) = \mathbf{s}^d$. If the origin is one of the quadrature points, it can be omitted from the quadrature point set because the origin does not contribute to the quadrature rule $\sum_{j=1}^{N_g} \bar{W}_j \boldsymbol{g}(\bar{\boldsymbol{\gamma}}_j)$ with $\boldsymbol{g}(\mathbf{x}) = \mathbf{x}^d$.

Now we show how the spherical rule can be constructed from a quadrature rule, for example, the SGQ rule, based on Eq. (4.108). Because all the points $\mathbf{s}_j$ of the spherical rule need to satisfy the constraint $\left\|\mathbf{s}_j\right\|_2 = 1$, they can be obtained by projecting the SGQ points $\bar{\boldsymbol{\gamma}}_j$ in Euclidean space onto the surface of the hypersphere, i.e.,

$$\mathbf{s}_j = \frac{\bar{\boldsymbol{\gamma}}_j}{\left\|\bar{\boldsymbol{\gamma}}_j\right\|_2} \tag{4.109}$$

where $\|\cdot\|_2$ is the L_2-norm of a vector.

Remark 4.14: There may exist duplicated points of $\mathbf{s}_j$ obtained from Eq. (4.109). For example, if $\bar{\boldsymbol{\gamma}}_{j_1} = [p_t, 0, \cdots, 0]^T$ and $\bar{\boldsymbol{\gamma}}_{j_2} = [2p_t, 0, \cdots, 0]^T$, the points $\mathbf{s}_{j_1}$ and $\mathbf{s}_{j_2}$ obtained from the projection of $\bar{\boldsymbol{\gamma}}_{j_1}$ and $\bar{\boldsymbol{\gamma}}_{j_2}$ by Eq. (4.109) are identical.

Using the quadrature rule and Eq. (4.109) into Eq. (4.108) yields

$$\int_{U_n} \mathbf{s}^d d\sigma(\mathbf{s}) = \sum_{j=1}^{N_p} \bar{W}_j (\bar{\boldsymbol{\gamma}}_j)^d \Big/ \int_0^\infty r^{n-1+d} \exp(-r^2) dr$$
$$= \sum_{j=1}^{N_p} \bar{W}_j (\mathbf{s}_j)^d \|\bar{\boldsymbol{\gamma}}_j\|_2^d \Big/ \int_0^\infty r^{n-1+d} \exp(-r^2) dr \quad (4.110)$$
$$= \sum_{j=1}^{N_p} w_{s,j} (\mathbf{s}_j)^d$$

where N_p is the number of projected quadrature points. Note that $N_p \leq N_g$. The weights $w_{s,j}$ of the spherical rule are

$$w_{s,j} = \bar{W}_j \|\bar{\boldsymbol{\gamma}}_j\|_2^d \Big/ \int_0^\infty r^{n-1+d} \exp(-r^2) dr = \bar{W}_j \|\bar{\boldsymbol{\gamma}}_j\|_2^d \Big/ \left(\frac{\Gamma(n/2 + d/2)}{2} \right) \quad (4.111)$$

Remark 4.15: Equation (4.111) is automatically satisfied if d is odd and the quadrature rule is fully symmetric because in this case both $I(\mathbf{x}^d)$ and $\int_{U_n} \mathbf{s}^d d\sigma(\mathbf{s})$ are 0 as a result of full symmetry of the quadrature rule.

Theorem 4.6: If the dth degree (d is even) fully symmetric quadrature rule is used in Eq. (4.108), the spherical rule constructed in Eq. (4.110) with the points given by Eq. (4.109) and the weights given by Eq. (4.111) is also a dth degree rule.

Proof: The spherical rule in Eq. (4.110) constructed from the dth degree quadrature rule, is exact for $\int_{U_n} \mathbf{s}^d d\sigma(\mathbf{s})$. To show that it is a dth degree spherical rule, it suffices to show that it is exact for any spherical rules $\int_{U_n} \mathbf{s}^\beta d\sigma(\mathbf{s})$ with a degree $0 \leq \beta \leq d$.

$$\int_{U_n} \mathbf{s}^\beta d\sigma(\mathbf{s}) = \frac{\int_0^\infty r^d r^{n-1} \exp(-r^2) dr}{\int_0^\infty r^d r^{n-1} \exp(-r^2) dr} \int_{U_n} \mathbf{s}^\beta d\sigma(\mathbf{s})$$
$$= \frac{\int_0^\infty r^{d-\beta} r^\beta r^{n-1} \exp(-r^2) dr \int_{U_n} \mathbf{s}^\beta d\sigma(\mathbf{s})}{\left(\frac{\Gamma(n/2 + d/2)}{2} \right)} \quad (4.112)$$

Let $\mathbf{x} = r\mathbf{s}$ with $\mathbf{s}^T\mathbf{s} = 1$ and $r = \sqrt{\mathbf{x}^T \mathbf{x}}$. The numerator in the last equality of Eq. (4.112) can be rewritten as

$$\int_{\mathbb{R}^n} \left[\left(\sqrt{x_1^2 + \cdots + x_n^2} \right)^{d-\beta} \mathbf{x}^\beta \right] \exp(-\mathbf{x}^T \mathbf{x}) d\mathbf{x} \quad (4.113)$$

If β is odd, $\left[\left(\sqrt{x_1^2+\cdots+x_n^2}\right)^{d-\beta}\mathbf{x}^\beta\right]$ is not a polynomial but still an odd function. Because of the symmetry of the integration region and the property of the odd function, the integral in Eq. (4.113) vanishes.

So, for any odd $\beta < d$,

$$\int_{U_n}\mathbf{s}^\beta d\sigma(\mathbf{s}) = 0 = \sum_{j=1}^{N_p} w_{s,j}\left(\mathbf{s}_j\right)^\beta \tag{4.114}$$

The second equality holds because the spherical rule is fully symmetric.

If β is even, then $d-\beta$ is even and $\left[\left(\sqrt{x_1^2+\cdots+x_n^2}\right)^{d-\beta}\mathbf{x}^\beta\right]$ is a polynomial with a degree $\frac{d-\beta}{2}+\beta=\frac{d+\beta}{2}\leq\frac{d+d}{2}=d$. Therefore, the integral in Eq. (4.113) can be exactly calculated by the dth degree quadrature rule as follows

$$\int_{\mathbb{R}^n}\left[\left(\sqrt{x_1^2+\cdots+x_n^2}\right)^{d-\beta}\mathbf{x}^\beta\right]\exp\left(-\mathbf{x}^T\mathbf{x}\right)d\mathbf{x} = \sum_{j=1}^{N_p}\bar{W}_j\left\|\bar{\boldsymbol{\gamma}}_j\right\|_2^{d-\beta}\bar{\boldsymbol{\gamma}}_j^\beta \tag{4.115}$$

Combining Eq. (4.112) and Eq. (4.115) leads to

$$\begin{aligned}\int_{U_n}\mathbf{s}^\beta d\sigma(\mathbf{s}) &= \sum_{j=1}^{N_p}\bar{W}_j\left\|\bar{\boldsymbol{\gamma}}_j\right\|_2^{d-\beta}\bar{\boldsymbol{\gamma}}_j^\beta\Bigg/\left(\frac{\Gamma(n/2+d/2)}{2}\right)\\ &= \sum_{j=1}^{N_p}\bar{W}_j\left\|\bar{\boldsymbol{\gamma}}_j\right\|_2^{d}\left(\frac{\bar{\boldsymbol{\gamma}}_j}{\left\|\bar{\boldsymbol{\gamma}}_j\right\|_2}\right)^\beta\Bigg/\left(\frac{\Gamma(n/2+d/2)}{2}\right)\\ &= \sum_{j=1}^{N_p}w_{s,j}\left(\mathbf{s}_j\right)^\beta\end{aligned} \tag{4.116}$$

where Eqs. (4.109) and (4.111) are used to arrive at the last equality.

By Eq. (4.114) and Eq. (4.116), it can be seen that any spherical rules with a degree $\beta \leq d$ can be exactly calculated by Eq. (4.110). ■

Proposition 4.2: The third-degree ($L = 2$) spherical rule in Eq. (4.97) and the fifth-degree ($L = 3$) spherical rule in Eq. (4.98) can be obtained by the projection of the sparse-grid quadrature rule.

Proof: It has been proved that the UT is a subclass of the SGQ rules at the accuracy level $L = 2$ in Theorem 4.1. Thus, it suffices to prove that the third-degree spherical rule in Eq. (4.97) is identical to the projected UT.

The quadrature rule using the UT is given by

$$\int_{\mathbb{R}^n} g(\mathbf{x}) N(\mathbf{x};\mathbf{0},\mathbf{I}) d\mathbf{x} \approx \frac{\kappa}{(n+\kappa)} g(\mathbf{0}) + \frac{1}{2(n+\kappa)} \sum_{j=1}^{n} \left[g\left(\sqrt{n+\kappa}\,\mathbf{e}_j\right) + g\left(-\sqrt{n+\kappa}\,\mathbf{e}_j\right) \right] \tag{4.117}$$

where κ is the tunable scaling parameter. Hence,

$$I(g) = \int_{\mathbb{R}^n} g(\mathbf{x}) \exp\left(-\mathbf{x}^T \mathbf{x}\right) d\mathbf{x} = \pi^{n/2} \int_{\mathbb{R}^n} g\left(\frac{\mathbf{x}}{\sqrt{2}}\right) N(\mathbf{x};\mathbf{0},\mathbf{I}) d\mathbf{x} \approx \sum_{j=1}^{2n+1} \bar{W}_j g(\bar{\boldsymbol{\gamma}}_j)$$
$$= \frac{\kappa \pi^{n/2}}{(n+\kappa)} g(\mathbf{0}) + \frac{\pi^{n/2}}{2(n+\kappa)} \sum_{j=1}^{n} \left(g\left(\sqrt{\frac{n+\kappa}{2}}\mathbf{e}_j\right) + g\left(-\sqrt{\frac{n+\kappa}{2}}\mathbf{e}_j\right)\right) \tag{4.118}$$

where the sigma points $\bar{\boldsymbol{\gamma}}_j$ and weights $\bar{W}_j$ are given by

$$\bar{\boldsymbol{\gamma}}_j = \begin{cases} \mathbf{0} & j = 1 \\ \sqrt{\frac{n+\kappa}{2}}\mathbf{e}_{j-1} & j = 2,\cdots,n+1 \\ -\sqrt{\frac{n+\kappa}{2}}\mathbf{e}_{j-n-1} & j = n+2,\cdots,2n+1 \end{cases} \tag{4.119}$$

and

$$\bar{W}_j = \begin{cases} \frac{\kappa}{(n+\kappa)} \pi^{n/2} & j = 1 \\ \frac{1}{2(n+\kappa)} \pi^{n/2} & j = 2,\cdots,2n+1 \end{cases} \tag{4.120}$$

respectively. By Eq. (4.109), projecting the sigma points $\bar{\boldsymbol{\gamma}}_j$ leads to the points $\mathbf{s}_j$ of the spherical rule

$$\mathbf{s}_j = \frac{\bar{\boldsymbol{\gamma}}_j}{\left\|\bar{\boldsymbol{\gamma}}_j\right\|_2} = \begin{cases} \mathbf{e}_j & j = 1,\cdots,n \\ -\mathbf{e}_{j-n} & j = n+1,\cdots,2n \end{cases} \tag{4.121}$$

Note that the point at the origin has been omitted.

By Remark 4.15 and Theorem 4.6, for the third-degree rule, d is chosen to be 2. The weights $w_{s,j}$ of the spherical rule can be obtained using (4.111):

$$w_{s,j} = \bar{W}_j \left\|\bar{\boldsymbol{\gamma}}_j\right\|_2^d \Big/ \left(\frac{\Gamma(n/2+d/2)}{2}\right) = \frac{1}{2(n+\kappa)} \pi^{n/2} \cdot \frac{(n+\kappa)}{2} \Big/ \left(\frac{n/2\,\Gamma(n/2)}{2}\right) = A_n/(2n) \tag{4.122}$$

Hence, using (4.121) and (4.122) in (4.110) yields the spherical rule

$$I_{U_n,2}(g_s)=\frac{A_n}{2n}\sum_{j=1}^{n}\left[g_s(\mathbf{e}_j)+g_s(-\mathbf{e}_j)\right] \tag{4.123}$$

Equation (4.123) is identical to the third-degree spherical rule and also identical to the third-degree spherical rule of (4.97) by Genz's method (Genz 2003) in Theorem 4.4.

For the fifth-degree (accuracy level $L = 3$) SGQ, (4.64) is repeated here.

$$I_{n,3}(g)=C_0 g(\mathbf{0})+C_1\sum_{j=1}^{2n} g(\mathbf{P}_{j,4})+C_2\sum_{j=1}^{2n(n-1)} g(\mathbf{P}_{j,5})$$

For convenience, $\bar{\boldsymbol{\gamma}}_j$ is divided into two sets: $\bar{\boldsymbol{\gamma}}_{j,1}$ and $\bar{\boldsymbol{\gamma}}_{j,2}$ where $\bar{\boldsymbol{\gamma}}_{j,1}=\mathbf{P}_{j,4}/\sqrt{2}$ and $\bar{\boldsymbol{\gamma}}_{j,2}=\mathbf{P}_{j,5}/\sqrt{2}$. The weights of $\bar{\boldsymbol{\gamma}}_{j,1}$ and $\bar{\boldsymbol{\gamma}}_{j,2}$ are $\bar{W}_{j,1}=C_1\cdot\pi^{n/2}$ and $\bar{W}_{j,2}=C_2\cdot\pi^{n/2}$, respectively. Note that the point at the origin is omitted in the construction of the spherical rule by projection.

The point set $\{\mathbf{s}_{j,1}\}$ projected from $\bar{\boldsymbol{\gamma}}_{j,1}$ are given by

$$\mathbf{s}_{j,1}=\frac{\bar{\boldsymbol{\gamma}}_{j,1}}{\left\|\bar{\boldsymbol{\gamma}}_{j,1}\right\|_2}=\begin{cases}\mathbf{e}_j & j=1,\cdots,n\\ -\mathbf{e}_{j-n} & j=n+1,\cdots,2n\end{cases} \tag{4.124}$$

The point set $\{\mathbf{s}_{j,2}\}$ projected from $\bar{\boldsymbol{\gamma}}_{j,2}$ are given by

$$\{\mathbf{s}_{j,2}\}=\frac{\bar{\boldsymbol{\gamma}}_{j,2}}{\left\|\bar{\boldsymbol{\gamma}}_{j,2}\right\|_2}=\begin{cases}\sqrt{\frac{1}{2}}(\mathbf{e}_k+\mathbf{e}_l)=\mathbf{s}_j^+ & k<l,\ k,l=1,2,\cdots,n\\ -\sqrt{\frac{1}{2}}(\mathbf{e}_k+\mathbf{e}_l)=-\mathbf{s}_j^+ & k<l,\ k,l=1,2,\cdots,n\\ \sqrt{\frac{1}{2}}(\mathbf{e}_k-\mathbf{e}_l)=\mathbf{s}_j^- & k<l,\ k,l=1,2,\cdots,n\\ -\sqrt{\frac{1}{2}}(\mathbf{e}_k-\mathbf{e}_l)=-\mathbf{s}_j^- & k<l,\ k,l=1,2,\cdots,n\end{cases} \tag{4.125}$$

For the fifth-degree rule, d is chosen to be 4 as discussed in Remark 4.15 and Theorem 4.6. Hence, the weights $w_{s,j}$ of the projected spherical points using (4.111) are given by

$$
\begin{aligned}
w_{s,j} &= \bar{W}_j \|\boldsymbol{\gamma}_j\|_2^4 \Big/ \left(\frac{\Gamma(n/2+d/2)}{2}\right) = \bar{W}_j \|\boldsymbol{\gamma}_j\|_2^4 \Big/ \left(\frac{\Gamma(n/2)n(n+2)}{8}\right) \\
&= \begin{cases} \bar{W}_{j,1} \|\boldsymbol{\gamma}_{j,1}\|_2^4 \Big/ \left(\frac{\Gamma(n/2)n(n+2)}{8}\right) & j=1,\cdots,2n \\ \bar{W}_{j,2} \|\boldsymbol{\gamma}_{j,2}\|_2^4 \Big/ \left(\frac{\Gamma(n/2)n(n+2)}{8}\right) & j=2n+1,\cdots,2n^2 \end{cases} \\
&= \begin{cases} \left(\frac{4-n}{18}\right) \Big/ \left(\frac{\Gamma(n/2)n(n+2)}{8}\right) \pi^{n/2} \cdot \left(\frac{\sqrt{3}}{\sqrt{2}}\right)^4 & j=1,\cdots,2n \\ \frac{1}{36} \Big/ \left(\frac{\Gamma(n/2)n(n+2)}{8}\right) \pi^{n/2} \left(\frac{\sqrt{6}}{\sqrt{2}}\right)^4 & j=2n+1,\cdots,2n^2 \end{cases} \\
&= \begin{cases} A_n(4-n)/(2n(n+2)) & j=1,\cdots,2n \\ A_n/(n(n+2)) & j=2n+1,\cdots,2n^2 \end{cases}
\end{aligned}
\tag{4.126}
$$

Hence,

$$
\begin{aligned}
I_{U_n,3}(\boldsymbol{g}_s) &= \frac{(4-n)A_n}{2n(n+2)} \sum_{j=1}^{n} \left[\boldsymbol{g}_s(\mathbf{e}_j) + \boldsymbol{g}_s(-\mathbf{e}_j)\right] + \\
&+ \frac{A_n}{n(n+2)} \sum_{j=1}^{n(n-1)/2} \left[\boldsymbol{g}_s(\mathbf{s}_j^+) + \boldsymbol{g}_s(-\mathbf{s}_j^+) + \boldsymbol{g}_s(\mathbf{s}_j^-) + \boldsymbol{g}_s(-\mathbf{s}_j^-)\right]
\end{aligned}
\tag{4.127}
$$

Equation (4.127) is the same as the fifth-degree spherical rule in (4.98) given by Genz's method. ■

The high-degree (> 5) spherical rule obtained by Genz's method and the rule obtained by the projection of the SGQ rule are not identical in general. For example, the high-degree SGQ (4.44) always contains the point $[\hat{p}, \hat{p}, 0, \cdots, 0]^T$, where $\hat{p}$ is a nonzero point that belongs to the univariate level-2 quadrature point set. The corresponding point of the projected SGQ obtained by (4.109) is $\left[\sqrt{2}/2, \sqrt{2}/2, 0, \cdots, 0\right]^T$. However, such a point does not necessarily belong to the point sets of the spherical rules by Genz's method. For example, for the accuracy level $L = 4$, the possible values of the components of the points in (4.96) are 0, $\pm\sqrt{1/3}$, $\pm\sqrt{2/3}$, and 1 only.

4.8.3 The Relations Between the Sparse-Grid Quadrature Rule and the Spherical-Radial Cubature Rule

In this section, two ways of obtaining the spherical-radial cubature rule from the SGQ rule are shown.

Spherical-radial cubature rule constructed from the SGQ rule by projection

If the spherical rule is obtained by the projection of the SGQ rule instead of using Genz's method, a new spherical-radial cubature rule can be constructed by combining this new spherical rule with the radial rule. We name this new cubature rule the projected sparse-grid quadrature (PSGQ) rule.

The following proposition reveals the relation between the PSGQ rule and the third-degree and fifth-degree spherical-radial cubature rules (Jia et al. 2015).

***Proposition 4.3*:** The third-degree spherical-radial cubature rule of Eq. (4.100) and the fifth-degree spherical-radial cubature rule of Eq. (4.101) can be obtained by the PSGQ rule.

Proof: The proposition is true because: (1) the same radial rules of Eq. (4.100) and Eq. (4.101) can be used by the PSGQ rule; (2) the third-degree and the fifth-degree spherical rules of Eq. (4.100) and Eq. (4.101) can be obtained by the projection of the third- and fifth-degree SGQ rule as proved in Proposition 4.2. ■

Spherical-radial cubature rule constructed directly from the SGQ rule without projection

The third-degree spherical-radial cubature rule can be viewed as a special case of the UT when the UT parameter $\kappa = 0$, and the UT is a subclass of the level-2 SGQ (Jia et al. 2012b). As shown in the previous section, the spherical-radial cubature rule can be obtained from the PSGQ rule. A natural question, in turn, is whether the spherical-radial cubature rules can be directly generated by the SGQ without using projection. In the following, we will show that the third-degree cubature rule and the fifth-degree cubature rule shown in (4.100) and (4.101) can be constructed from SGQ rules with a specific choice of univariate quadrature points and weights. Nevertheless, more general fifth-degree and higher degree (seven or higher, or equivalently $L \geq 4$) cubature rules cannot.

***Theorem 4.7*:** The spherical-radial cubature rule using the third-degree spherical rule in (4.100) and the arbitrary-degree radial rule can be constructed by a third-degree sparse-grid quadrature rule.

Proof: The integral (4.84) can be approximated by the third-degree spherical rule (4.97) and the arbitrary-degree radial rule (4.88)

$$\int_{\mathbb{R}^n} \mathbf{g}(\mathbf{x}) N(\mathbf{x};\mathbf{0},\mathbf{I}) d\mathbf{x}$$

$$\approx \begin{cases} \dfrac{A_n w_{r,1}}{2n\pi^{n/2}} \mathbf{g}(\mathbf{0}) + \sum_{i=2}^{N_r} \sum_{j=1}^{n} \dfrac{A_n w_{r,i}}{2n\pi^{n/2}} \left(\mathbf{g}\left(\sqrt{2} r_i \mathbf{e}_j\right) + \mathbf{g}\left(-\sqrt{2} r_i \mathbf{e}_j\right) \right), & r_1 = 0 \\ \sum_{i=1}^{N_r} \sum_{j=1}^{n} \dfrac{A_n w_{r,i}}{2n\pi^{n/2}} \left(\mathbf{g}\left(\sqrt{2} r_i \mathbf{e}_j\right) + \mathbf{g}\left(-\sqrt{2} r_i \mathbf{e}_j\right) \right), & r_1 \neq 0 \end{cases}$$

$$= \begin{cases} \hat{W}_{c,0} \mathbf{g}(\mathbf{0}) + \sum_{i=2}^{N_r} \sum_{j=1}^{n} \hat{W}_{c,i} \left(\mathbf{g}\left(\hat{r}_i \mathbf{e}_j\right) + \mathbf{g}\left(-\hat{r}_i \mathbf{e}_j\right) \right), & r_1 = 0 \\ \sum_{i=1}^{N_r} \sum_{j=1}^{n} \hat{W}_{c,i} \left(\mathbf{g}\left(\hat{r}_i \mathbf{e}_j\right) + \mathbf{g}\left(-\hat{r}_i \mathbf{e}_j\right) \right), & r_1 \neq 0 \end{cases} \tag{4.128}$$

where $\hat{W}_{c,0} = \dfrac{A_n w_{r,1}}{2n\pi^{n/2}}$, $\hat{W}_{c,i} = \dfrac{A_n w_{r,i}}{2n\pi^{n/2}}$, and $\hat{r}_i = \sqrt{2} r_i$. Note that r_1 is chosen to be zero without loss of generality and other r_i can be set to zero as well.

By the algorithm shown in Table 4.1, the third-degree SGQ using the univariate quadrature point set $\{-\hat{p}_l, \cdots, -\hat{p}_1, 0, \hat{p}_1, \cdots, \hat{p}_l\}$ with the weights $(\hat{w}_{l+1}, \cdots, \hat{w}_2, \hat{w}_1, \hat{w}_2, \cdots, \hat{w}_{l+1})$ for I_2 is given by

$$I_{n,2} = (1 - n + n\hat{w}_1) \mathbf{g}(\mathbf{0}) + \sum_{i=1}^{l} \hat{w}_{i+1} \sum_{j=1}^{n} \left[\mathbf{g}\left(\hat{p}_i \mathbf{e}_j\right) + \mathbf{g}\left(-\hat{p}_i \mathbf{e}_j\right) \right] \tag{4.129}$$

For $r_1 = 0$, Eq. (4.129) to be equivalent to (4.128), we may choose

$$\begin{cases} 1 - n + n\hat{w}_1 = \hat{W}_{c,0} \\ l = N_r - 1 \\ \hat{w}_{i+1} = \hat{W}_{c,i} \\ \hat{p}_i = \hat{r}_i \end{cases} \tag{4.130}$$

For $r_1 \neq 0$, Eq. (4.129) to be equivalent to (4.128), we may choose

$$\begin{cases} 1 - n + n\hat{w}_1 = 0 \\ l = N_r \\ \hat{w}_{i+1} = \hat{W}_{c,i} \\ \hat{p}_i = \hat{r}_i \end{cases} \tag{4.131}$$

Now, we only need to show that with the above choices, the points and weights are qualified as the univariate points and weights of I_2. That is, they match the first two moments:

$$\hat{w}_1 + 2\sum_{i=1}^{l} \hat{w}_{i+1} = 1$$

$$2\sum_{i=1}^{l} \hat{w}_{i+1}\hat{p}_i^2 = 1$$

Since (4.129) is equivalent to (4.128), it is a third-degree rule and exact for $g(\mathbf{x}) = 1$ and $g(\mathbf{x}) = \mathbf{x}^T\mathbf{x}$, of which $I_{n,2}$ are 1 and n respectively. So,

$$\left(1 - n + n\hat{w}_1\right) + 2n\sum_{i=1}^{l} \hat{w}_{i+1} = 1 \quad \text{or} \quad \hat{w}_1 + 2\sum_{i=1}^{l} \hat{w}_{i+1} = 1 \tag{4.132a}$$

$$2n\sum_{i=1}^{l} \hat{w}_{i+1}\hat{p}_i^2 = n \quad \text{or} \quad 2\sum_{i=1}^{l} \hat{w}_{i+1}\hat{p}_i^2 = 1 \tag{4.132b}$$

Clearly, the first two moments are matched. Therefore, the spherical-radial cubature rule using the third-degree spherical rule in (4.97) and the arbitrary-degree radial rule can be constructed by the sparse-grid quadrature rule directly. ■

Proposition 4.4: The fifth-degree spherical-radial cubature rule shown in (4.98) can be directly constructed from the sparse-grid quadrature.

Proof: When the fifth-degree or level-3 (L-3) SGQ (4.62) is constructed, the univariate quadrature rule I_1 using the univariate point $\{0\}$ with the weight of 1, the univariate quadrature rule I_2 using three symmetric univariate point set $\{-\hat{p}_1, 0, \hat{p}_1\}$ with the weight $\left(\hat{w}_2, \hat{w}_1, \hat{w}_2\right)$, and the univariate quadrature rule I_3 using five symmetric univariate point set $\{-\hat{p}_2, -\hat{p}_1, 0, \hat{p}_1, \hat{p}_2\}$ with the weight $\left(\hat{w}_5, \hat{w}_4, \hat{w}_3, \hat{w}_4, \hat{w}_5\right)$, are chosen as the sparse-grid constituents. $\hat{w}_1, \cdots, \hat{w}_5$ can be obtained by the moment matching method when $\hat{p}_1$ and $\hat{p}_2$ are given.

If $\hat{p}_1 = \sqrt{(n+2)/2}$ and $\hat{p}_2 = \sqrt{n+2}$, the univariate quadrature should satisfy the following moment matching equations:

$$\begin{cases} \hat{w}_1 + 2\hat{w}_2 = M_0 = 1 \\ 2\hat{w}_2\hat{p}_1^2 = M_2 = 1 \end{cases} \tag{4.133a}$$

$$\begin{cases} \hat{w}_3 + 2\hat{w}_4 + 2\hat{w}_5 = M_0 = 1 \\ 2\hat{w}_4\hat{p}_1^2 + 2\hat{w}_5\hat{p}_2^2 = M_2 = 1 \\ 2\hat{w}_4\hat{p}_1^4 + 2\hat{w}_5\hat{p}_2^4 = M_4 = 3 \end{cases} \tag{4.133b}$$

where M_j denotes the jth-order moment. After some algebra, the univariate quadrature weights can be solved from (4.133) as

$$\hat{w}_1=\frac{n}{n+2}, \hat{w}_2=\frac{1}{n+2}, \hat{w}_3=\frac{n^2+n+4}{(n+2)^2}, \hat{w}_4=\frac{2(n-1)}{(n+2)^2}, \hat{w}_5=\frac{(4-n)}{2(n+2)^2}$$

Using these values, the coefficients in (4.62) can be obtained.

$$A_1=\frac{2}{n+2}, B_1=0, C_1=\frac{(4-n)}{2(n+2)^2}, D_1=\frac{1}{(n+2)^2}.$$

Comparing (4.62) having the above coefficients to the fifth-degree spherical-radial cubature rule (4.101), they are identical. Thus, the fifth-degree spherical-radial cubature rule (4.101) can be directly constructed from the sparse-grid quadrature. ■

It should be emphasized that in general, the spherical-radial cubature rule using the fifth-degree spherical rule in (4.98) and the *arbitrary-degree* radial rule cannot always be constructed from a fifth-degree SGQ rule, which can be explained as follows.

When $L = 3$, as shown in (4.98), there are $2^2\mathbf{C}_n^2$ spherical points that are not on the axes $\mathbf{e}_j (j=1,\cdots,n)$. If there are m ($m \geq 1$, $m \in \mathbb{N}$) non-zero points used for the fifth-degree radial rule, the number of the spherical-radial points that are not on the axes is $m\cdot 2^2\mathbf{C}_n^2$.

For the SGQ with accuracy level $L = 3$, there is only one group of accuracy level sequences $(2,2,1,\cdots,1)$ with 2 elements greater than 1. In addition, there are $\mathbf{C}_n^2$ combinations of such accuracy level sequences with two elements being 2 and others being 1. Hence, if there are $\tilde{m}$ ($\tilde{m} \geq 2, \tilde{m} \in \mathbb{N}$) nonzero quadrature points used for the univariate quadrature rule with level $l = 2$, there are $\tilde{m}^2\mathbf{C}_n^2$ SGQ points that are not on the axes.

Because $m \cdot 2^2$ is not always equal to $\tilde{m}^2$, the fifth-degree spherical rule in (4.98) and the *arbitrary-degree* radial rule cannot be always constructed by the fifth-degree SGQ rule.

The next question is whether an arbitrary degree spherical-radial cubature rule given in Table 4.2 can be directly constructed by the SGQ in Table 4.1. In fact, the higher-degree spherical-radial cubature rule cannot be directly obtained by the SGQ when $L \geq 4$. It can be shown using the seventh-degree ($L = 4$) cubature rule and the SGQ rule as an example.

When $L = 4$, $I_{U_n,L}(g_s)$ is a seventh-degree spherical rule and $i_1+i_2+\quad+i\quad=n+L-1=n+3$. There is only one group of accuracy level sequences such as $\Xi=(2,2,2,1,\cdots,1)$ with 3 elements greater than 1. Because there are $\mathbf{C}_n^3$ combinations of Ξ, by the algorithm shown in Table 4.2, there are $2^3\mathbf{C}_n^3$ spherical points corresponding to all Ξ with 3 elements greater than 1.

If there are p ($p \geq 2$) non-zero points used for the seventh-degree radial rule, the number of the spherical-radial cubature points corresponding to all Ξ with 3 elements greater than 1 is $p \cdot 2^3 \mathbf{C}_n^3$.

For the SGQ with accuracy level $L = 4$, there is also only one group of accuracy level sequences such as $(2,2,2,1,\cdots,1)$ with 3 elements greater than 1. If there are q ($q \geq 2$) nonzero quadrature points used for the univariate quadrature rule with level $l = 2$, there are $q^3 \mathbf{C}_n^3$ SGQ points corresponding to all Ξ with 3 elements greater than 1.

When $p \cdot 2^3 = q^3$, the number of SGQ points corresponding to all Ξ with 3 elements greater than 1 is identical to that of the spherical-radial cubature rule. For the spherical-radial cubature rule, the location of points for all Ξ with 3 elements greater than 1 only depends on the radial rule. If the radial rule has m points and they are $[r_1,\cdots,r_m]$, the spherical-radial cubature points would be $r_i[\pm s,\pm s,\pm s,\cdots,0]^T$, $i=1,\cdots,m$, where the value of s can be obtained by the spherical rule. For each point, the values of all nonzero elements are the same. For the SGQ rule, however, the location of points depends on the univariate quadrature rule. If the $\tilde{m}$ nonzero points of the level-2 univariate quadrature are assumed to be $[s_1,\cdots,s_{\tilde{m}}]$, there exist such SGQ points $[\pm s_1,\pm s_{\tilde{m}},\pm s_1,0,\cdots,0]^T$ corresponding to the accuracy level sequence $(2,2,2,1,\cdots,1)$. Note that the values of the non-zero elements can be different.

Hence, the seventh-degree spherical-radial cubature rule cannot be directly obtained by the SGQ rule. The same conclusion follows for the higher-degree spherical-radial cubature rule.

With the discussions of the relations among these numerical integration rules, the relations among the filters constructed from these rules can be given likewise. Relations among UKF, CKF using the algorithms shown in Table 4.2, and SGQF using the algorithm shown in Table 4.1 can be briefly summarized in Fig. 4.3. Specifically, (1) the UKF can be constructed from the third-degree CKF using specific cubature (spherical-radial) points and weights; (2) the third-degree and the fifth-degree CKFs can be constructed from the SGQF by the projection; (3) the third-degree CKFs and the fifth-degree CKF shown in Eqs. (4.100) and (4.101) can also be constructed directly from the SGQF using specific SGQ points and weights. However, the general fifth-degree CKFs and higher-degree (> 5) CKFs cannot always be directly constructed by the SGQF and vice versa. Note that the CKF used to show the relations with UKF and SGQF in Fig. 4.3, is based on the Genz's spherical rule. Arbitrary-degree CKFs not using Genz's spherical rule can also be obtained by using the PSGQ rule.

Remark 4.16: Given the accuracy level L, the number of points of both the SGQ rule and the spherical-radial cubature rule increases polynomially with the dimension in the same order of $L - 1$ (Jia et al. 2013).

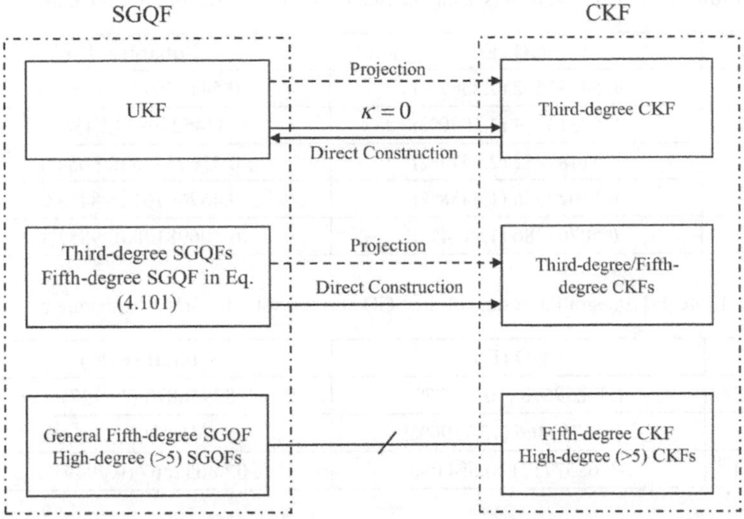

Fig. 4.3 Relations among UKF, CKF, and SGQF.

Next, two numerical integration problems are used to compare the performance of the SGQ rule and the spherical-radial cubature rule. The performance of their corresponding filters including the UKF, the SGQF, and the CKF will be presented in the later chapters.
The first integration problem is given by

$$\int_{\mathbb{R}^3} \sqrt{\mathbf{x}^T A^{-1}\mathbf{x}}\, N(\mathbf{x};\mathbf{0},\mathbf{I})\,d\mathbf{x} \tag{4.134}$$

where $A = \begin{bmatrix} 140 & -1 & -340 \\ -1 & 10 & -2 \\ -340 & -2 & 860 \end{bmatrix}$.

The SGQ rule and the spherical-radial cubature rule are compared with the Monte Carlo method using 1,000,000 points. The Monte Carlo result is 0.50203311 and is considered as the true value. For the SGQ rule, $2l$–1 univariate GHQ points are used for the univariate quadrature rule with accuracy l. Note that the UT is a subclass of the level-2 ($L = 2$) SGQ (Jia et al. 2012b) and that this level-2 SGQ has zero weight for the point at the origin and is equivalent to the UT with $\kappa = 3 - n = 3 - 3 = 0$ as well as the level-2 spherical-radial cubature rule. The results using the SGQ rule and the cubature rule are shown in Table 4.3, where E_r stands for the relative absolute error. Both rules approach the true value as the accuracy level increases. The SGQ improves the integration accuracy consistently whereas the spherical-radial cubature rule shows some fluctuations in the result.

Table 4.3 Integration results using the SGQ and the spherical-radial cubature rule.

L	SGQ (E_r)	Cubature (E_r)
2	0.54137592 (7.8367%)	0.54137592 (7.8367%)
3	0.52313364 (4.2030%)	0.42452798 (15.4383%)
4	0.51487232 (2.5574%)	0.51693525 (2.9684%)
5	0.51029556 (1.6458%)	0.46766761 (6.8453%)
6	0.50763786 (1.1164%)	0.50698178 (0.9857%)

Table 4.4 Integration results of the SGQ and the spherical-radial cubature rule.

L	SGQ (E_r)	Cubature (E_r)
2	–1.37259598 (101.5227%)	–0.87956873 (29.1371%)
3	–0.58753466 (13.7390%)	–0.60741880 (10.8196%)
4	–0.68577171 (0.6841%)	–0.68652667 (0.7949%)

The second integration problem is given by

$$\int_{\mathbb{R}^6} \cos\left(\sqrt{1+\mathbf{x}^T\mathbf{x}}\right) N(\mathbf{x};\mathbf{0},\mathbf{I})\,d\mathbf{x} \tag{4.135}$$

The Monte Carlo result using 10,000,000 points is –0.68111242 and is considered as the true value. The results using the SGQ and spherical-radial cubature rules are shown in Table 4.4.

In this case, it can be seen that the spherical-radial cubature rule is more accurate than the SGQ when $L \leq 3$ but they have a very close performance when $L = 4$.

Hence, it is hard to claim which, the SGQ or the spherical-radial cubature rule, is better since their performance depends on the specific nonlinear function and they are equally good if the nonlinear function is polynomial. However, it is true that the UT and the conventional third-degree spherical-radial cubature rule cannot provide accurate enough results and the integral accuracy can be improved by increasing the accuracy level. This is also true when these rules are applied in the nonlinear filtering problems, which will be shown in the subsequent chapters.

4.9 Positive Weighted Quadrature

Although the high-degree spherical-radial cubature rule (Jia et al. 2013) can provide higher than the third degree of accuracy, the cubature weight may not be always positive, which can cause the negative definite covariance matrix in challenging scenarios. The GHQ rule always provides positive quadrature weights but suffers the curse-of-dimensionality problem. In this section, we introduce a class of compact quadrature rules (CQR) that achieve a higher degree of accuracy as well as positive quadrature weights (Jia and Xin 2017). The CQR is constructed based on the following two propositions.

Proposition 4.5: The point set $\mathbf{p}=\left\{\left[p_1,p_2,\cdots,p_j,\cdots,p_n\right]^T: p_j=\pm u, 1\le j\le n-1, p_n=\prod_{j=1}^{n-1}\operatorname{sgn}\left(p_j\right)\cdot u\right\}$ is symmetric where "sgn" is the signum function.

Proof: Define $\mathbf{p}_s=\left\{\left[p_1,p_2,\cdots,p_j,\cdots,p_{n-1}\right]^T: p_j=\pm u, 1\le j\le n-1\right\}$. It is easy to check that $\mathbf{p}_s$ is symmetric. If $\left[p_1,p_2,\cdots,p_j,\cdots,p_n\right]^T\in\mathbf{p}$, there exists $\left[\tilde{p}_1,\tilde{p}_2,\cdots,\tilde{p}_j,\cdots,\tilde{p}_{n-1}\right]^T\in\mathbf{p}_s$ and $\tilde{p}_n=\prod_{j=1}^{n-1}\operatorname{sgn}\left(\tilde{p}_j\right)\cdot u=-\prod_{j=1}^{n-1}\operatorname{sgn}\left(p_j\right)\cdot u=-p_n$. Hence, $\left[p_1,p_2,\cdots,p_j,\cdots,-p_n\right]^T\in\mathbf{p}$ and the point set $\mathbf{p}$ is symmetric. ■

Proposition 4.6: Define two point sets

$$\boldsymbol{\gamma}_i=\left\{\left[\gamma_1,\gamma_2,\cdots,\gamma_j,\cdots,\gamma_n\right]^T:\gamma_j=\pm u, 1\le j\le n\right\}$$

and $\mathbf{p}=\left\{\left[p_1,p_2,\cdots,p_j,\cdots,p_n\right]^T: p_j=\pm u, 1\le j\le n-1, p_n=\prod_{j=1}^{n-1}\operatorname{sgn}\left(p_j\right)\cdot u\right\}$,

then $\sum_{i=1}^{2^n} w_i\cdot F\left(\boldsymbol{\gamma}_i,\boldsymbol{\nu}\right)=\sum_{i-1}^{2^{n-1}} 2w_i\cdot F\left(\mathbf{p}_i,\boldsymbol{\nu}\right)$ where $\boldsymbol{\gamma}_i\in\boldsymbol{\gamma}$, $\mathbf{p}_i\in\mathbf{p}$, $\boldsymbol{\nu}=\left[\nu_1,\nu_2,\cdots,\nu_i,\cdots,\nu_n\right]$, $F\left(\mathbf{x},\boldsymbol{\nu}\right)$ is a polynomial function of the form $\prod_i x_i^{\nu_i}$ and ν_i is a whole number, $w_i>0$ is the weight that has the same value for the point $\boldsymbol{\gamma}_i$ (or $\mathbf{p}_i$) as the one for $-\boldsymbol{\gamma}_i$ (or $-\mathbf{p}_i$).

Proof: By definition,

$$\sum_{i=1}^{2^n} w_i\cdot F\left(\boldsymbol{\gamma}_i,\boldsymbol{\nu}\right)=\sum_{i=1}^{2^n} w_i\cdot\left(\prod_{j=1}^{n-1}\left(\gamma_j\right)^{\nu_j}\cdot\left(\gamma_n\right)^{\nu_n}\right) \tag{4.136}$$

$$\sum_{i=1}^{2^{n-1}} 2w_i\cdot F\left(\mathbf{p}_i,\boldsymbol{\nu}\right)=2\sum_{i=1}^{2^{n-1}} w_i\cdot\left(\prod_{j=1}^{n-1}\left(p_j\right)^{\nu_j}\cdot\left(\prod_{j=1}^{n-1}\operatorname{sgn}\left(p_j\right)\cdot u\right)^{\nu_n}\right) \tag{4.137}$$

If there exists any ν_j, $1\le j\le n$ that is an odd number, $\sum_{i=1}^{2^n} w_i\cdot F\left(\boldsymbol{\gamma}_i,\boldsymbol{\nu}\right)=\sum_{i=1}^{2^{n-1}} 2w_i\cdot F\left(\mathbf{p}_i,\boldsymbol{\nu}\right)=0$ due to the symmetry of both the point set $\boldsymbol{\gamma}$ and the point set $\mathbf{p}$. If all ν_j, $1\le j\le n$ are even numbers, we have

$$\prod_{j=1}^{n-1}(\gamma_j)^{v_j}\cdot(\gamma_n)^{v_n}+\prod_{j=1}^{n-1}(\gamma_j)^{v_j}\cdot(-\gamma_n)^{v_n}=2\prod_{j=1}^{n-1}(\gamma_j)^{v_j}\cdot(\gamma_n)^{v_n}$$

$$=2\prod_{j=1}^{n-1}(\gamma_j)^{v_j}\cdot\left(\prod_{j=1}^{n-1}\operatorname{sgn}(\gamma_j)\cdot u\right)^{v_n}$$

$$=2\prod_{j=1}^{n-1}(p_j)^{v_j}\cdot\left(\prod_{j=1}^{n-1}\operatorname{sgn}(p_j)\cdot u\right)^{v_n}$$

$$=2\prod_{j=1}^{n-1}(p_j)^{v_j}\cdot(p_n)^{v_n} \tag{4.138}$$

Therefore, one can conclude that $\sum_{i=1}^{2^n} w_i\cdot F(\boldsymbol{\gamma}_i,\boldsymbol{v})=\sum_{i=1}^{2^{n-1}} 2w_i\cdot F(\mathbf{p}_i,\boldsymbol{v})$. ■

The point set $\mathbf{p}$ in the above two propositions are used to construct the CQR. Specifically, two high-degree CQRs are constructed from their respective conventional quadrature rules.

Fifth-degree rule ($n \geq 3$)

A fifth-degree CQR with all positive weights is given by (Stroud 1971)

$$\int_{\mathbb{R}^n} g(\mathbf{x})N(\mathbf{x};\mathbf{0},\mathbf{I})=\sum_{i=1}^{n}W_1\left(g(\boldsymbol{\gamma}_i)+g(-\boldsymbol{\gamma}_i)\right)+\sum_{j=1}^{2^n}W_2 g(\bar{\boldsymbol{\gamma}}_j) \tag{4.139}$$

where the quadrature points and weights are given by

$$\gamma_i=\sqrt{\frac{n+2}{2}}\,\mathbf{e}_i,\ W_1=\frac{4}{(n+2)^2} \tag{4.140a}$$

$$\bar{\gamma}_j=\left\{\left[\pm\sqrt{\frac{n+2}{n-2}}\right],\left[\pm\sqrt{\frac{n+2}{n-2}}\right],\dots,\left[\pm\sqrt{\frac{n+2}{n-2}}\right]\right\},\ W_2=\frac{(n-2)^2}{2^n(n+2)^2} \tag{4.140b}$$

The fifth-degree CQR is then constructed by applying the Proposition 4.6:

$$\int_{\mathbb{R}^n} g(\mathbf{x})N(\mathbf{x};\mathbf{0},\mathbf{I})=\sum_{i=1}^{n}W_1\left(g(\boldsymbol{\gamma}_i)+g(-\boldsymbol{\gamma}_i)\right)+\sum_{j=1}^{2^{n-1}}2W_2 g(\mathbf{p}_j) \tag{4.141}$$

where

$$\mathbf{p}_j = \left\{ \left[p_1, p_2, \cdots, p_m, \cdots, p_n \right]^T : p_m = \pm\sqrt{\frac{n+2}{n-2}}, 1 \le m \le n-1,\ p_n = \prod_{m=1}^{n-1} \operatorname{sgn}(p_m) \cdot \sqrt{\frac{n+2}{n-2}} \right\} \tag{4.142}$$

Seventh-degree rule ($n \le 6$)

A seventh-degree CQR with all positive weights is given by (Adurthi 2013)

$$\int_{\mathbb{R}^n} \mathbf{g}(\mathbf{x}) N(\mathbf{x}; \mathbf{0}, \mathbf{I}) = W_0 \mathbf{g}(\mathbf{0}) + \sum_{i=1}^{n} W_1 \left[\mathbf{g}(\boldsymbol{\gamma}_{i,1}) + \mathbf{g}(-\boldsymbol{\gamma}_{i,1}) \right] + \sum_{i=1}^{2^n} W_2 \mathbf{g}(\boldsymbol{\gamma}_{i,2}) + \sum_{i=1}^{2n(n-1)} W_3 \mathbf{g}(\boldsymbol{\gamma}_{i,3}) \tag{4.143}$$

where

$$\boldsymbol{\gamma}_{i,1} = r_1 \mathbf{e}_i,\ \boldsymbol{\gamma}_{i,2} = \left\{ \left[\pm r_2, \pm r_2, \cdots, \pm r_2 \right] \right\},\ \boldsymbol{\gamma}_{i,3} = \left\{ \left[0, \cdots, 0, \pm r_3, 0, \cdots, 0, \pm r_3, 0, \cdots, 0 \right] \right\} \tag{4.144a}$$

r_1, r_2, and r_3 are given by solving the following equations (Adurthi 2013)

$$\begin{cases} 2(8-n) r_1^{-4} + r_2^{-4} + 2(n-1) r_3^{-4} = 1 \\ 2(8-n) r_1^{-2} + r_2^{-2} + 2(n-1) r_3^{-2} = 3 \\ r_2^{-2} + 2 r_3^{-2} = 1 \end{cases} \tag{4.144b}$$

with the weights

$$\begin{cases} W_0 = 1 - 2nW_1 - 2^n W_2 - 2n(n-1) W_3 \\ W_1 = (8-n) / r_1^6 \\ W_2 = 1 / \left(2^n r_2^6 \right) \\ W_3 = 1 / \left(2 r_3^6 \right) \end{cases} \tag{4.145}$$

Using the Proposition 4.6, the seventh-degree CQR can be constructed by

$$\int_{\mathbb{R}^n} \mathbf{g}(\mathbf{x}) N(\mathbf{x}; \mathbf{0}, \mathbf{I}) = W_0 \mathbf{g}(\mathbf{0}) + \sum_{i=1}^{n} W_1 \left[\mathbf{g}(\boldsymbol{\gamma}_{i,1}) + \mathbf{g}(-\boldsymbol{\gamma}_{i,1}) \right] + \sum_{i=1}^{2^{n-1}} 2W_2 \mathbf{g}(\mathbf{p}_{i,2}) + \sum_{i=1}^{2n(n-1)} W_3 \mathbf{g}(\boldsymbol{\gamma}_{i,3}) \tag{4.146}$$

where

$$\mathbf{p}_{i,2} = \left\{ \left[p_1, p_2, \cdots, p_j, \cdots, p_n \right]^T : p_j = \pm r_2, 1 \le j \le n-1,\ p_n = \prod_{j=1}^{n-1} \operatorname{sgn}(p_j) \cdot r_2 \right\} \tag{4.147}$$

The fifth-degree and seventh-degree CQRs have several unique benefits if it is used in the filtering problems: (1) they are more accurate than the widely used UT and the third-degree spherical-radial cubature rule; (2) all the CQR weights are positive, which ensures the correct computation of the covariance matrix; (3) CQR requires fewer quadrature points than the original fifth- and seventh-degree quadrature rules and the GHQ. In addition, for the problems with the dimension less than or equal to six ($n \leq 6$), the number of CQR points is less than the number of the high-degree spherical-radial cubature rule (Jia et al. 2013) and the SGQ rule (Jia et al. 2011).

Note that other CQRs can be constructed similarly but it is hard to find a general form for higher degree CQRs with all positive weights.

Remark 4.17: The CQR is suitable for low or medium dimensional problems. It will become computationally intensive for high dimensional problems. Note that the number of points of the high-degree spherical-radial cubature rule (Jia et al. 2013) and the SGQ (Jia et al. 2011, 2012b) increases polynomially with the dimensional *n*. Hence, for the high dimensional ($n > 6$) problems, the high-degree spherical-radial cubature rule and the SGQ are preferred from the computation perspective. However, the quadrature weights for these two rules may become negative. In this regard, the CQR possesses the unique advantages over other numerical rules.

4.10 Adaptive Quadrature

We have introduced several different quadrature rules that can be used in the nonlinear filtering algorithm. For estimation problems with different complexity and accuracy requirements, we have the freedom of selecting those numerical rules with various levels of accuracy and computation cost. To improve the estimation efficiency of the filtering algorithm, the selection of the numerical rules can be done adaptively according to the measure of nonlinearity of a stochastic system. In this section, the adaptive SGQF algorithm incorporating the measure of nonlinearity (*MoN*) is given.

4.10.1 Global Measure of Nonlinearity for Stochastic Systems

In order to consider the nonlinearity of the entire system, including both the dynamic model and the measurement model, the joint system is written as follows (Li 2012).

$$\mathbf{z}_k = \begin{bmatrix} \mathbf{x}_{k+1} \\ \mathbf{y}_k \end{bmatrix} = \boldsymbol{\beta}_k(\mathbf{x}_k) + \begin{bmatrix} \mathbf{v}_k \\ \mathbf{n}_k \end{bmatrix} \tag{4.148}$$

where $\boldsymbol{\beta}_k(\mathbf{x}_k) = \begin{bmatrix} \boldsymbol{f}(\mathbf{x}_k) \\ \boldsymbol{h}(\mathbf{x}_k) \end{bmatrix}$. Only the nonlinearity of $\boldsymbol{\beta}_k(\mathbf{x}_k)$ needs to be considered because the joint system is linear in the white Gaussian noise.

Measuring the deviation of a nonlinear function from the set of all linear functions is different from the traditional approach that measures the deviation of a certain nonlinear function from a specific linear function (e.g., a linear one approximating that nonlinear function), which ignores the random effect of the variable $\mathbf{x}_k$. The proposed measure is given by (Li 2012):

$$J_k = \inf_{\boldsymbol{\alpha}_k \in \zeta} J(\boldsymbol{\alpha}_k - \boldsymbol{\beta}_k(\mathbf{x})) = \inf_{\boldsymbol{\alpha}_k \in \zeta} \sqrt{E(\| \boldsymbol{\alpha}_k(\mathbf{x}) - \boldsymbol{\beta}_k(\mathbf{x}) \|^2)} \tag{4.149}$$

where $\boldsymbol{\alpha}_k \in \zeta$, ζ is the set of all linear functions with the same dimension as the variable $\mathbf{x}$. It can be interpreted as the greatest lower bound of the distances from $\boldsymbol{\beta}_k$. To constrain the measure J_k in the range of [0, 1], the measure can be normalized by (Li 2012, Sun et al. 2015)

$$v_k = \frac{J_k}{\left[tr(\mathbf{C}_{\boldsymbol{\beta}_k}) \right]^{1/2}} \tag{4.150}$$

where $\mathbf{C}_{\boldsymbol{\beta}_k}$ is the covariance matrix of $\boldsymbol{\beta}_k$ and $\mathrm{tr}(\mathbf{C}_{\boldsymbol{\beta}_k})$ is the trace. Analytical solutions to Eqs. (4.149) and (4.150) are available and given by Eqs. (4.151) and (4.152) respectively (Li 2012).

$$J_k = \left[tr(\mathbf{C}_{\boldsymbol{\beta}_k} - \mathbf{C}_{\boldsymbol{\beta}_k \mathbf{x}_k} \mathbf{C}_{\mathbf{x}_k}^{-1} \mathbf{C}_{\boldsymbol{\beta}_k \mathbf{x}_k}^T) \right]^{1/2} \tag{4.151}$$

$$v_k = \sqrt{1 - \frac{tr(\mathbf{C}_{\boldsymbol{\beta}_k \mathbf{x}_k} \mathbf{C}_{\mathbf{x}_k}^{-1} \mathbf{C}_{\boldsymbol{\beta}_k \mathbf{x}_k}^T)}{tr(\mathbf{C}_{\boldsymbol{\beta}_k})}} \tag{4.152}$$

where $\mathbf{C}_{\mathbf{x}_h}$ and $\mathbf{C}_{\boldsymbol{\beta}_h \mathbf{x}_h}$ are the covariance of $\mathbf{x}_k$ and cross-covariance between $\mathbf{x}_k$ and $\boldsymbol{\beta}_k$, respectively. Due to the nonlinearity of the dynamic equation and measurement equation, the covariance matrices $\{\mathbf{C}_{\boldsymbol{\beta}_k \mathbf{x}_k}, \mathbf{C}_{\boldsymbol{\beta}_k}\}$ are commonly approximated by Gaussian integration techniques, such as the UT (Sun et al. 2015).

4.10.2 Local Measure of Nonlinearity for Stochastic Systems

The local measure of the nonlinearity of the dynamic systems is confined in the neighborhood of the current operating point. Here we briefly introduce the weighted least square method (WLSM) (Duník et al. 2013b).

If a nonlinear function is $f(\mathbf{x})$, $\mathbb{R}^n \rightarrow \mathbb{R}$, some selected points, $\boldsymbol{\varphi}_i$, $i = 0, 1 \ldots, r$, can be generated by UT or the SGQ rule according to the current estimate, i.e., mean and covariance. Then these points are transformed through the nonlinear function given by $\chi_i = f(\boldsymbol{\varphi}_i)$. The objective is to find a vector $\boldsymbol{\theta} \in \mathbb{R}^n$ so as to minimize the total Euclidean distance between χ_i and $\boldsymbol{\varphi}_i^T \boldsymbol{\theta}$. The cost function can be defined as:

$$J = (\mathbf{Z} - \mathbf{X}\boldsymbol{\theta})^T \mathbf{W}(\mathbf{Z} - \mathbf{X}\boldsymbol{\theta}) \tag{4.153}$$

where $\mathbf{X} = [\boldsymbol{\varphi}_0, \boldsymbol{\varphi}_1, ..., \boldsymbol{\varphi}_r]^T$ and $\mathbf{Z} = [\chi_0, \chi_1, ..., \chi_r]^T$. W is a weighting matrix used to mitigate the residue difference caused by the distance between different sigma points. It is a matrix with the distance $\mathbf{D}_i = \|\boldsymbol{\varphi}_i - \boldsymbol{\varphi}_0\|^2$ as the diagonal element, i.e., $\mathbf{W} = \text{diag}([\mathbf{D}_c, \mathbf{D}_1, ..., \mathbf{D}_r])$. $\mathbf{D}_c$ is the weight on $\boldsymbol{\varphi}_0$ and takes a value larger than other $\mathbf{D}_i$.
The optimal $\boldsymbol{\theta}^*$ is given by (Duník et al. 2013b):

$$\boldsymbol{\theta}^* = \arg\min_{\boldsymbol{\theta}} J(\boldsymbol{\theta}) = (\mathbf{X}^T \mathbf{W}^{-1} \mathbf{X})^{-1} \mathbf{X}^T \mathbf{W}^{-1} \mathbf{Z} \tag{4.154}$$

The *MoN* can be defined as $MoN = J(\boldsymbol{\theta}^*)$. In addition, if the nonlinear transformation is $\mathbb{R}^n \rightarrow \mathbb{R}^m$, i.e., $\chi_i \in \mathbb{R}^m$, $i = 0,1...,r$, for each dimension of χ_i, MoN_j, $j = 1,2,\ldots,m$, can be calculated as (4.154). Then the MoN for the m-dimensional nonlinear function is defined as:

$$MoN = \max_{j=1,2,...,m} MoN_j \tag{4.155}$$

***Remark 4.18*:** The accuracy level of the SGQ or the spherical-radial cubature rules can be selected according to the value of *MoN*, and when used in the filtering framework, it leads to the adaptive quadrature filter.

4.11 Summary

There is no optimal solution for the general nonlinear estimation problem. The grid-based Gaussian approximation filter is a competitive alternative of the EKF, which uses the linearization technique and the classical Kalman filter. Note that the grid-based Gaussian approximation filter has the same performance as the Kalman filter for linear systems. We have discussed several different rules such as GHQ, SGQ, and spherical-radial cubature rules, to generate the grid points and weights. From the numerical integration perspective, these rules are only distinct by different numerical techniques to calculate the Gaussian weighted integrals, which is the core of the nonlinear Gaussian approximation filters.

The authors believe that one of the future directions is to design the "smart" grid so that it can adaptively change according to the nonlinearity and uncertainty of the system. Some promising progress has been seen in the design of the grid. In (Duník et al. 2013a), the randomized UKF is proposed. In (Steinbring et al. 2016), the smart sampling method, which obtains the grid/samples using the criterion that minimizes the distance between the Gaussian distribution and the samples, is proposed. The Chi distribution and uniform sampling on the surface of hypersphere are integrated to give samples of multivariate Gaussian distribution in (Kurz and Hanebeck 2017).

References

Adurthi, N. 2013. The Conjugate Unscented Transform—A Method to Evaluate Multidimensional Expectation Integrals. M.S. Thesis, State University of New York at Buffalo, Buffalo, New York.

Arasaratnam, I., S. Haykin and R.J. Elliott. 2007. Discrete-time nonlinear filtering algorithms using Gauss-Hermite quadrature. Proceedings of the IEEE 95: 953–977.

Arasaratnam, I. and S. Haykin. 2009. Cubature Kalman filters. IEEE Transactions on Automatic Control 54: 1254–1269.

Duník, J., O. Straka and M. Šimandl. 2013a. Stochastic integration filter. IEEE Transactions on Automatic Control 58: 1561–1566.

Duník, J., O. Straka and M. Šimandl. 2013b. Nonlinearity and non-Gaussianity measures for stochastic dynamic systems. 16th International Conference on Information Fusion. Turkey.

Genz, A. 2003. Fully symmetric interpolatory rules for multiple integrals over hyper-spherical surfaces. Journal of Computational and Applied Mathematics 157: 187–195.

Heiss, F. and V. Winschel. 2008. Likelihood approximation by numerical integration on sparse grids. Journal of Econometrics 14: 62–80.

Ito, K. and K. Xiong. 2000. Gaussian filters for nonlinear filtering problems. IEEE Transactions on Automatic Control 45: 910–927.

Jia, B., M. Xin and Y. Cheng. 2011. Sparse Gauss-Hermite quadrature filter with application to spacecraft attitude estimation. Journal of Guidance, Control, and Dynamics 34: 367–379.

Jia, B., M. Xin and Y. Cheng. 2012a. Anisotropic sparse Gauss-Hermite quadrature filter. Journal of Guidance, Control, and Dynamics 35: 1014–1022.

Jia, B., M. Xin and Y. Cheng. 2012b. Sparse-grid quadrature nonlinear filtering. Automatica 48: 327–341.

Jia, B., M. Xin and Y. Cheng. 2013. High-degree cubature Kalman filter. Automatica 49: 510–518.

Jia, B., M. Xin and Y. Cheng. 2015. Relations between sparse-grid quadrature rule and spherical-radial cubature rule in nonlinear Gaussian estimation. IEEE Transactions on Automatic Control 60: 199–204.

Jia, B. and M. Xin. 2017. Orbital uncertainty propagation using positive weighted compact quadrature rule. Journal of Spacecraft and Rockets 54: 683–697.

Julier, S.J. and J.K. Uhlmann. 2002. Reduced sigma point filters for the propagation of means and covariances through nonlinear transformations. Proceedings of the 2002 American Control Conference. USA, 887–892.

Julier, S.J. 2003. The spherical simplex unscented transformation. Proceedings of the 2003 American Control Conference. USA, 2430–2434.

Julier, S.J. and J.K. Uhlmann. 2004. Unscented filtering and nonlinear estimation. Proceedings of the IEEE 92: 401–422.

Kurz, G. and U.D. Hanebeck. 2017. Linear regression Kalman filtering based on hyperspherical deterministic sampling. 2017 IEEE 56th Annual Conference on Decision and Control. Australia, 977–983.

Li, X. 2012. Measure of nonlinearity for stochastic systems. 15th International Conference on Information Fusion. Singapore, 1073–1080.

Macon, N. and A. Spitzbart. 1958. Inverses of vandermonde matrices. The American Mathematical Monthly 65: 95–100.

Nobile, F., R. Tempone and C.G. Webster. 2008a. An anisotropic sparse grid stochastic collocation method for partial differential equations with random input data. SIAM Journal on Numerical Analysis 46: 2411–2442.

Nobile, F., R. Tempone and C.G. Webster. 2008b. A sparse grid stochastic collocation method for partial differential equations with random input data. SIAM Journal on Numerical Analysis 46: 2309–2345.

Särkkä, S.. 2008. Unscented Rauch–Tung–Striebel smoother. IEEE Transactions on Automatic Control 53: 845–849.

Steinbring, J., M. Pander and U.D. Hanebeck. 2016. The smart sampling Kalman filter with symmetric samples. Journal of Advances in Information Fusion 11: 71–90.

Stroud, A.H. 1971. Approximate Calculation of Multiple Integrals. Prentice-Hall, Inc., Englewood Cliffs.

Sun, T., M. Xin and B. Jia. 2015. Nonlinearity-based adaptive sparse-grid quadrature filter. Proceedings of the American Control Conference. USA, 2499–2504.

Tenne, D. and T. Singh. 2003. The higher order unscented filter. Proceedings of the American Control Conference. USA, 2441–2446.

5 Nonlinear Estimation Extensions

The conventional grid-based Gaussian approximation filters with different grid selection strategies are introduced in Chapter 4. In this chapter, we present some extensions of the grid-based Gaussian approximation filters via various forms to handle the computational stability, noise uncertainty, model uncertainty, and non-Gaussian noises. Specifically, the continuous-discrete form is introduced to solve the continuous-discrete nonlinear estimation problem considering that many real applications are described by the continuous dynamic equations. To improve the numerical stability, the square root form of the grid-based Gaussian approximation filters has been developed in the literature. In this chapter, we summarize the generalized form that considers possible negative weights. In many real applications, the process noise and measurement noise contain uncertainty. With the imperfect noise statistics, we introduce the robust grid-based Gaussian approximation filter.

It is also widely known that many nonlinear systems contain non-Gaussian noise. It is very hard to directly solve the nonlinear non-Gaussian estimation problem. Fortunately, the common non-Gaussian noise can be approximated by the finite sum of Gaussian noise. The grid-based Gaussian mixture filter is introduced. Parameters in the Gaussian mixture filter can be flexibly chosen. The connection between the Gaussian mixture filter and other grid-based filters is discussed. In a real application, the uncertainty of the system may vary with time. Hence, the adaptive structure filter, which represents the uncertainty by Gaussian distribution or Gaussian mixture distribution, adaptively, can be effective in providing robust estimation performance.

5.1 Grid-based Continuous-Discrete Gaussian Approximation Filter

The nonlinear continuous-time mathematical model can be described by

$$d\mathbf{x} = \boldsymbol{f}\left(\mathbf{x}(t)\right)dt + \sqrt{\mathbf{Q}}d\mathbf{v} \tag{5.1}$$

where $\mathbf{x}(t) \in \mathbb{R}^n$, $\mathbf{v}$ is the standard Brownian process. $\mathbf{Q}$ is the spectral density matrix of the process noise. Note the measurement equation is the same as Eq. (4.2).

For the continuous-time dynamic system with discrete measurement, the measurement update step is the same as the conventional discrete-time grid-based Gaussian approximation filter. However, the prediction of the mean and covariance should be rewritten.

Based on Eq. (5.1), the time evolution of an arbitrary function of the state $\phi(\mathbf{x})$ is governed by the following equation (Särkkä and Solin 2012)

$$d\phi = \sum_j \frac{\partial \phi}{\partial x_j}\left(\boldsymbol{f}(\mathbf{x},t)dt + \sqrt{\mathbf{Q}}d\mathbf{v}\right)_j + \frac{1}{2}\sum_{ij}\mathbf{Q}_{ij}\frac{\partial^2\phi}{\partial x_i \partial x_j}dt \tag{5.2}$$

The expectation of the function of the state is then given by (Särkkä and Solin 2012)

$$E\left[\frac{d\phi}{dt}\right] = E\left[\sum_j \frac{\partial \phi}{\partial x_j} f_j(\mathbf{x},t)\right] + \frac{1}{2}E\left[\sum_{ij}\mathbf{Q}_{ij}\frac{\partial^2\phi}{\partial x_i \partial x_j}\right] \tag{5.3}$$

By selecting $\phi(\mathbf{x}) = x_i$ and then $\phi(\mathbf{x}) = \left(x_i - E[x_i]\right)\left(x_j - E[x_j]\right)$, Eq. (5.3) leads to the prediction equations of the state and covariance

$$\frac{d\mathbf{x}}{dt} = E\left[\boldsymbol{f}\left(\mathbf{x}(t)\right)\right] \tag{5.4}$$

$$\frac{d\mathbf{P}}{dt} = E\left[(\mathbf{x}-\hat{\mathbf{x}})\boldsymbol{f}^T(\mathbf{x},t)\right] + E\left[\boldsymbol{f}(\mathbf{x},t)(\mathbf{x}-\hat{\mathbf{x}})^T\right] + \mathbf{Q} \tag{5.5}$$

The mean and covariance of the state are difficult to obtain unless the Fokker-Planck-Kolmogorov equation is solved. For real applications, the mean and covariance of the state are often approximated. The propagation of the mean and the covariance can be approximated by the moment differential equations.

$$\frac{d\hat{\mathbf{x}}}{dt} = \boldsymbol{f}(\hat{\mathbf{x}}) \tag{5.6}$$

$$\frac{d\mathbf{P}(\hat{\mathbf{x}})}{dt} = \mathbf{F}(\hat{\mathbf{x}})\mathbf{P}(\hat{\mathbf{x}}) + \mathbf{P}(\hat{\mathbf{x}})\mathbf{F}^T(\hat{\mathbf{x}}) + \mathbf{Q} \tag{5.7}$$

where $\mathbf{F}(\hat{\mathbf{x}})$ is the Jacobian matrix evaluated at $\hat{\mathbf{x}}$.

Based on Eqs. (5.6) and (5.7), as derived in (Särkkä 2007), the mean and covariance can be propagated using the grid-based integration technique.

$$\frac{d\hat{\mathbf{x}}}{dt} = \sum_i W_i \boldsymbol{f}(\hat{\mathbf{x}} + \mathbf{S}\boldsymbol{\gamma}_i) \tag{5.8}$$

$$\frac{d\mathbf{P}}{dt} = \sum_i W_i \boldsymbol{f}(\hat{\mathbf{x}} + \mathbf{S}\boldsymbol{\gamma}_i)(\mathbf{S}\boldsymbol{\gamma}_i)^T + \sum_i W_i (\mathbf{S}\boldsymbol{\gamma}_i)\left(\boldsymbol{f}(\hat{\mathbf{x}} + \mathbf{S}\boldsymbol{\gamma}_i)\right)^T + \mathbf{Q} \tag{5.9}$$

where $\boldsymbol{\gamma}_i$ and W_i are quadrature points and weights, respectively, defined in Chapter 4.

Besides the form described above, the 1.5 order strong Itô-Taylor scheme is used in the continuous-discrete Kalman filter (Arasaratnam et al. 2010). Given the time interval $[t, t+\delta]$, the new discretized model is given by (Särkkä and Solin 2012)

$$\mathbf{x}(t+\delta) = \mathbf{x}(t) + \delta \boldsymbol{f}(\mathbf{x}(t)) + \frac{\delta^2}{2} L_0 \boldsymbol{f}(\mathbf{x}(t)) + \sqrt{\mathbf{Q}}\boldsymbol{\eta} + \left(L\boldsymbol{f}(\mathbf{x}(t))\right)\boldsymbol{\omega} \tag{5.10}$$

where $\boldsymbol{\eta}$ and $\boldsymbol{\omega}$ are a suitable pair of zero mean Gaussian random vectors,

$$L_0 = \frac{\partial}{\partial t} + \sum_i f_i \frac{\partial}{\partial x_i} + \frac{1}{2}\sum_{p,q,j} \sqrt{\mathbf{Q}}_{pj} \sqrt{\mathbf{Q}}_{qj} \frac{\partial^2}{\partial x_p \partial x_q} \tag{5.11}$$

$$L_i = \sum_j \sqrt{\mathbf{Q}}_{ij} \frac{\partial}{\partial x_j} \tag{5.12}$$

Note that Lf is a matrix with its ith row and jth column element being $L_i f_j (i, j = 1, \ldots, n)$.

Based on Eqs. (5.10)–(5.12), the mean and the covariance of the state can be described by

$$\hat{\mathbf{x}}(t+\delta) = E\left[\boldsymbol{f}_d(\mathbf{x}, t)\right] \tag{5.13}$$

$$\begin{aligned} \mathbf{P}(t+\delta) = {} & E\left[\boldsymbol{f}_d(\mathbf{x})\left(\boldsymbol{f}_d(\mathbf{x})\right)^T\right] - \hat{\mathbf{x}}(t+\delta)\left(\hat{\mathbf{x}}(t+\delta)\right)^T \\ & + \frac{\delta^3}{3} E\left[L\boldsymbol{f}(\mathbf{x})\left(L\boldsymbol{f}(\mathbf{x})\right)^T\right] + \frac{\delta^2}{2}\sqrt{\mathbf{Q}} E\left[\left(L\boldsymbol{f}(\mathbf{x})\right)^T\right] \\ & + \frac{\delta^2}{2} E\left[L\boldsymbol{f}(\mathbf{x})\right]\sqrt{\mathbf{Q}}^T + \delta\mathbf{Q} \end{aligned} \tag{5.14}$$

$$\boldsymbol{f}_d(\mathbf{x}, t) = \mathbf{x} + \delta \boldsymbol{f}(\mathbf{x}, t) + \frac{\delta^2}{2} L_0 \boldsymbol{f}(\mathbf{x}, t) \tag{5.15}$$

$$\delta = T/M \tag{5.16}$$

where T is the sampling interval and M is the steps.

Note that for the methods introduced above, it is not guaranteed that one performs better than the other in all applications. How to choose the filter is problem dependent. The detailed comparison is shown in (Särkkä and Solin 2012).

Remark 5.1: Solvers are required for the differential equation (Kulikov and Kulikova 2017). The most common solver uses the Runge-Kutta integrator. High accuracy integrators are also available (Chapter 7).

Besides the above algorithms, alternatively, the continuous-time dynamic system can be discretized. Compared to the system discretization algorithm, the continuous-discrete filtering algorithm is more suitable for intermittent measurements so that the sampling periods can be flexibly chosen.

5.2 Augmented Grid-based Gaussian Approximation Filter

Roughly speaking, the grid-based Gaussian approximation filters can be classified into two forms, augmented and non-augmented forms. The non-augmented form has been introduced in Chapter 4. For conciseness, the dynamic equation and measurement equation are not repeated here. They are given in Eq. (4.1) and Eq. (4.2), respectively. In the following, the augmented form of the filter is presented. The augmented form can provide more accurate results with a slight increase of the computational complexity.

The augmented state and covariance are defined as

$$\mathbf{x}^a = \begin{bmatrix} \mathbf{x} \\ \mathbf{0} \\ \mathbf{0} \end{bmatrix} \tag{5.17}$$

$$\mathbf{P}^a = \begin{bmatrix} \mathbf{P} & \mathbf{0} & \mathbf{0} \\ \mathbf{0} & \mathbf{Q} & \mathbf{0} \\ \mathbf{0} & \mathbf{0} & \mathbf{R} \end{bmatrix} \tag{5.18}$$

Note that we assume the means of the process noise and the measurement noise are $\mathbf{0}$.

Given the initial state $\mathbf{x}^a_{k-1|k-1}$ and covariance $\mathbf{P}^a_{k-1|k-1}$, the filtering algorithm can be rewritten as follows.

Prediction:

$$\hat{\mathbf{x}}^a_{k|k-1} = \sum_{i=1}^{N_p^a} W_i^a \chi^a_{k|k-1,i} \tag{5.19}$$

$$\boldsymbol{\chi}^{a}_{k|k-1,i,x} = \boldsymbol{f}\left(\boldsymbol{\xi}^{a}_{k-1|k-1,i,x}\right) + \boldsymbol{\xi}^{a}_{k-1|k-1,i,v} \tag{5.20}$$

$$\mathbf{P}^{a}_{k|k-1} = \sum_{i=1}^{N_p^a} W_i^a \left(\boldsymbol{\chi}^{a}_{k|k-1,i} - \hat{\mathbf{x}}^{a}_{k|k-1}\right)\left(\boldsymbol{\chi}^{a}_{k|k-1,i} - \hat{\mathbf{x}}^{a}_{k|k-1}\right)^T \tag{5.21}$$

where N_p^a is the total number of quadrature points; $\boldsymbol{\chi}^{a}_{k|k-1,i,v} = \boldsymbol{\xi}^{a}_{k-1|k-1,i,v}$ and $\boldsymbol{\chi}^{a}_{k|k-1,i,n} = \boldsymbol{\xi}^{a}_{k-1|k-1,i,n}$; the subscripts '$x$', '$v$', and '$n$' denote the components associated with the state, the process noise, and measurement noise respectively; $\boldsymbol{\xi}^{a}_{k-1|k-1,i}$ is the transformed quadrature point obtained by

$$\boldsymbol{\xi}^{a}_{k-1|k-1,i} = \mathbf{S}^{a}_{k-1}\boldsymbol{\gamma}^{a}_{i} + \hat{\mathbf{x}}^{a}_{k-1|k-1} \;\; ; \;\; \mathbf{P}^{a}_{k-1|k-1} = \mathbf{S}^{a}_{k-1}\left(\mathbf{S}^{a}_{k-1}\right)^T \tag{5.22}$$

Note that $\boldsymbol{\gamma}_i^a$ and W_i^a are respectively the quadrature points and weights corresponding to the augmented state and standard Gaussian distribution.

Update:

$$\hat{\mathbf{x}}^{a}_{k|k} = \hat{\mathbf{x}}^{a}_{k|k-1} + \mathbf{K}_k\left(\mathbf{y}_k - \hat{\mathbf{y}}_k\right) \tag{5.23}$$

$$\mathbf{P}^{a}_{k|k} = \mathbf{P}^{a}_{k|k-1} - \mathbf{K}_k\mathbf{P}^{T}_{xy} \tag{5.24}$$

$$\mathbf{K}_k = \mathbf{P}_{xy}\left(\mathbf{P}_{yy}\right)^{-1} \tag{5.25}$$

where

$$\hat{\mathbf{y}}_k - \sum_{i=1}^{N_p^a} W_i^a \boldsymbol{\chi}^{a}_{k|k,i} \tag{5.26}$$

$$\boldsymbol{\chi}^{a}_{k|k,i,x} = \boldsymbol{h}\left(\boldsymbol{\chi}^{a}_{k|k-1,i,x}\right) + \boldsymbol{\chi}^{a}_{k|k-1,i,n} \tag{5.27}$$

$$\mathbf{P}_{xy} = \sum_{i=1}^{N_p^a} W_i^a \left(\boldsymbol{\chi}^{a}_{k|k-1,i} - \hat{\mathbf{x}}^{a}_{k|k-1}\right)\left(\boldsymbol{\chi}^{a}_{k|k,i} - \hat{\mathbf{y}}_k\right)^T \tag{5.28}$$

$$\mathbf{P}_{yy} = \sum_{i=1}^{N_p^a} W_i^a \left(\boldsymbol{\chi}^{a}_{k|k,i} - \hat{\mathbf{y}}_k\right)\left(\boldsymbol{\chi}^{a}_{k|k,i} - \hat{\mathbf{y}}_k\right)^T \tag{5.29}$$

Remark 5.2: The obvious difference between the augmented and non-augmented Gaussian approximation filters is that the augmented one only draws points and weights once per estimation circle while the non-augmented Gaussian approximation filter draws points and weights twice per estimation circle. In addition, the augmented Gaussian approximation filter can achieve better performance than the non-augmented Gaussian approximation filters

(Wu et al. 2005). The number of points and weights of the former, however, is larger than that of the latter.

5.3 Square-root Grid-based Gaussian Approximation Filter

The square-root form of the Gaussian approximation filter can potentially improve the numerical stability (Arasaratnam and Haykin 2008, Arasaratnam and Haykin 2011) or improve the efficiency (Van Der Merwe and Wan 2001) of the original grid-based Gaussian approximation filters. The term **S** (factor of the covariance matrix **P**, $\mathbf{P} = \mathbf{S}\mathbf{S}^T$) is propagated directly without factorization. To avoid factorization of the covariance matrix, many linear algebra techniques can be leveraged, such as the QR decomposition and Cholesky factor updating. In the following, the typical form of the square-root grid-based Gaussian approximation filter is summarized (Van Der Merwe and Wan 2001).

When the initial covariance is given, the Cholesky factorization or singular value decomposition (SVD) is used to obtain $\mathbf{S}_{0|0}$. The initial grid set is then given by

$$\xi_{0|0} = \left[\hat{\mathbf{x}}_{0|0} + \mathbf{S}_{0|0}\boldsymbol{\gamma}_1, \quad \hat{\mathbf{x}}_{0|0} + \mathbf{S}_{0|0}\boldsymbol{\gamma}_2, \quad \cdots, \quad \hat{\mathbf{x}}_{0|0} + \mathbf{S}_{0|0}\boldsymbol{\gamma}_{N_p} \right] \tag{5.30}$$

γ_i are the quadrature points.

Prediction:

Each point in the grid set can be propagated by

$$\boldsymbol{\chi}_{k|k-1,i} = \boldsymbol{f}\left(\xi_{k-1|k-1,i}\right) \quad i = 1, \cdots, N_p \tag{5.31}$$

The predicted state value is given by

$$\hat{\mathbf{x}}_{k|k-1} = \sum_{i=1}^{N_p} W_i \boldsymbol{\chi}_{k|k-1,i} \tag{5.32}$$

where N_p is the number of points, and W_i are the quadrature weights.

The factor $\mathbf{S}_{k-1|k-1}$ of the covariance $\mathbf{P}_{k-1|k-1}$ can be propagated directly using the QR decomposition with a subsequent Cholesky update (Van Der Merwe and Wan 2001).

$$\mathbf{S}_{k|k-1} = \mathrm{QR}\left\{\sqrt{W_1^p}\left(\boldsymbol{\chi}_{k|k-1,1}^p - \hat{\mathbf{x}}_{k|k-1}\right), \cdots \sqrt{W_{N_1}^p}\left(\boldsymbol{\chi}_{k|k-1,N_1}^p - \hat{\mathbf{x}}_{k|k-1}\right) \ \sqrt{\mathbf{Q}}\right\} \tag{5.33}$$

$$\mathbf{S}_{k|k-1} = \mathrm{cholupdate}\left\{\mathbf{S}_{k|k-1}, \left(\boldsymbol{\chi}_{k|k-1,j}^n - \hat{\mathbf{x}}_{k-1|k-1}\right), W_j^n\right\} \quad j = 1, \cdots, N_2 \tag{5.34}$$

Note that N_1 and N_2 are the number of points with positive and negative weights, respectively. The superscripts 'p' and 'n' denote the point with positive and negative weights, respectively. In addition, $N_1 + N_2 = N_p$. Note that $\mathbf{S}$ is a lower triangular matrix.

Update:

Each point in the grid set is used to predict the measurement

$$\hat{\mathbf{y}}_k = \sum_{i=1}^{N_p} W_i \boldsymbol{\chi}_{k|k,i} = \sum_{i=1}^{N_p} W_i \boldsymbol{h}\left(\xi_{k|k-1,i}\right) \quad i = 1, \cdots, N_p \tag{5.35}$$

where the grid $\xi_{k|k-1}$ is given by

$$\xi_{k|k-1} = \left[\hat{\mathbf{x}}_{k|k-1} + \mathbf{S}_{k|k-1}\boldsymbol{\gamma}_1, \quad \hat{\mathbf{x}}_{k|k-1} + \mathbf{S}_{k|k-1}\boldsymbol{\gamma}_2, \quad \cdots, \quad \hat{\mathbf{x}}_{k|k-1} + \mathbf{S}_{k|k-1}\boldsymbol{\gamma}_{N_p}\right] \tag{5.36}$$

Similarly, the factor $\mathbf{S}_{k|k,y}$ of the covariance $\mathbf{P}_{yy}$ can be updated by Cholesky update (Van Der Merwe and Wan 2001).

$$\mathbf{S}_{k|k} = \text{cholupdate}\left\{\mathbf{S}_{k|k-1}, \mathbf{U}, -1\right\} \tag{5.37}$$

where $\mathbf{U} = \mathbf{K}_k \mathbf{S}_{k|k,y}$. Note '$-1$' is the weight used in 'cholupdate'. The factor $\mathbf{S}_{k|k,y}$ of the covariance $\mathbf{P}_{yy}$ can be obtained by QR decomposition and Cholesky update.

$$\mathbf{S}_{k|k,y} = \text{QR}\left\{\sqrt{W_1^p}\left(\boldsymbol{\chi}_{k|k,1}^p - \hat{\mathbf{y}}_k\right), \cdots \sqrt{W_{N_1}^p}\left(\boldsymbol{\chi}_{k|k,N_1}^p - \hat{\mathbf{y}}_k\right) \quad \sqrt{\mathbf{R}}\right\} \tag{5.38}$$

$$\mathbf{S}_{k|k,y} = \text{cholupdate}\left\{\mathbf{S}_{k|k,y}, \left(\boldsymbol{\chi}_{k|k,j}^n - \hat{\mathbf{y}}_k\right), W_j^n\right\} \quad j = 1, \cdots, N_2 \tag{5.39}$$

The state estimation is updated by

$$\hat{\mathbf{x}}_{k|k} = \hat{\mathbf{x}}_{k|k-1} + \mathbf{K}_k\left(\mathbf{y}_k - \hat{\mathbf{y}}_k\right) \tag{5.40}$$

where the Kalman gain $\mathbf{K}_k$ and covariance $\mathbf{P}_{xy}$ are given by

$$\mathbf{K}_k = \mathbf{P}_{xy}\left(\left(\mathbf{S}_{k|k,y}\right)^T\right)^{-1}\left(\mathbf{S}_{k|k,y}\right)^{-1} \tag{5.41}$$

$$\mathbf{P}_{xy} = \sum_{i=1}^{N_p} W_i\left(\boldsymbol{\chi}_{k|k-1,i} - \hat{\mathbf{x}}_{k|k-1}\right)\left(\boldsymbol{\chi}_{k|k,i} - \hat{\mathbf{y}}_k\right)^T \tag{5.42}$$

Remark 5.3: If all weights are positive, the Cholesky update step is not necessary. In addition, although the square-root form of the grid-based Gaussian approximation filter can improve the numerical properties over the conventional

Gaussian approximation filter, it cannot fully mitigate the effect of the negative weights of the quadrature rules.

5.4 Constrained Grid-based Gaussian Approximation Filter

The constrained nonlinear filter is very useful to incorporate extra information available to the system. Typical constraints include equality and inequality constraints. They can be described by Eq. (5.43) or (5.44), respectively.

$$\mathbf{D}\mathbf{x}_k = \mathbf{d} \tag{5.43}$$

$$\mathbf{D}\mathbf{x}_k \leq \mathbf{d} \tag{5.44}$$

Note that $\mathbf{D}$ and $\mathbf{d}$ are given. The state estimate $\hat{\mathbf{x}}_k$ should satisfy the constraints.

There are many methods to incorporate such constraints in the filtering algorithm. For the equality constraint, it can be viewed as a perfect measurement. The measurement equation is then augmented by

$$\begin{bmatrix} \mathbf{y}_k \\ \mathbf{d} \end{bmatrix} = \begin{bmatrix} h(\mathbf{x}_k) \\ \mathbf{D}\mathbf{x}_k \end{bmatrix} + \begin{bmatrix} \mathbf{n}_k \\ 0 \end{bmatrix} \tag{5.45}$$

Then, the standard grid-based Gaussian approximation filter can be applied.

For the inequality constraint, a constrained estimation is formulated as follows

$$\begin{aligned} \tilde{\mathbf{x}}_{k|k} &= \arg\min_{\mathbf{x}} \left(\mathbf{x} - \hat{\mathbf{x}}_{k|k}\right)^T \mathbf{P}_{k|k} \left(\mathbf{x} - \hat{\mathbf{x}}_{k|k}\right) \\ &\text{subject to } \mathbf{D}\mathbf{x} \leq \mathbf{d} \end{aligned} \tag{5.46}$$

The estimate of the state can be obtained by solving the optimization problem, which enforces the inequality constraint. The probability density function truncation algorithm (Simon 2010) is another method to solve such a problem. In addition, other constraints can be included, such as the sparsity of the state or parameters (Jia and Wang 2013, 2014). The state estimation can be improved greatly if adequate constraints can be formulated.

Because of the importance of interval constraints of the state estimation, two typical algorithms are introduced in the following (Teixeira et al. 2010).

5.4.1 Interval-constrained Unscented Transformation

For convenience, let's assume the initial state estimate and covariance matrix are $\hat{\mathbf{x}}$ and $\mathbf{P}$, respectively. In addition, it is assumed that $\underline{\mathbf{d}} \leq \hat{\mathbf{x}} \leq \overline{\mathbf{d}}$. The points and weights for the interval-constrained unscented transformation with standard normal distribution is given by (Teixeira et al. 2010)

$$\tilde{\boldsymbol{\gamma}}_i = \begin{cases} \hat{\mathbf{x}} & i=1 \\ \hat{\mathbf{x}} + \theta_{i-1}\mathbf{S}\boldsymbol{\gamma}_{i-1} & i=2\cdots n+1 \\ \hat{\mathbf{x}} + \theta_{i-n-1}\mathbf{S}\boldsymbol{\gamma}_{i-n-1} & i=n+2\cdots 2n+1 \end{cases} \tag{5.47}$$

where $\theta_i = \min\left(\mathrm{col}_i\left(\Theta\right)\right)$, Θ is a matrix and the element of the *i*th row and *j*th column of Θ is given by

$$\Theta_{ij} = \begin{cases} \sqrt{n+\kappa} & \breve{\mathrm{S}}_{ij} = 0 \\ \min\left(\sqrt{n+\kappa}, \left(\overline{\mathrm{d}}_i - \hat{x}_i\right)/\breve{\mathrm{S}}_{ij}\right) & \breve{\mathrm{S}}_{ij} > 0 \\ \min\left(\sqrt{n+\kappa}, \left(\underline{\mathrm{d}}_i - \hat{x}_i\right)/\breve{\mathrm{S}}_{ij}\right) & \breve{\mathrm{S}}_{ij} < 0 \end{cases} \tag{5.48}$$

where $\breve{\mathbf{S}} = \left[\mathbf{S} \quad -\mathbf{S}\right]$ and $\mathbf{P} = \mathbf{S}\mathbf{S}^T$.
The weights should be revised as (Vachhani et al. 2006)

$$\tilde{W}_i = \begin{cases} b & j=1 \\ a\theta + b & j=2,\cdots,2n+1 \end{cases} \tag{5.49}$$

$$a = \frac{2\kappa - 1}{2(n+\kappa)\left(\sum_{i=1}^{n}\theta_i - (2n+1)\sqrt{n+\kappa}\right)} \tag{5.50}$$

$$b = \frac{1}{2(n+\kappa)} - \frac{2\kappa - 1}{\left(2\sqrt{n+\kappa}\right)\left(\sum_{i=1}^{n}\theta_i - (2n+1)\sqrt{n+\kappa}\right)} \tag{5.51}$$

The derivation of weights and points are given in (Vachhani et al. 2006). By using the interval-constrained unscented transformation, the interval constraint can be enforced in the prediction step. However, after the update, the constraints may not be kept. Hence, the estimation projection should be performed.

5.4.2 Estimation Projection and Constrained Update

If $\hat{\mathbf{x}}$ does not satisfy the interval constraint, it is necessary to project it onto the constraint boundary using the quadratic programming method (Teixeira et al. 2009). It can be described by

$$\begin{gathered} \hat{\mathbf{x}}_p = \arg\min_{\mathbf{x}} \left(\mathbf{x} - \hat{\mathbf{x}}\right)^T \mathbf{P}\left(\mathbf{x} - \hat{\mathbf{x}}\right) \\ \text{subject to } \underline{\mathbf{d}} \le \mathbf{x} \le \overline{\mathbf{d}} \end{gathered} \tag{5.52}$$

Besides the direct estimation projection after the conventional update step of the grid-based Gaussian approximation filters, the constrained update step can be obtained via solving the optimization as follows.

$$\hat{\mathbf{x}}_p = \arg\min_{\mathbf{x}} \left((\mathbf{y} - \boldsymbol{h}(\mathbf{x}))^T \mathbf{R}^{-1} (\mathbf{y} - \boldsymbol{h}(\mathbf{x})) + (\mathbf{x} - \hat{\mathbf{x}})^T \mathbf{P}^{-1} (\mathbf{x} - \hat{\mathbf{x}}) \right) \quad (5.53)$$
$$\text{subject to } \underline{\mathbf{d}} \le \mathbf{x} \le \overline{\mathbf{d}}$$

Note that each point can be projected via the above optimization. In addition, the solution of the classical Kalman filter is identical to the solution of Eq. (5.53) for the linear system (Jazwinski 1970). We only summarized one of constrained Kalman update algorithms. There are other algorithms available. Selection of the constrained filter is problem dependent.

5.5 Robust Grid-based Gaussian Approximation Filter

In this section, two robust Kalman filters, Huber based robust filter and H_∞ filter, are introduced.

5.5.1 Huber-based Filter

The estimator using Huber's technique has a bounded and continuous influence function for outliers (Gandhi and Mili 2010, Li et al. 2015). The effect of outliers is reduced by the bounded influence function. The estimation can be represented by minimizing the cost function

$$\hat{\mathbf{x}}_{k|k} = \arg\min_{\mathbf{x}} \sum_{i=1}^{m+n} \rho(\mathbf{e}_{k,i}) \quad (5.54)$$

where $\rho(\tau) = \begin{cases} 1/2\tau^2 & |\tau| < \gamma \\ \gamma|\tau| - 1/2\gamma^2 & |\tau| \ge \gamma \end{cases}$. n and m are the dimension of the system state and measurement, respectively. $\mathbf{e}_{k,i}$ is the ith component of the vector $\mathbf{e}_k$, which is the error term of the estimation. It can be seen that $\rho(\cdot)$ is a quadratic function of the error when the estimation error is small, while it is a linear function of the error if the estimation error is large. Hence, it reduces the sensitivity to large error terms, such as outlier values. The robustness is thus expected. The function ρ is shown in Fig. 5.1 for different γ. Different levels of performance can be achieved for different γ parameters.

To derive the error term of the estimation, we combine the estimate and the measurement for the system as follows.

$$\begin{bmatrix} \mathbf{y}_k \\ \hat{\mathbf{x}}_{k|k} \end{bmatrix} = \begin{bmatrix} \boldsymbol{h}(\mathbf{x}_k) \\ \mathbf{x}_k \end{bmatrix} + \begin{bmatrix} \mathbf{n}_k \\ \delta\mathbf{x}_k \end{bmatrix} \quad (5.55)$$

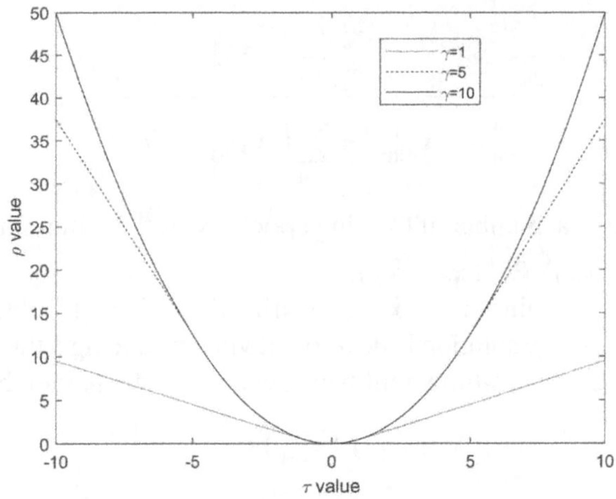

Fig. 5.1 ρ function value with different parameters.

where $\delta \mathbf{x}_k$ denotes error between the true state value and the estimated value.

Define $\boldsymbol{\varepsilon} = \begin{bmatrix} \mathbf{n}_k \\ \delta \mathbf{x}_k \end{bmatrix}$, $E\left(\boldsymbol{\varepsilon} \cdot \boldsymbol{\varepsilon}^T\right) = \begin{bmatrix} \mathbf{R}_k & \mathbf{0} \\ \mathbf{0} & \mathbf{P}_{k|k-1} \end{bmatrix} \triangleq \bar{\mathbf{R}}_k$, the error term of the estimation can be described by (Li et al. 2015)

$$\mathbf{e}_k = \bar{\mathbf{R}}_k^{-1/2} \begin{bmatrix} \mathbf{n}_k \\ \delta \mathbf{x}_k \end{bmatrix} = \bar{\mathbf{R}}_k^{-1/2} \begin{bmatrix} \mathbf{y}_k \\ \hat{\mathbf{x}}_{k|k} \end{bmatrix} - \bar{\mathbf{R}}_k^{-1/2} \begin{bmatrix} h(\mathbf{x}_k) \\ \mathbf{x}_k \end{bmatrix} \tag{5.56}$$

Equation (5.54) can be solved via the following equation.

$$\sum_{i=1}^{m+n} \phi\left(\mathbf{e}_{k,i}\right) \frac{\partial \mathbf{e}_{k,i}}{\partial \mathbf{x}_k} = 0 \tag{5.57}$$

where ϕ is the derivative of the function $\rho(\cdot)$, i.e., $\phi = \rho'$ with respect to error $\mathbf{e}_{k,i}$.

The Huber based Kalman filter (Li et al. 2015) can use various grid-based Gaussian integral techniques. There are other research works based on the Huber estimation framework, such as the square-root version of the Huber based filter. In addition, other weighting functions can be used, such as the Hampel based weight function (Chang and Li 2017).

5.5.2 H_∞ Filter

The principle of the H_∞ filter is to guarantee the smallest energy error for all fixed energy disturbances (Jiang et al. 2016). It is achieved by optimizing $\hat{\mathbf{x}}_k = \arg\min \|J\|_\infty$ with the cost function

$$J = \frac{\sum_{k=1}^{N} \left\| \mathbf{x}_k - \hat{\mathbf{x}}_k \right\|^2}{\left\| \mathbf{x}_0 - \hat{\mathbf{x}}_0 \right\|_{\mathbf{P}_0^{-1}}^2 + \sum_{k=1}^{N} \left(\left\| \mathbf{v}_k \right\|_{\mathbf{Q}_k^{-1}}^2 + \left\| \mathbf{n}_k \right\|_{\mathbf{R}_k^{-1}}^2 \right)} \tag{5.58}$$

where N is the total number of filtering epochs, $\mathbf{x}_0$ is the initial state. $\left\| \mathbf{x}_0 - \hat{\mathbf{x}}_0 \right\|_{\mathbf{P}_0^{-1}}^2$ denotes $(\mathbf{x}_0 - \hat{\mathbf{x}}_0)^T \mathbf{P}_0^{-1} (\mathbf{x}_0 - \hat{\mathbf{x}}_0)$.

The analytic solution of $\hat{\mathbf{x}}_k = \arg\min \|J\|_\infty$ is hard to obtain. Hence, a suboptimal recursive solution is developed via introducing a threshold value γ.

Consider a discrete-time nonlinear dynamic system given by

$$\mathbf{x}_k = \boldsymbol{f}(\mathbf{x}_{k-1}) + \mathbf{v}_{k-1} \tag{5.59}$$

$$\mathbf{y}_k = \boldsymbol{h}(\mathbf{x}_k) + \mathbf{n}_k \tag{5.60}$$

$$\mathbf{z}_k = \mathbf{L}_k \mathbf{x}_k \tag{5.61}$$

where $\mathbf{x}_{k-1} \in \mathbb{R}^n$ is the state vector at the time $k-1$; $\mathbf{y}_k \in \mathbb{R}^m$ is the measurement at the time k; $\mathbf{z}_k$ is the signal to be estimated and $\mathbf{L}_k$ is a given matrix. If only the states are to be estimated, $\mathbf{L}_k = \boldsymbol{I}$. $\boldsymbol{f}(\cdot)$ and $\boldsymbol{h}(\cdot)$ are the nonlinear system dynamics and the observation function, respectively. $\mathbf{v}_{k-1}$ and $\mathbf{n}_k$ are the process noise and the measurement noise, respectively, and are assumed to be bounded signal with unknown statistics, i.e., $\sum_{k=0}^{\infty} \mathbf{v}_k^T \mathbf{v}_k < \infty$ and $\sum_{k=0}^{\infty} \mathbf{n}_k^T \mathbf{n}_k < \infty$.

The aim of the H_∞ filtering is to find $\hat{\mathbf{z}}_k$ that minimizes the estimation error $\left\| \hat{\mathbf{z}}_k - \mathbf{z}_k \right\|$ for any $\hat{\mathbf{x}}_0$, $\mathbf{v}_{k-1}$, and $\mathbf{n}_k$. To avoid using the infinite magnitudes for $\hat{\mathbf{x}}_0$, $\mathbf{v}_{k-1}$, and $\mathbf{n}_k$, a commonly used cost function is given from the game theory (Simon 2006)

$$J = \frac{\sum_{j=0}^{k} \left\| \hat{\mathbf{z}}_j - \mathbf{L}_j \mathbf{x}_j \right\|^2}{\left\| \mathbf{x}_0 - \hat{\mathbf{x}}_0 \right\|_{\mathbf{P}_0^{-1}}^2 + \sum_{j=0}^{k-1} \left\| \mathbf{v}_j \right\|_{\mathbf{Q}_j^{-1}}^2 + \sum_{j=0}^{k} \left\| \mathbf{n}_j \right\|_{\mathbf{R}_j^{-1}}^2} \tag{5.62}$$

where $\mathbf{P}_0$ is the initial covariance matrix that indicates how close the initial estimate $\hat{\mathbf{x}}_0$ is to the true initial states; $\mathbf{Q}_j$ and $\mathbf{R}_j$ are positive definite matrices that are selected by the designer and can be chosen as the estimates of the covariance matrices of the noise. The notation of $\|\mathbf{x}\|_{\mathbf{P}}^2$ is defined as the square of the weighted l_2 norm of $\mathbf{x}$, i.e., $\|\mathbf{x}\|_{\mathbf{P}}^2 = \mathbf{x}^T \mathbf{P} \mathbf{x}$. Note that Eq. (5.62) defines the mapping from the unknown disturbances, including the initial estimation error $\mathbf{x}_0 - \hat{\mathbf{x}}_0$, the process noise $\{\mathbf{v}_j\}_{j=1}^{k-1}$, and the measurement noise $\{\mathbf{n}_j\}_{j=1}^{k}$, to the estimation error $\{e_j\}_{j=0}^{k} = \hat{\mathbf{z}}_j - \mathbf{z}_j$.

For the H_∞ filtering problem, $\hat{\mathbf{z}}_j$ is calculated to satisfy the following condition

$$\sup J < \gamma^2 \tag{5.63}$$

where 'sup' represents supremum and γ^2 is a given scalar. The nonlinear H_∞ filter can be designed to achieve the desired accuracy by iterating γ to arrive at a suboptimal solution. One such sub-optimal solution can be obtained by the extended H_∞ filter.

The extended H_∞ filter (EH_∞F) has been well studied and can be summarized as follows (Einicke and White 1999, Li and Jia 2010). Similarly to the EKF, EH_∞F contains two steps: prediction and update.

Prediction:

$$\hat{\mathbf{x}}_{k|k-1} = \boldsymbol{f}\left(\hat{\mathbf{x}}_{k-1|k-1}\right) \tag{5.64}$$

$$\mathbf{P}_{k|k-1} = \mathbf{F}_{k-1}\mathbf{P}_{k-1|k-1}\mathbf{F}_{k-1}^T + \mathbf{Q}_{k-1} \tag{5.65}$$

where $\mathbf{F}_{k-1}$ is the Jacobian matrix of $\boldsymbol{f}(\cdot)$ evaluated at $\hat{\mathbf{x}}_{k-1|k-1}$.

Update:

$$\hat{\mathbf{x}}_{k|k} = \hat{\mathbf{x}}_{k|k-1} + \mathbf{K}_k\left[\mathbf{y}_k - \boldsymbol{h}\left(\hat{\mathbf{x}}_{k|k-1}\right)\right] \tag{5.66}$$

$$\mathbf{P}_{k|k} = \mathbf{P}_{k|k-1} - \mathbf{P}_{k|k-1}\begin{bmatrix}\mathbf{H}_k^T & \mathbf{L}_k^T\end{bmatrix}\mathbf{R}_{e,k}^{-1}\begin{bmatrix}\mathbf{H}_k^T & \mathbf{L}_k^T\end{bmatrix}^T\mathbf{P}_{k|k-1} \tag{5.67}$$

where
$$\mathbf{K}_k = \mathbf{P}_{k|k-1}\mathbf{H}_k^T\left(\mathbf{H}_k\mathbf{P}_{k|k-1}\mathbf{H}_k^T + \mathbf{R}_k\right)^{-1} \tag{5.68}$$

$$\mathbf{R}_{e,k} = \begin{bmatrix}\mathbf{R}_k & 0\\ 0 & -\gamma^2 I\end{bmatrix} + \begin{bmatrix}\mathbf{H}_k^T & \mathbf{L}_k^T\end{bmatrix}^T\mathbf{P}_{k|k-1}\begin{bmatrix}\mathbf{H}_k^T & \mathbf{L}_k^T\end{bmatrix} \tag{5.69}$$

and $\mathbf{H}_k$ is the Jacobian matrix of $\boldsymbol{h}(\cdot)$ evaluated at $\hat{\mathbf{x}}_{k|k-1}$.

The EH_∞F is based on the first-order linearization of the nonlinear model and is thus not accurate enough for many applications when the nonlinearity is high or the initial estimation error is large. Many quadrature point-based and derivative-free numerical rules can be used to improve the accuracy of the EH_∞F such as the UT, the spherical-radial cubature rule, the GHQ rule, and the SGQ rule introduced in Chapter 4.

With the Gaussian assumption, the predicted mean $\hat{\mathbf{x}}_{k|k-1}$ and covariance $\mathbf{P}_{k|k-1}$ can be obtained (Jia and Xin 2013, Jia et al. 2012a,b) by the quadrature approximations (5.70) and (5.71).

$$\hat{\mathbf{x}}_{k|k-1} = \sum_{i=1}^{N_p} W_i \boldsymbol{f}\left(\xi_{k-1,i}\right) \tag{5.70}$$

$$\mathbf{P}_{k|k-1} = \sum_{i=1}^{N_p} W_i \left(\boldsymbol{f}\left(\xi_{k-1,i}\right) - \hat{\mathbf{x}}_{k|k-1}\right)\left(\boldsymbol{f}\left(\xi_{k-1,i}\right) - \hat{\mathbf{x}}_{k|k-1}\right)^T + \mathbf{Q}_{k-1} \tag{5.71}$$

where $$\xi_{k-1,i} = \mathbf{S}_{k-1}\boldsymbol{r}_i + \hat{\mathbf{x}}_{k-1|k-1} \quad \mathbf{P}_{k-1|k-1} = \mathbf{S}_{k-1}\mathbf{S}_{k-1}^T \tag{5.72}$$

$\boldsymbol{r}_i$ and W_i are the quadrature points and weights respectively, which can be generated by the numcrical rule shown in Chapter 4 and will be given afterward; N_p is the number of points.

The grid-based nonlinear H_∞ filters can be designed by embedding the numerical rule into the extended H_∞ filtering framework (5.64)–(5.69). Since the prediction step in (5.64) and (5.65) can be readily replaced by the quadrature approximations (5.70) and (5.71), it remains to replace the Jacobian matrix $\mathbf{H}_k$ by the quadrature approximation in the update step. This can be done by the statistical linear error propagation method (Lee 2008, Arasaratnam et al. 2007).

This method is based on the following two approximations (Arasaratnam et al. 2007, Lee 2008).

$$\mathbf{P}_{k|k-1}^{yy} \triangleq E\left[\left(\mathbf{y}_k - \hat{\mathbf{y}}_{k|k-1}\right)\left(\mathbf{y}_k - \hat{\mathbf{y}}_{k|k-1}\right)^T\right] \approx \mathbf{H}_k \mathbf{P}_{k|k-1} \mathbf{H}_k^T \tag{5.73}$$

$$\mathbf{P}_{k|k-1}^{xy} \triangleq E\left[\left(\mathbf{x}_k - \hat{\mathbf{x}}_{k|k-1}\right)\left(\mathbf{y}_k - \hat{\mathbf{y}}_{k|k-1}\right)^T\right] \approx \mathbf{P}_{k|k-1} \mathbf{H}_k^T \tag{5.74}$$

where $\mathbf{P}_{k|k-1}^{yy}$ is the measurement covariance and $\mathbf{P}_{k|k-1}^{xy}$is the cross-correlation covariance.
Based on Eq. (5.74), we have,

$$\mathbf{H}_k = \left(\left(\mathbf{P}_{k|k-1}\right)^{-1} \mathbf{P}_{k|k-1}^{xy}\right)^T. \tag{5.75}$$

Substituting Eqs. (5.73) and (5.74) into Eq. (5.68) yields

$$\mathbf{K}_k = \mathbf{P}_{k|k-1}\mathbf{H}_k^T\left(\mathbf{H}_k \mathbf{P}_{k|k-1}\mathbf{H}_k^T + \mathbf{R}_k\right)^{-1} \approx \mathbf{P}_{k|k-1}^{xy}\left(\mathbf{P}_{k|k-1}^{yy} + \mathbf{R}_k\right)^{-1} \tag{5.76}$$

Hence, the update Eq. (5.66) can be rewritten as

$$\hat{\mathbf{x}}_{k|k} = \hat{\mathbf{x}}_{k|k-1} + \mathbf{P}_{k|k-1}^{xy}\left(\mathbf{P}_{k|k-1}^{yy} + \mathbf{R}_k\right)^{-1}\left(\mathbf{y}_k - \hat{\mathbf{y}}_{k|k-1}\right) \tag{5.77}$$

where $\hat{\mathbf{y}}_{k|k-1}$ is the computed observation vector.
Substituting Eqs. (5.73) and (5.74) into the update Eq. (5.67) yields

$$\begin{aligned}\mathbf{P}_{k|k} &= \mathbf{P}_{k|k-1} - \mathbf{P}_{k|k-1}\begin{bmatrix}\mathbf{H}_k^T & \mathbf{L}_k^T\end{bmatrix}\mathbf{R}_{e,k}^{-1}\begin{bmatrix}\mathbf{H}_k^T & \mathbf{L}_k^T\end{bmatrix}^T\mathbf{P}_{k|k-1} \\ &= \mathbf{P}_{k|k-1} - \begin{bmatrix}\mathbf{P}_{k|k-1}\mathbf{H}_k^T & \mathbf{P}_{k|k-1}\end{bmatrix}\mathbf{R}_{e,k}^{-1}\begin{bmatrix}\mathbf{H}_k\mathbf{P}_{k|k-1} \\ \mathbf{P}_{k|k-1}\end{bmatrix} \\ &\approx \mathbf{P}_{k|k-1} - \begin{bmatrix}\mathbf{P}_{k|k-1}^{xy} & \mathbf{P}_{k|k-1}\end{bmatrix}\mathbf{R}_{e,k}^{-1}\begin{bmatrix}\begin{bmatrix}\mathbf{P}_{k|k-1}^{xy}\end{bmatrix}^T \\ \mathbf{P}_{k|k-1}\end{bmatrix}\end{aligned} \tag{5.78}$$

We assume $\mathbf{L}_k = \mathbf{I}$ here since the state estimation problem is considered. Substituting Eqs. (5.73) and (5.74) into Eq. (5.69) yields

$$\begin{aligned}\mathbf{R}_{e,k} &= \begin{bmatrix}\mathbf{R}_k & 0 \\ 0 & -\gamma^2\mathbf{I}\end{bmatrix} + \begin{bmatrix}\mathbf{H}_k^T & \mathbf{L}_k^T\end{bmatrix}^T\mathbf{P}_{k|k-1}\begin{bmatrix}\mathbf{H}_k^T & \mathbf{L}_k^T\end{bmatrix} \\ &= \begin{bmatrix}\mathbf{R}_k & 0 \\ 0 & -\gamma^2\mathbf{I}\end{bmatrix} + \begin{bmatrix}\mathbf{H}_k^T & \mathbf{L}_k^T\end{bmatrix}^T\begin{bmatrix}\mathbf{P}_{k|k-1}\mathbf{H}_k^T & \mathbf{P}_{k|k-1}\end{bmatrix} \\ &= \begin{bmatrix}\mathbf{R}_k & 0 \\ 0 & -\gamma^2\mathbf{I}\end{bmatrix} + \begin{bmatrix}\mathbf{H}_k\mathbf{P}_{k|k-1}\mathbf{H}_k^T & \mathbf{H}_k\mathbf{P}_{k|k-1} \\ \mathbf{P}_{k|k-1}\mathbf{H}_k^T & \mathbf{P}_{k|k-1}\end{bmatrix} \\ &\approx \begin{bmatrix}\mathbf{R}_k + \mathbf{P}_{k|k-1}^{yy} & \begin{bmatrix}\mathbf{P}_{k|k-1}^{xy}\end{bmatrix}^T \\ \mathbf{P}_{k|k-1}^{xy} & -\gamma^2\mathbf{I} + \mathbf{P}_{k|k-1}\end{bmatrix}\end{aligned} \tag{5.79}$$

Since $\hat{\mathbf{y}}_{k|k-1}$, $\mathbf{P}_{k|k-1}^{yy}$, and $\mathbf{P}_{k|k-1}^{xy}$ involve computing Gaussian weighted integrals, they can be calculated by using the previous quadrature point-based integral approximations, i.e.,

$$\hat{\mathbf{y}}_{k|k-1} = \sum_{i=1}^{N_p} W_i \boldsymbol{h}\left(\xi_{k,i}\right) \tag{5.80}$$

$$\mathbf{P}_{k|k-1}^{xy} = \sum_{i=1}^{N_p} W_i\left(\tilde{\xi}_{k,i} - \hat{\mathbf{x}}_{k|k-1}\right)\left(\boldsymbol{h}\left(\tilde{\xi}_{k,i}\right) - \hat{\mathbf{y}}_{k|k-1}\right)^T \tag{5.81}$$

$$\mathbf{P}_{k|k-1}^{yy} = \sum_{i=1}^{N_p} W_i\left(\boldsymbol{h}\left(\tilde{\xi}_{k,i}\right) - \hat{\mathbf{y}}_{k|k-1}\right)\left(\boldsymbol{h}\left(\tilde{\xi}_{k,i}\right) - \hat{\mathbf{y}}_{k|k-1}\right)^T \tag{5.82}$$

where $$\tilde{\xi}_{k,i} = \tilde{\mathbf{S}}_k \boldsymbol{r}_i + \hat{\mathbf{x}}_{k|k-1} \quad \mathbf{P}_{k|k-1} = \tilde{\mathbf{S}}_k\tilde{\mathbf{S}}_k^T \tag{5.83}$$

Note that the value γ in Eq. (5.79) should be chosen such that the covariance $\mathbf{P}_{k|k}$ is positive definite. A way to choose γ is given by (Li and Jia 2010)

$$\gamma^2 = \alpha \max\left\{\mathrm{eig}\left(\mathbf{P}_{k|k-1}^{-1} + \mathbf{P}_{k|k-1}^{-1}\mathbf{P}_{k|k-1}^{xy}\mathbf{R}_k^{-1}\left[\mathbf{P}_{k|k-1}^{-1}\mathbf{P}_{k|k-1}^{xy}\right]^T\right)^{-1}\right\} \tag{5.84}$$

where α is a scalar greater than one and $\max\left\{\text{eig}(*)^{-1}\right\}$ denotes the maximum eigenvalue of $(*)^{-1}$.

Many numerical integration rules given in Chapter 4 can be utilized to provide quadrature points $\boldsymbol{r}_i$ and weights W_i in Eqs. (5.72) and (5.83) such as the UT (Julier and Uhlmann 2004), the spherical-radial cubature rule (Arasaratnam and Haykin 2009, Jia et al. 2013), the GHQ rule (Arasaratnam et al. 2007, Ito and Xiong 2000), and the SGQ rule. These derivative-free numerical rules will lead to different grid-based H_∞ filters.

5.6 Gaussian Mixture Filter

We consider a class of nonlinear discrete-time dynamic systems described by

$$\mathbf{x}_k = \boldsymbol{f}(\mathbf{x}_{k-1}) + \mathbf{v}_{k-1} \tag{5.85}$$

$$\mathbf{y}_k = \boldsymbol{h}(\mathbf{x}_k) + \mathbf{n}_k \tag{5.86}$$

where $\mathbf{x}_k \in \mathbb{R}^n$ is the state vector and $\mathbf{y}_k \in \mathbb{R}^m$ is the measurement. $\mathbf{v}_{k-1}$ and $\mathbf{n}_k$ are the process noise and measurement noise, respectively, and their PDFs are represented by the Gaussian mixtures (GM) $p(\mathbf{v}_{k-1}) = \sum_{p=1}^{N_p} \beta^p N\left(\mathbf{v}_{k-1}^p; \bar{\mathbf{v}}_{k-1}^p, \mathbf{Q}_{k-1}^p\right)$ and $p(\mathbf{n}_k) = \sum_{q=1}^{N_q} \upsilon^q N\left(\mathbf{n}_k^q; \bar{\mathbf{n}}_k^q, \mathbf{R}_k^q\right)$, respectively, where $N\left(\mathbf{n}_k^q; \bar{\mathbf{n}}_k^q, \mathbf{R}_k^q\right)$ denotes a normal distribution with mean $\bar{\mathbf{n}}_k^q$ and covariance $\mathbf{R}_k^q$, β^p and υ^q are the weights of the Gaussian component. The superscripts 'p' and 'q' denote the pth and qth component of the GM; 'N_p' and 'N_q' denote the number of Gaussian components. Due to the non-Gaussian noise and nonlinear dynamics, the estimated state will have a non-Gaussian PDF, which can be modeled as the GM as well.

Grid-based Gaussian Mixture Filter

Assume that the initial state PDF at the beginning of each filtering cycle can be represented by the GM $p(\mathbf{x}) = \sum_{l=1}^{N_l} \alpha^l N\left(\mathbf{x}; \hat{\mathbf{x}}^l, \mathbf{P}^l\right)$. In Fig. 5.2 one cycle of the grid-based GM filter is illustrated using the CKF to perform the prediction and update as an example. The grid-based Gaussian approximation filter runs on each component of the GM to predict and update the component's mean and covariance. The prediction step of the grid-based Gaussian approximation filter is first used for each of the N_l GM components. Note that after the prediction step, there are $N_l \times N_p$ Gaussian components contributed from the GM of the initial state PDF and the GM of the process noise. After that, the update step

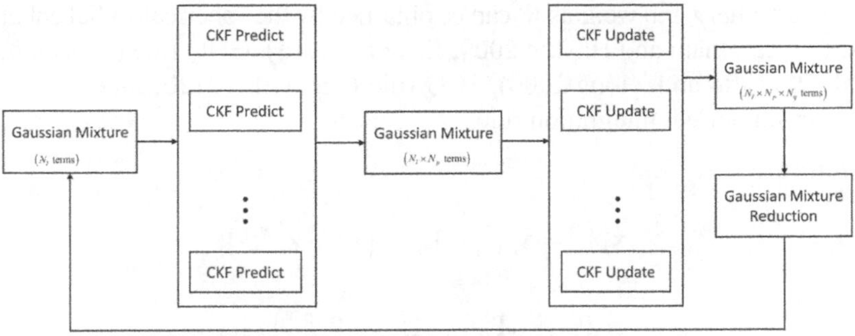

Fig. 5.2 One filtering cycle of the cubature GM filter.

of the grid-based Gaussian approximation filter is used for each Gaussian component and leads to $N_l \times N_p \times N_q$ Gaussian components added by the GM of the measurement noise. It can be seen that the number of Gaussian components increases after each filtering cycle. To limit the computational complexity, the number of Gaussian components has to be reduced after the update step. In the following, the prediction step and update step for each Gaussian component using the grid-based Gaussian approximation filter framework are introduced.

Remark 5.4: It is flexible to choose the number of Gaussian components in the Gaussian mixture filter at different stages, such as N_l.

Prediction step:

Given the initial estimate of the mean $\hat{\mathbf{x}}^l_{k-1|k-1}$ and covariance $\mathbf{P}^l_{k-1|k-1}$ at the time $k-1$ for the lth Gaussian component, the predicted mean and covariance can be computed by the quadrature approximation (Ito and Xiong 2000, Arasaratnam et al. 2007)

$$\hat{\mathbf{x}}^{l,p}_{k|k-1} = \sum_{i=1}^{N_u} W_i \boldsymbol{f}\left(\xi^l_{k-1,i}\right) + \overline{\mathbf{v}}^p_{k-1} \tag{5.87}$$

$$\mathbf{P}^{l,p}_{k|k-1} = \sum_{i=1}^{N_u} W_i \left[\boldsymbol{f}\left(\xi^l_{k-1,i}\right) - \left(\hat{\mathbf{x}}^{l,p}_{k|k-1} - \overline{\mathbf{v}}^p_{k-1}\right)\right]\left[\boldsymbol{f}\left(\xi^l_{k-1,i}\right) - \left(\mathbf{x}^{l,p}_{k|k-1} - \overline{\mathbf{v}}^p_{k-1}\right)\right]^T + \mathbf{Q}^p_k \tag{5.88}$$

where N_u is the total number of quadrature points, $l = 1, \cdots, N_l$, $p = 1, \cdots, N_p$; The superscript "l,p" denotes the value using the lth Gaussian component of the GM of the initial state PDF and the pth component of the GM of the process noise. $\overline{\mathbf{v}}^p_{k-1}$ is the mean of the pth Gaussian component of the GM representation of the process noise; $\xi^l_{k-1,i}$ is the transformed quadrature point given by

$$\xi^l_{k-1,i} = \mathbf{S}^l_{k-1}\boldsymbol{\gamma}_i + \hat{\mathbf{x}}^l_{k-1|k-1}, \qquad \mathbf{P}^l_{k-1|k-1} = \mathbf{S}^l_{k-1}\left(\mathbf{S}^l_{k-1}\right)^T \tag{5.89}$$

The points γ_i and weights W_i can be obtained by the spherical-radial cubature rule (Arasaratnam and Haykin 2009, Jia et al. 2013), GHQ rule (Arasaratnam et al. 2007, Ito and Xiong 2000), SGQ rule (Jia et al. 2012b, Jia et al. 2011), or other numerical integration rules.

Update Step:

$$\hat{\mathbf{x}}_{k|k}^{l,p,q} = \hat{\mathbf{x}}_{k|k-1}^{l,p} + \mathbf{L}_k^{l,p,q}\left(\mathbf{y}_k - \mathbf{z}_k^{l,p,q}\right) \tag{5.90}$$

$$\mathbf{P}_{k|k}^{l,p,q} = \mathbf{P}_{k|k-1}^{l,p} - \mathbf{L}_k^{l,p,q}\left(\mathbf{P}_{xz}^{l,p,q}\right)^T \tag{5.91}$$

where
$$\mathbf{L}_k^{l,p,q} = \mathbf{P}_{xz}^{l,p,q}\left(\mathbf{R}_k^q + \mathbf{P}_{zz}^{l,p,q}\right)^{-1} \tag{5.92}$$

$$\mathbf{z}_k^{l,p,q} = \sum_{i=1}^{N_u} W_i \boldsymbol{h}\left(\tilde{\xi}_{k,i}^{l,p}\right) + \overline{\mathbf{n}}_k^q \tag{5.93}$$

$$\mathbf{P}_{xz}^{l,p,q} = \sum_{i=1}^{N_u} W_i \left[\tilde{\xi}_{k,i}^{l,p} - \hat{\mathbf{x}}_{k|k-1}^{l,p}\right]\left[\boldsymbol{h}\left(\tilde{\xi}_{k,i}^{l,p}\right) - \left(\mathbf{z}_k^{l,p,q} - \overline{\mathbf{n}}_k^q\right)\right]^T \tag{5.94}$$

$$\mathbf{P}_{zz}^{l,p,q} = \sum_{i=1}^{N_u} W_i \left[\boldsymbol{h}\left(\tilde{\xi}_{k,i}^{l,p}\right) - \left(\mathbf{z}_k^{l,p,q} - \overline{\mathbf{n}}_k^q\right)\right]\left[\boldsymbol{h}\left(\tilde{\xi}_{k,i}^{l,p}\right) - \left(\mathbf{z}_k^{l,p,q} - \overline{\mathbf{n}}_k^q\right)\right]^T \tag{5.95}$$

$\overline{\mathbf{n}}_k^q$ is the mean of the qth Gaussian component of the GM representation of the measurement noise; $\xi_{k,i}^{l,p}$ is the transformed quadrature point given by

$$\xi_{k,i}^{l,p} = \tilde{\mathbf{S}}_k^{l,p}\boldsymbol{\gamma}_i + \hat{\mathbf{x}}_{k|k-1}^{l,p}, \quad \mathbf{P}_{k|k-1}^{l,p} = \tilde{\mathbf{S}}_k^{l,p}\left(\tilde{\mathbf{S}}_k^{l,p}\right)^T \tag{5.96}$$

Remark 5.5: The weight of the Gaussian component $N\left(\mathbf{x}; \hat{\mathbf{x}}_{k|k}^{l,p,q}, \mathbf{P}_{k|k}^{l,p,q}\right)$ is $\alpha^{l,p,q}$. The final Gaussian mixture can be represented by $\sum_{l=1}^{N_l}\sum_{p=1}^{N_p}\sum_{q=1}^{N_q} \alpha^{l,p,q} N\left(\mathbf{x}; \hat{\mathbf{x}}_{k|k}^{l,p,q}, \mathbf{P}_{k|k}^{l,p,q}\right)$. Note that the weight $\alpha^{l,p,q}$ is given by (Arasaratnam et al. 2007)

$$\alpha^{l,p,q} = \frac{\alpha^l \beta^p \upsilon^q \Omega^{l,p,q}}{\sum_{q=1}^{N_q}\sum_{l=1}^{N_l}\sum_{p=1}^{N_p} \alpha^l \beta^p \upsilon^q \Omega^{l,p,q}} \tag{5.97}$$

with $\Omega^{l,p,q} = N\left(\mathbf{y}; \mathbf{z}^{l,p,q}, \mathbf{P}_{zz}^{l,p,q}\right)$.

The number of Gaussian components increases significantly as the time evolves. In order to avoid excessive computation load, some Gaussian components can be removed or merged. There are many GM reduction algorithms (Runnalls 2007, Williams 2003), such as pruning Gaussian components with negligible weights, joining near Gaussian components, and regeneration of GM via Kullback-Leibler approach. Note that to keep the estimation accuracy, the GM reduction procedure is not necessary if the number of Gaussian components is less than a specified threshold.

5.7 Simplified Grid-based Gaussian Mixture Filter

The simplified grid-based GM filter integrates the Gaussian approximation filter framework and the GM filter (Kottakki et al. 2014). Specifically, for each filtering cycle, the predicted PDF is represented by the Gaussian mixture and the updated PDF is represented by the Gaussian distribution.

For the conventional grid-based filter, the predicted PDF from time instant k–1 to k is represented by the Gaussian distribution. For the simplified grid Gaussian mixture filter, the PDF of the predicted state is represented by a Gaussian mixture as follows.

$$p\left(\hat{\mathbf{x}}_{k|k-1}, \mathbf{P}_{k|k-1}\right) = \sum_{i=1}^{N_s} w^{(i)} p^{(i)}\left(\hat{\mathbf{x}}_{k|k-1}^{(i)}, \mathbf{P}_{k|k-1}^{(i)}\right) \tag{5.98}$$

If we let $\hat{\mathbf{x}}_{k|k-1}^{(i)} = \boldsymbol{f}\left(\xi_{k-1,i}\right)$, $\mathbf{P}_{k|k-1}^{(i)} = \mathbf{Q}_{k-1}$, and $w^{(i)} = W_i$, the state estimate of the Gaussian mixture is given by

$$\hat{\mathbf{x}}_{k|k-1,GM} = \sum_{i=1}^{N_s} w^{(i)} \hat{\mathbf{x}}_{k|k-1}^{(i)} = \sum_{i=1}^{N_s} W_i \boldsymbol{f}\left(\xi_{k-1,i}\right) = \hat{\mathbf{x}}_{k|k-1} \tag{5.99}$$

$$\begin{aligned}
\mathbf{P}_{k|k-1,GM} &= \sum_{i=1}^{N_s} w^{(i)} \left(\mathbf{Q}_{k-1} + \left(\hat{\mathbf{x}}_{k|k-1}^{(i)} - \hat{\mathbf{x}}_{k|k-1}\right)\left(\hat{\mathbf{x}}_{k|k-1}^{(i)} - \hat{\mathbf{x}}_{k|k-1}\right)^T \right) \\
&= \sum_{i=1}^{N_s} W_i \left(\mathbf{Q}_{k-1} + \left(\boldsymbol{f}\left(\xi_{k-1,i}\right) - \hat{\mathbf{x}}_{k|k-1}\right)\left(\boldsymbol{f}\left(\xi_{k-1,i}\right) - \hat{\mathbf{x}}_{k|k-1}\right)^T \right) \\
&= \mathbf{Q}_{k-1} + \sum_{i=1}^{N_s} W_i \left(\boldsymbol{f}\left(\xi_{k-1,i}\right) - \hat{\mathbf{x}}_{k|k-1}\right)\left(\boldsymbol{f}\left(\xi_{k-1,i}\right) - \hat{\mathbf{x}}_{k|k-1}\right)^T = \mathbf{P}_{k|k-1}
\end{aligned} \tag{5.100}$$

Note that $\xi_{k-1,i}$ and W_i are the same as those used in Section 4.1. The subscript '*GM*' denotes the value calculated using the Gaussian mixture in Eq. (5.98). By Eqs. (5.99) and (5.100), it can be seen that the mean and covariance calculated by the grid-based Gaussian approximation filter in the prediction step has the same values as those using the GM representation in Eq. (5.98).

For each Gaussian component, the state estimation $\hat{\mathbf{x}}_{k|k-1}^{(i)}$ and covariance $\mathbf{P}_{k|k-1}^{(i)}$ are updated by the update step of the grid-based Gaussian approximation filter. The weights can be updated in the same way used in the GM filter (Kottakki et al. 2014). Specifically, the weight of the ith Gaussian component is given by

$$\tilde{w}_k^{(i)} = w^{(i)} \exp\left\{-0.5\left(\mathbf{y}_k - \boldsymbol{h}\left(\hat{\mathbf{x}}_{k|k-1}^{(i)}\right)\right)^T \mathbf{P}_{yy}^{-1}\left(\mathbf{y}_k - \boldsymbol{h}\left(\hat{\mathbf{x}}_{k|k-1}^{(i)}\right)\right)\right\} \quad (5.101)$$

To keep the constraint of the PDF, the weights should be normalized.

$$\bar{w}_k^{(i)} = \frac{\tilde{w}_k^{(i)}}{\sum_i \tilde{w}_k^{(i)}} \quad (5.102)$$

After the update step, as discussed in the general GM filter section, it needs to merge the Gaussian components. As a special case, all Gaussian components are merged into one Gaussian distribution. The final state estimation and covariance matrix are given by

$$\hat{\mathbf{x}}_{k|k} = \sum_{i=1}^{N_s} \bar{w}^{(i)} \hat{\mathbf{x}}_{k|k}^{(i)} \quad (5.103)$$

$$\mathbf{P}_{k|k} = \sum_{i=1}^{N_s} \bar{w}^{(i)} \left(\mathbf{P}_{k|k}^{(i)} + \left(\hat{\mathbf{x}}_{k|k}^{(i)} - \hat{\mathbf{x}}_{k|k}\right)\left(\hat{\mathbf{x}}_{k|k}^{(i)} - \hat{\mathbf{x}}_{k|k}\right)^T\right). \quad (5.104)$$

The prediction step uses the same equations as those in the conventional grid-based Gaussian approximation filter. However, the predicted PDF is represented by the Gaussian mixture, as shown in Eq. (5.98). Note that the GM is only used after the prediction.

***Remark 5.6*:** The weights used in Eq. (5.98) should be positive. However, it is not always true for the sparse-grid quadrature or high-degree spherical-radial cubature rules.

5.8 Adaptive Gaussian Mixture Filter

As we mentioned in Section 5.6, the number of Gaussian components for the GM filter should be controlled. In general, if the number of Gaussian components is greater than a threshold, extra Gaussian components are merged. There will be accuracy loss when the number of Gaussian components is forced to be an exact number. The accuracy can be improved by choosing the Gaussian components adaptively. In this section, we review the existing design strategies in order to inspire readers to investigate better strategies in the future.

The first strategy uses the Fokker-Planck equation or Chapman-Kolmogorov equation to evaluate the GM approximation. Instead of updating the number of Gaussian components, the weights of different Gaussian components are dynamically updated (Terejanu et al. 2011).

The idea is to approximate the PDF via the GM with unknown weights. The unknown weights are then computed via the optimization of the cost function.

The cost function is defined as the L_2-norm of the difference between the true PDF propagated via the Chapman-Kolmogorov equation $p\left(\mathbf{x}_{k|k-1} \mid \mathbf{y}_{k-1}\right)$ and the approximated PDF via the GM $\hat{p}\left(\mathbf{x}_{k|k-1} \mid \mathbf{y}_{k-1}\right)$.
The weights are updated via solving the following optimization problem

$$\begin{aligned} &\min_{w_k^{(i)}} \frac{1}{2} \int \left(p\left(\mathbf{x}_{k|k-1} \mid \mathbf{y}_{k-1}\right) - \hat{p}\left(\mathbf{x}_{k|k-1} \mid \mathbf{y}_{k-1}\right)\right)^2 \\ &\text{s.t.} \sum_{i=1}^{N_s} w_k^{(i)} = 1, \quad w_k^{(i)} \geq 0 \end{aligned} \tag{5.105}$$

where the PDF is represented by the GM.

$$\hat{p}\left(\mathbf{x}_{k|k-1} \mid \mathbf{y}_{k-1}\right) = \sum_{i=1}^{N_s} w_k^{(i)} N\left(\mathbf{x}_k ; \hat{\mathbf{x}}_{k|k-1}^{(i)}, \mathbf{P}_{k|k-1}^{(i)}\right) \tag{5.106}$$

and $p\left(\mathbf{x}_{k|k-1} \mid \mathbf{y}_{k-1}\right)$ can be solved via the Chapman-Kolmogorov equation.

$$\begin{aligned} p\left(\mathbf{x}_{k|k-1} \mid \mathbf{y}_{k-1}\right) &= \int p\left(\mathbf{x}_k \mid \mathbf{x}_{k-1}\right) p\left(\mathbf{x}_{k-1} \mid \mathbf{y}_{k-1}\right) d\mathbf{x}_{k-1} \\ &\approx \int p\left(\mathbf{x}_k \mid \mathbf{x}_{k-1}\right) \sum_{i=1}^{N_s} w_{k-1}^{(i)} N\left(\mathbf{x}_{k-1} ; \hat{\mathbf{x}}_{k-1|k-1}^{(i)}, \mathbf{P}_{k-1|k-1}^{(i)}\right) d\mathbf{x}_{k-1} \end{aligned} \tag{5.107}$$

After some algebra, the optimization problem can be rewritten as (Terejanu et al. 2011)

$$\begin{aligned} &\min_{w_k^{(i)}} \frac{1}{2} \mathbf{w}_k^T \mathbf{M} \mathbf{w}_k - \mathbf{w}_k^T \mathbf{z} \\ &\text{s.t.} \sum_{i=1}^{N_s} w_k^{(i)} = 1, \quad w_k^{(i)} \geq 0 \end{aligned} \tag{5.108}$$

where $\mathbf{w}_k = \left[w_k^{(1)}, \cdots, w_k^{(N_s)} \right]^T$, the element $\mathbf{M}_{ij}$ (*i*th row and *j*th column) of $\mathbf{M}$ is described by $\mathbf{M}_{ij} = N\left(\mathbf{x}_k ; \hat{\mathbf{x}}_{k|k-1}^{(i)} - \hat{\mathbf{x}}_{k|k-1}^{(j)}, \mathbf{P}_{k|k-1}^{(i)} + \mathbf{P}_{k|k-1}^{(j)}\right)$. The vector $\mathbf{z}$ is given by

$$\mathbf{z} = \left[z^{(1)}, \cdots, z^{(N_s)} \right]^T \tag{5.109}$$

$$z^{(i)} = \sum_{j=1}^{N_s} w_{k-1}^{(j)} \int N\left(f\left(\mathbf{x}_{k-1}\right); \hat{\mathbf{x}}_{k|k-1}^{(i)}, \mathbf{P}_{k|k-1}^{(i)}\right) N\left(\mathbf{x}_{k-1}; \hat{\mathbf{x}}_{k-1|k-1}^{(j)}, \mathbf{P}_{k-1|k-1}^{(j)}\right) d\mathbf{x}_{k-1} \tag{5.110}$$

The weights are then updated when the solution of Eq. (5.108) is obtained. The update step is the same as the update step of the conventional GM filter.

The second strategy is to explore the intermediate area of the ensemble Kalman filter and the particle filter (Stordal et al. 2011). This strategy is initialized by sampling N particles from the initial PDF. The weight of each particle is the same. After that, the GM is initialized via constructing N Gaussian kernels. When the new measurement is available, each Gaussian component is updated via the classical Kalman filter and the weight is adjusted based on the prior and likelihood. The adaptive Gaussian mixture filter using this strategy is equivalent to the sequential importance resampling filter (SIR) when specific parameters are chosen (Stordal et al. 2011).

Another Gaussian mixture filter is motivated by the particle filter and the Gaussian sum re-approximation. The particles in the particle filter are replaced by the Gaussian distribution. The Gaussian sum re-approximation technique aims to reproduce the original probability distribution while bounds the covariance of each Gaussian component and uses fewer Gaussian components (Psiaki 2016).

5.9 Interacting Multiple Model Filter

Interacting Multiple Model Filter (IMMF) (Bar-Shalom et al. 2004) works for the dynamic systems whose model changes over time. For example, the dynamics of a maneuvering target can be different at different times. In IMMF, the change of the model is assumed to follow a finite-state, discrete-time Markov chain. Besides providing the estimated state, the IMMF can estimate the model probability from a set of candidate models, which can be very useful for the maneuver detection of the target. One filtering cycle of the IMMF includes state interaction, filtering, and state estimate combination, as shown in Fig. 5.3 (Liland 2017). Each component is explained as follows (Hartikainen et al. 2011).

Interaction:

The mixing probabilities $\mu_k^{i|j}$ for the ith model and the jth model are calculated by

$$\mu_k^{i|j} = \frac{1}{\overline{c}_j} p_{ij} \mu_{k-1}^i \tag{5.111}$$

$$\overline{c}_j = \sum_{i=1}^{N_M} p_{ij} \mu_{k-1}^i \tag{5.112}$$

where μ_{k-1}^i is the probability of the model i at time k–1 and $\overline{c}_j$ is a normalization factor. p_{ij} is the Markov model transfer probability from model j to model i. N_M is the number of models.

The mixed mean $\hat{\mathbf{x}}_{m,k-1|k-1}^j$ and covariance $\hat{\mathbf{P}}_{m,k-1|k-1}^j$ for each model j are given by

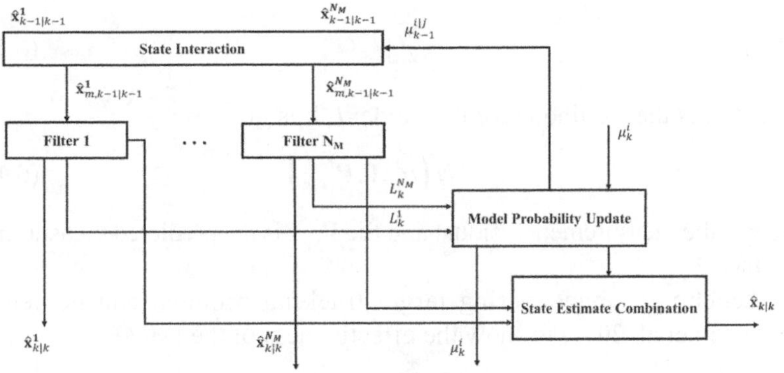

Fig. 5.3 The Diagram of the IMMF algorithm (Genovese 2001).

$$\hat{\mathbf{x}}^{j}_{m,k-1|k-1} = \sum_{i=1}^{N_M} \mu_k^{i|j} \hat{\mathbf{x}}^{i}_{k-1|k-1} \tag{5.113}$$

$$\mathbf{P}^{j}_{m,k-1|k-1} = \sum_{i=1}^{N_M} \mu_k^{i|j} \left(\mathbf{P}^{i}_{k-1|k-1} + \left(\hat{\mathbf{x}}^{i}_{k-1|k-1} - \hat{\mathbf{x}}^{j}_{m,k-1|k-1} \right) \left(\hat{\mathbf{x}}^{i}_{k-1|k-1} - \hat{\mathbf{x}}^{j}_{m,k-1|k-1} \right)^T \right) \tag{5.114}$$

where $\hat{\mathbf{x}}^{i}_{k-1|k-1}$ and $\mathbf{P}^{i}_{k-1|k-1}$ are the updated mean and covariance for model i at time $k-1$.

Filtering:

For convenience, each model uses the nonlinear filtering method.

$$\left(\hat{\mathbf{x}}^{j}_{k|k}, \mathbf{P}^{j}_{k|k} \right) = \text{NF}\left(\hat{\mathbf{x}}^{j}_{m,k-1|k-1}, \mathbf{P}^{j}_{m,k-1|k-1}, \mathbf{y}_k, \mathbf{Q}, \mathbf{R} \right) \tag{5.115}$$

Note that the 'NF' denotes the nonlinear filter. Many nonlinear filters can be applied, for example, the extended Kalman filter and grid-based Gaussian approximation filters.

Combination:

The final estimate is given by

$$\hat{\mathbf{x}}_{k|k} = \sum_{i=1}^{N_M} \mu_k^{i} \hat{\mathbf{x}}^{i}_{k|k} \tag{5.116}$$

$$\mathbf{P}_{k|k} = \sum_{i=1}^{N_M} \mu_k^{i} \left(\mathbf{P}^{i}_{k|k} + \left(\hat{\mathbf{x}}^{i}_{k|k} - \hat{\mathbf{x}}_{k|k} \right) \left(\hat{\mathbf{x}}^{i}_{k|k} - \hat{\mathbf{x}}_{k|k} \right)^T \right) \tag{5.117}$$

where μ_k^i is the updated model probability of the ith model at time k. It is given by

$$\mu_k^i = \frac{1}{c} L_k^i \overline{c}_i \tag{5.118}$$

$$c = \sum_{i=1}^{N_M} L_k^i \bar{c}_i \tag{5.119}$$

Note that L_k^i is the likelihood for the model i. It is given by

$$L_k^i = N\left(\eta_k^i; \mathbf{0}, \mathbf{P}_{yy,k}^i\right) \tag{5.120}$$

where η_k^i is the measurement residual and the $\mathbf{P}_{yy,k}^i$ is the predicted measurement covariance.

A benchmark maneuvering target tracking problem can be seen in (Hartikainen et al. 2011) to show the effectiveness of the IMMF.

5.10 Summary

Although many different extensions of the grid-based Gaussian approximation filters are introduced, a comprehensive summary is difficult. There are many other extensions in the literature, such as the Gaussian processing grid-based filter (Deisenroth et al. 2009), the reduced state estimators (Moireau and Chapelle 2011), nonlinear estimators for correlated noises (Wang et al. 2012), and moving horizon estimation (Gao et al. 2018). The extension forms introduced in this chapter can be used jointly. It is necessary to choose an adequate estimator based on different applications. Novel filters can be developed by choosing suitable forms in a complementary way.

References

Arasaratnam, I., S. Haykin and R.J. Elliott. 2007. Discrete-time nonlinear filtering algorithms using Gauss-Hermite quadrature. Proceedings of the IEEE 95: 953–977.

Arasaratnam, I. and S. Haykin. 2008. Square-root quadrature Kalman filtering. IEEE Transactions on Signal Processing 56: 2589–2593.

Arasaratnam, I. and S. Haykin. 2009. Cubature Kalman filters. IEEE Transactions on Automatic Control 54: 1254–1269.

Arasaratnam, I., S. Haykin and T.R. Hurd. 2010. Cubature Kalman filtering for continuous-discrete systems: theory and simulations. IEEE Transactions on Signal Processing 58: 4977–4993.

Arasaratnam, I. and S. Haykin. 2011. Cubature Kalman smoothers. Automatica 47: 2245–2250.

Bar-Shalom, Y., X. Li and T. Kirubarajan. 2004. Estimation with Applications to Tracking and Navigation: Theory Algorithms and Software. John Wiley & Sons, Inc., New York.

Chang, L. and K. Li. 2017. Unified form for the robust Gaussian information filtering based on M-estimate. IEEE Signal Processing Letters 24: 412–416.

Deisenroth, M.P., M.F. Huber and U.D. Hanebeck. 2009. Analytic moment-based Gaussian process filtering. Proceedings of the 26th Annual International Conference on Machine Learning. Canada, 225–232.

Einicke, G.A. and L.B. White. 1999. Robust extended Kalman filtering. IEEE Transactions on Signal Processing 47: 2596–2599.

Gandhi, M.A. and L. Mili. 2010. Robust Kalman filter based on a generalized maximum-likelihood-type estimator. IEEE Transactions on Signal Processing 58: 2509–2520.

Gao, B., S. Gao, G. Hu, Y. Zhong and C. Gu. 2018. Maximum likelihood principle and moving horizon estimation based adaptive unscented Kalman filter. Aerospace Science and Technology 73: 184–196.

Genovese, A.F. 2001. The interacting multiple model algorithm for accurate state estimation of maneuvering targets. Johns Hopkins APL Technical Digest 22: 614–623.

Hartikainen, J., A. Solin and S. Särkkä. 2011. Optimal filtering with Kalman filters and smoothers a Manual for the Matlab toolbox EKF/UKF. Aalto University School of Science, Espoo, Finland.

Ito, K. and K. Xiong. 2000. Gaussian filters for nonlinear filtering problems. IEEE Transactions on Automatic Control 45: 910–927.

Jazwinski, A. 1970. Stochastic Processing and Filtering Theory. Academic Press, New York.

Jia, B., M. Xin and Y. Cheng. 2011. Sparse Gauss-Hermite quadrature filter with application to spacecraft attitude estimation. Journal of Guidance, Control, and Dynamics 34: 367–379.

Jia, B., M. Xin and Y. Cheng. 2012a. Anisotropic sparse Gauss-Hermite quadrature filter. Journal of Guidance, Control, and Dynamics 35: 1014–1022.

Jia, B., M. Xin and Y. Cheng. 2012b. Sparse-grid quadrature nonlinear filtering. Automatica 48: 327–341.

Jia, B., M. Xin and Y. Cheng. 2013. High-degree cubature Kalman filter. Automatica 49: 510–518.

Jia, B. and X. Wang. 2013. Gene regulatory network inference by point-based Gaussian approximation filters incorporating the prior information. EURASIP Journal on Bioinformatics and Systems Biology 1: 16.

Jia, B. and M. Xin. 2013. Sparse-grid quadrature H-infinity filter for discrete-time systems with uncertain noise statistics. IEEE Transactions on Aerospace and Electronic Systems 49: 1626–1636.

Jia, B. and X. Wang. 2014. Regularized EM algorithm for sparse parameter estimation in nonlinear dynamic systems with application to gene regulatory network inference. EURASIP Journal on Bioinformatics and Systems Biology 1: 5.

Jiang, C., S. Zhang and Q. Zhang. 2016. A new adaptive H-Infinity filtering algorithm for the GPS/INS integrated navigation. Sensors 16: 2127.

Julier, S.J. and J.K. Uhlmann. 2004. Unscented filtering and nonlinear estimation. Proceedings of the IEEE 92: 401–422.

Kottakki, K.K., S. Bhartiya and M. Bhushan. 2014. State estimation of nonlinear dynamical systems using nonlinear update based unscented Gaussian sum filter. Journal of Process Control 24: 1425–1443.

Kulikov, G. Yu and M.V. Kulikova. 2017. Accurate continuous–discrete unscented Kalman filtering for estimation of nonlinear continuous-time stochastic models in radar tracking. Signal Processing 139: 25–35.

Lee, D. 2008. Nonlinear estimation and multiple sensor fusion using unscented information filtering. IEEE Signal Processing Letters 15: 861–864.

Li, K., B. Hu, L. Chang and Y. Li. 2015. Robust square-root cubature Kalman filter based on Huber's M-estimation methodology. Proceedings of the Institution of Mechanical Engineers, Part G: Journal of Aerospace Engineering 229: 1236–1245.

Li, W. and Y. Jia. 2010. H-infinity filtering for a class of nonlinear discrete-time systems based on unscented transform. Signal Processing 90: 3301–3307.

Liland, E. 2017. An ILP Approach to Multi Hypothesis Tracking. Retrieved from osf.io/jp4pp.

Moireau, P. and D. Chapelle. 2011. Reduced-order unscented Kalman filtering with application to parameter identification in large-dimensional systems. ESAIM: Control, Optimisation and Calculus of Variations 17: 380–405.

Psiaki, M.L. 2016. Gaussian mixture nonlinear filtering with resampling for mix and narrowing. IEEE Transactions on Signal Processing 64: 5499–5512.

Runnalls, A.R. 2007. Kullback-Leibler approach to Gaussian mixture reduction. IEEE Transactions on Aerospace and Electronic Systems 43: 989–999.

Särkkä, S. 2007. On unscented Kalman filtering for state estimation of continuous-time nonlinear systems. IEEE Transactions on Automatic Control 52: 1631–1641.

Särkkä, S. and A. Solin. 2012. On continuous-discrete cubature Kalman filtering. IFAC Proceedings 45: 1221–1226.

Simon, D. 2006. Optimal State Estimation: Kalman, H Infinity, and Nonlinear Approaches. John Wiley & Sons, Inc., Hoboken.

Simon, D. 2010. Kalman filtering with state constraints: a survey of linear and nonlinear algorithms. IET Control Theory & Applications 4: 1303–1318.

Stordal, A.S., H.A. Karlsen, G. Nævdal, H.J. Skaug and B. Vallès. 2011. Bridging the ensemble Kalman filter and particle filters: the adaptive Gaussian mixture filter. Computational Geosciences 15: 293–305.

Teixeira, B.O.S., J. Chandrasekar, L.A.B. Tôrres, L.A. Aguirre and D.S. Bernstein. 2009. State estimation for linear and non-linear equality-constrained systems. International Journal of Control 82: 918–936.

Teixeira, B.O.S., L.A.B. Tôrres, L.A. Aguirre and D.S. Bernstein. 2010. On unscented Kalman filtering with state interval constraints. Journal of Process Control 20: 45–57.

Terejanu, G., P. Singla, T. Singh and P.D. Scott. 2011. Adaptive Gaussian sum filter for nonlinear Bayesian estimation. IEEE Transactions on Automatic Control 56: 2151–2156.

Vachhani, P., S. Narasimhan and R. Rengaswamy. 2006. Robust and reliable estimation via unscented recursive nonlinear dynamic data reconciliation. Journal of Process Control 16: 1075–1086.

Van Der Merwe, R. and E.A. Wan. 2001. The square-root unscented Kalman filter for state and parameter-estimation. 2001 IEEE International Conference on Acoustics, Speech, and Signal Processing. USA, 3461–3464.

Wang, X., Y. Liang, Q. Pan and F. Yang. 2012. A Gaussian approximation recursive filter for nonlinear systems with correlated noises. Automatica 48: 2290–2297.

Williams, J.L. 2003. Gaussian mixture reduction for tracking multiple maneuvering targets in clutter. M.S. Thesis, Air Force Institute of Technology, Wright-Patterson Air Force Base, Ohio.

Wu, Y., D. Hu, M. Wu and X. Hu. 2005. Unscented Kalman filtering for additive noise case: augmented versus nonaugmented. IEEE Signal Processing Letters 12: 357–360.

Multiple Sensor Estimation 6

With the rapid development of the network and sensor technologies, multiple sensors have been widely used in many estimation applications, such as tracking and information fusion. In this chapter, we discuss the distributed nonlinear estimation using multiple sensors. The main difference between the multiple sensor estimation and single sensor estimation is the fusion of information from different sensors. Specifically, how to integrate diverse and uncertain measurements or other available information is the key for the multiple sensor estimation. There are different strategies to process the information from different sensors in centralized, hierarchical, or decentralized manners. With an increasing number of sensors, it would be hard to use the centralized estimation algorithm.

In this chapter, we focus on the distributed multiple sensor estimation algorithms and introduce the main results in a practical way. Multiple sensor estimation has many implementation considerations. It is not our objective to give a complete survey or cover all topics in this respect. For instance, issues such as sensor registration (Taghavi et al. 2016), data association, out of sequence measurement (Maskell et al. 2006), and time synchronization (Amundson et al. 2007) are not discussed. Some implementation details in multiple sensor estimation applications, such as high-level information fusion (Liggins II et al. 2008) and decision making (Klein 2004) are also left to the references. We focus on nonlinear estimation of the single object and three distributed multiple sensor estimation algorithms. Specifically, the consensus-based, the covariance intersection-based, and the diffusion-based estimation algorithms are introduced. The grid-based nonlinear estimation can be directly used in these algorithms. For the scenario of multiple target estimation, the data association should be designed. The distributed particle filter and the sensor allocation problem are briefly discussed. A good literature review of the distributed estimation is given in (Mahmoud and Khalid 2013).

6.1 Main Fusion Structures

There are three main fusion structures for the multiple sensor estimation: centralized fusion, hierarchical fusion, and distributed fusion.

For the centralized fusion, as shown in Fig. 6.1, all information or measurements are collected and processed by the fusion center. The conventional Kalman filter or the information filter is then used to fuse information or measurements of all sensors. It can be seen that the fusion center is critical in this structure. The computational and communication complexity and the load of the fusion center are high.

To alleviate these burdens of the fusion node, the hierarchical fusion strategy is commonly used, as shown in Fig. 6.2. The information of different

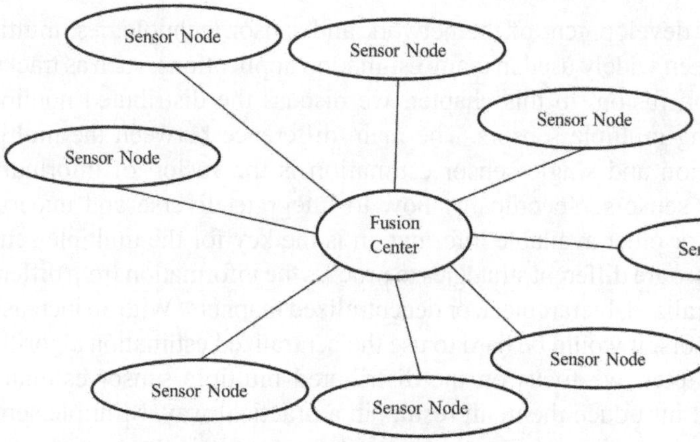

Fig. 6.1 Centralized fusion framework.

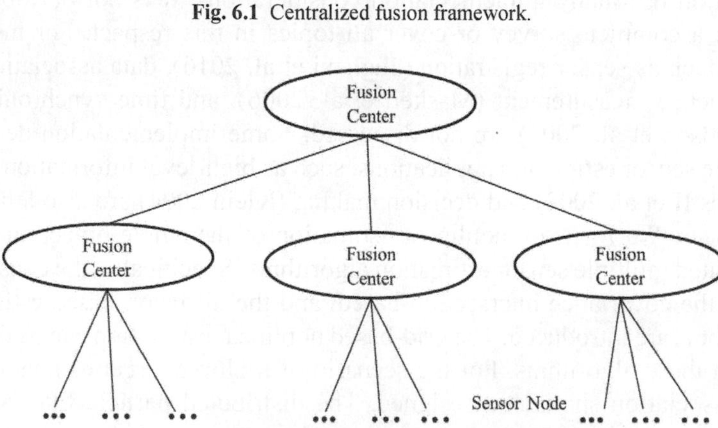

Fig. 6.2 Hierarchical fusion framework.

sensors is first collected by a local fusion center. Then, the information of local fusion centers is collected and processed by upper-level fusion centers. There are different fusion levels for different problems. In this case, the computational and communication load is shared by fusion centers at different levels. The structure of the network, however, is hard to change.

To further alleviate the computational and communication burden of fusion nodes, the fully distributed fusion strategy is proposed, as shown in Fig. 6.3. For this strategy, the sensor node is connected to some other nodes to send and receive information. The local estimation is based on the local information of the sensor and its directly connected sensor nodes (neighbors). A special case is that each sensor node connects to all other nodes in the network. In this special case, it can be viewed as a variant of the centralized fusion framework. For this strategy, the network of sensors can be flexibly designed. It is also robust to the sensor node failure or malicious attack.

How to choose the fusion structures is problem dependent. If powerful super sensor nodes are available, it is feasible to use the centralized or hierarchical fusion strategies. However, if sensor nodes have limited computational capability and the robustness of the network is the main concern, it is better to choose the distributed fusion strategy.

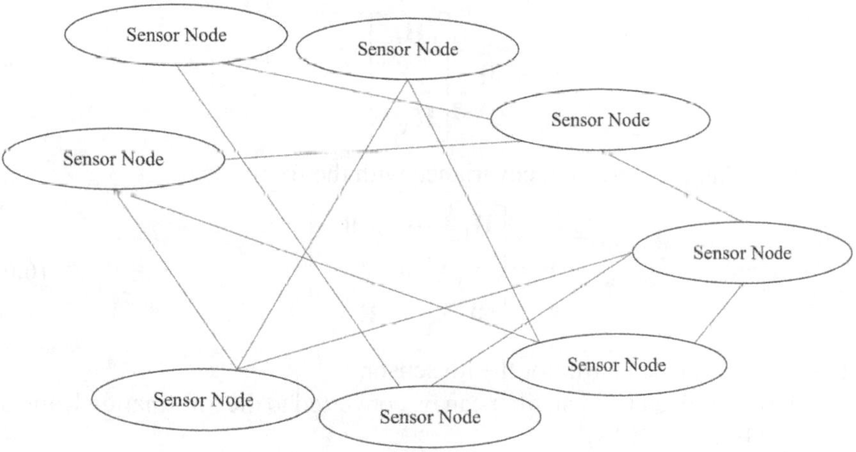

Fig. 6.3 Distributed fusion framework.

6.2 Grid-based Information Kalman Filters and Centralized Gaussian Nonlinear Estimation

For convenience, let's assume a linear measurement equation with the form

$$\mathbf{y}_{k,j} = \mathbf{H}_j \mathbf{x}_{k,j} + \mathbf{n}_j \quad 1 \le j \le N_{sn} \tag{6.1}$$

where '$\mathbf{H}$' denotes the Jacobian matrix of the measurement equation. $\mathbf{x}$ and $\mathbf{y}$ are the system state and observation, respectively. The subscript 'j' denotes the sensor index and '$\mathbf{n}$' denotes the sensor noise. The subscript 'k' denotes the time step. N_{sn} is the number of sensors.

For the centralized Kalman filter, the measurement from different sensors is concatenated into a single measurement vector. The update step of the centralized Kalman filter is given by

$$\mathbf{K}_k = \mathbf{P}_{k|k-1}\mathbf{H}^T\left(\mathbf{H}\mathbf{P}_{k|k-1}\mathbf{H}^T + \mathbf{R}\right)^{-1}, \tag{6.2}$$

$$\hat{\mathbf{x}}_k = \hat{\mathbf{x}}_{k|k-1} + \mathbf{K}_k\left(\mathbf{y}_k - \mathbf{H}\hat{\mathbf{x}}_{k|k-1}\right), \tag{6.3}$$

$$\mathbf{P}_{k|k} = \mathbf{P}_{k|k-1} - \mathbf{K}_k\mathbf{H}\mathbf{P}_{k|k-1}. \tag{6.4}$$

where $\mathbf{K}_k$ is the Kalman gain. $\mathbf{P}_{k|k-1}$ is the predicted covariance associated with the predicted state $\hat{\mathbf{x}}_{k|k-1}$. They can be obtained via the standard prediction step of the Kalman filter.

Note that $\mathbf{y}_k = \begin{bmatrix} \mathbf{y}_{k,1} \\ \vdots \\ \mathbf{y}_{k,N_{sn}} \end{bmatrix}$, $\mathbf{H}$ is the global measurement matrix with

$$\mathbf{H} = \begin{bmatrix} \mathbf{H}_1 \\ \vdots \\ \mathbf{H}_{N_{sn}} \end{bmatrix}. \tag{6.5}$$

$\mathbf{R}$ is the global measurement covariance with the form

$$\mathbf{R} = \begin{bmatrix} \mathbf{R}_1 & \cdots & \mathbf{0} \\ \vdots & \ddots & \vdots \\ \mathbf{0} & \cdots & \mathbf{R}_{N_{sn}} \end{bmatrix}, \tag{6.6}$$

$\mathbf{R}_i$ is the covariance matrix for the ith sensor.

The centralized Kalman filter can be converted to the information Kalman filter by (Terejanu 2013)

$$\left(\mathbf{P}_{k|k}\right)^{-1}\hat{\mathbf{x}}_{k|k} = \left(\mathbf{P}_{k|k-1}\right)^{-1}\hat{\mathbf{x}}_{k|k-1} + \mathbf{H}^T\mathbf{R}^{-1}\mathbf{y}_k \tag{6.7}$$

$$\left(\mathbf{P}_{k|k}\right)^{-1} = \left(\mathbf{P}_{k|k-1}\right)^{-1} + \mathbf{H}^T\mathbf{R}^{-1}\mathbf{H} \tag{6.8}$$

Define $\breve{\mathbf{y}} = \mathbf{P}^{-1}\mathbf{x}$, $\mathbf{i} = \mathbf{H}^T\mathbf{R}^{-1}\mathbf{y}$, $\breve{\mathbf{Y}} = \mathbf{P}^{-1}$, and $\mathbf{I} = \mathbf{H}^T\mathbf{R}^{-1}\mathbf{H}$. $\breve{\mathbf{y}}$ and $\breve{\mathbf{Y}}$ are called information state and information matrix, respectively. After some algebra, we have

$$\breve{\mathbf{y}}_{k|k} = \breve{\mathbf{y}}_{k|k-1} + \sum_{j=1}^{N_{sn}} \mathbf{i}_{k,j} \tag{6.9}$$

$$\breve{\mathbf{Y}}_{k|k} = \breve{\mathbf{Y}}_{k|k-1} + \sum_{j=1}^{N_{sn}} \mathbf{I}_{k,j} \tag{6.10}$$

As shown in Eqs. (6.9)–(6.10), the information filter uses the information state and the information matrix rather than the system state and covariance matrix in the conventional Kalman filter.

Let's consider a class of nonlinear discrete-time dynamical systems

$$\mathbf{x}_k = \boldsymbol{f}\left(\mathbf{x}_{k-1}\right) + \mathbf{v}_{k-1} \tag{6.11}$$

$$\mathbf{y}_{k,j} = \boldsymbol{h}_j\left(x_k\right) + \mathbf{n}_{k,j} \tag{6.12}$$

where $\mathbf{x}_k \in \mathbb{R}^n; \mathbf{y}_{k,j} \in \mathbb{R}^m$. $\mathbf{v}_{k-1}$ and $\mathbf{n}_{k,j}$ are independent white Gaussian process noise and measurement noise with covariance $\mathbf{Q}_{k-1}$ and $\mathbf{R}_{k,j}$, respectively. $\mathbf{y}_{k,j}$ is the measurement by the *j*th sensor.

The information state and the information matrix at the time $k-1$ are defined by $\breve{\mathbf{y}}_{k-1|k-1} = \mathbf{P}^{-1}_{k-1|k-1}\hat{\mathbf{x}}_{k-1|k-1}$ and $\breve{\mathbf{Y}}_{k-1|k-1} = \mathbf{P}^{-1}_{k-1|k-1}$, respectively. The system state $\hat{\mathbf{x}}_{k-1|k-1}$ and covariance $\mathbf{P}_{k-1|k-1}$ can be obtained by $\hat{\mathbf{x}}_{k-1|k-1} = \mathbf{P}_{k-1|k-1}\breve{\mathbf{y}}_{k-1|k-1}$ and $\mathbf{P}_{k-1|k-1} = \breve{\mathbf{Y}}^{-1}_{k-1|k-1}$, respectively. In the following, the Gaussian approximation information filter framework from which the grid-based information filter is derived is presented starting with the extended information filter.

The extended information filter (EIF) is derived from the EKF and contains the prediction and update steps:

Prediction:

The information state $\breve{\mathbf{y}}_{k|k-1}$ and the information matrix $\breve{\mathbf{Y}}_{k|k-1}$ can be predicted by

$$\breve{\mathbf{y}}_{k|k-1} = \mathbf{P}^{-1}_{k|k-1}\hat{\mathbf{x}}_{k|k-1} \tag{6.13}$$

$$\breve{\mathbf{Y}}_{k|k-1} = \mathbf{P}^{-1}_{k|k-1} \tag{6.14}$$

The predicted state and the associated covariance matrix at time *k* can be obtained by

$$\hat{\mathbf{x}}_{k|k-1} = \boldsymbol{f}\left(\hat{\mathbf{x}}_{k-1|k-1}\right) \tag{6.15}$$

$$\mathbf{P}_{k|k-1} = \mathbf{F}_{k|k-1}\mathbf{P}_{k-1|k-1}\mathbf{F}^T_{k|k-1} + \mathbf{Q}_{k-1} \tag{6.16}$$

where $\mathbf{F}_{k|k-1}$ is the Jacobian matrix of $\boldsymbol{f}$ with respect to $\hat{\mathbf{x}}_{k|k-1}$.

Update:

For multiple sensor estimation, the information state and the information matrix can be updated by Eqs. (6.9–6.10) (Lee 2008). But, the information state contribution $\mathbf{i}_{k,j}$ and the information matrix contribution $\mathbf{I}_{k,j}$ of the *j*th sensor are given by

$$\mathbf{i}_{k,j} = \mathbf{H}_{k,j}^T \mathbf{R}_{k,j}^{-1} \left[\left(\mathbf{y}_{k,j} - \boldsymbol{h}_j \left(\hat{\mathbf{x}}_{k|k-1} \right) \right) + \mathbf{H}_{k,j} \hat{\mathbf{x}}_{k|k-1} \right] \tag{6.17}$$

$$\mathbf{I}_{k,j} = \mathbf{H}_{k,j}^T \mathbf{R}_{k,j}^{-1} \mathbf{H}_{k,j} \tag{6.18}$$

where $\boldsymbol{h}_j$ and $\mathbf{H}_{k,j}$ are the *j*th measurement function and the associated Jacobian matrix at time *k*, respectively; $\mathbf{R}_{k,j}$ is the covariance of the measurement noise corresponding to the *j*th measurement equation.

Remark 6.1: From the above EIF filtering algorithm, it can be seen that the local information contributions of $\mathbf{i}_{k,j}$ and $\mathbf{I}_{k,j}$ are only computed at sensor *j* and the total information contribution including the information state and the information matrix are simply the sum of the local contributions. Therefore, the information filter is more suitable for decentralized sensor estimation than the conventional Kalman filter.

To embed grid-based filtering techniques into the EIF update structure, as shown in Eqs. (6.9), (6.10), (6.17), (6.18), the linearized measurement Jacobian matrix $\mathbf{H}_{k,j}$ needs to be replaced by the quadrature approximations. To this end, statistical linear error propagation approach is used (Lee 2008). For the *j*th sensor, the following equation can be obtained by the linear error propagation approach.

$$\mathbf{P}_{k|k-1,xy_j} \approx \mathbf{P}_{k|k-1} \mathbf{H}_{k,j}^T \tag{6.19}$$

where $\mathbf{P}_{k|k-1,xy_j}$ is the cross-correlation covariance that can be calculated by the quadrature approximation. Using the relation (6.19) into Eqs. (6.17) and (6.18), the measurement Jacobian matrix $\mathbf{H}_{k,j}$ in the update Eqs. (6.17) and (6.18) can be replaced by $\mathbf{P}_{k|k-1,xz_j}$ and $\mathbf{P}_{k|k-1}$, which is derived as follows.

$$\begin{aligned}
\mathbf{i}_{k,j} &= \mathbf{H}_{k,j}^T \mathbf{R}_{k,j}^{-1} \left[\left(\mathbf{y}_{k,j} - \hat{\mathbf{y}}_{k,j} \right) + \mathbf{H}_{k,j} \hat{\mathbf{x}}_{k|k-1} \right] \\
&= \left(\mathbf{P}_{k|k-1} \right)^{-1} \left(\mathbf{P}_{k|k-1} \right) \mathbf{H}_{k,j}^T \mathbf{R}_{k,j}^{-1} \left[\left(\mathbf{y}_{k,j} - \hat{\mathbf{y}}_{k,j} \right) + \mathbf{H}_{k,j} \left(\mathbf{P}_{k|k-1} \right)^T \left(\mathbf{P}_{k|k-1} \right)^{-T} \hat{\mathbf{x}}_{k|k-1} \right] \\
&\approx \left(\mathbf{P}_{k|k-1} \right)^{-1} \mathbf{P}_{k|k-1,xy_j} \mathbf{R}_{k,j}^{-1} \left[\left(\mathbf{y}_{k,j} - \hat{\mathbf{y}}_{k,j} \right) + \left(\mathbf{P}_{k|k-1,xy_j} \right)^T \left(\mathbf{P}_{k|k-1} \right)^{-T} \hat{\mathbf{x}}_{k|k-1} \right]
\end{aligned} \tag{6.20}$$

and

$$\begin{aligned}\mathbf{I}_{k,j} &= \mathbf{H}_{k,j}^T \mathbf{R}_{k,j}^{-1} \mathbf{H}_{k,j} \\ &= \left(\mathbf{P}_{k|k-1}\right)^{-1} \left(\mathbf{P}_{k|k-1}\right) \mathbf{H}_{k,j}^T \mathbf{R}_{k,j}^{-1} \mathbf{H}_{k,j} \left(\mathbf{P}_{k|k-1}\right)^T \left(\mathbf{P}_{k|k-1}\right)^{-T} \\ &\approx \left(\mathbf{P}_{k|k-1}\right)^{-1} \mathbf{P}_{k|k-1,xy_j} \mathbf{R}_{k,j}^{-1} \left(\mathbf{P}_{k|k-1,xy_j}\right)^T \left(\mathbf{P}_{k|k-1}\right)^{-T}\end{aligned} \tag{6.21}$$

Using different Gaussian integration techniques given in Chapter 4 into Eqs. (6.20), (6.21), different information filter can be obtained.

For the centralized information filter, the information of each node is collected by the fusion center. The estimation for each sensor is the same after the fusion. The hierarchical fusion strategy can also be directly used based on Eqs. (6.9), (6.10), (6.20), and (6.21) to obtain the hierarchical information filter. The distributed information filter for the fully connected network can be designed by using each sensor node as the fusion center. For an arbitrarily connected network, the redundant information should be removed in the fusion process. The redundant information is the common information among the estimates of different nodes. If the redundant information is used, it may lead to bias, over-confidence or divergence of the estimate (Durrant-Whyte and Henderson 2008).

The distributed fusion strategy, however, needs to be specially designed. For example, one direct method is to decompose Eqs. (6.9) and (6.10) using distributed computation strategies. In the following, we introduce such strategies.

6.3 Consensus-based Strategy

Consensus algorithms provide ways to achieve the agreement of the network, which can be used in the multiple sensor estimation.

For the distributed fusion algorithm, the network topology should first be provided and it can be represented by an undirected connected graph $G = (V, E)$, where V is the set of vertices or nodes of the graph and E is the set of edges or lines of the graph (Kamal et al. 2013). Before the introduction of the distributed fusion algorithm, the average consensus algorithm is briefly reviewed as follows since it is the fundamental component of the consensus-based estimation algorithms.

6.3.1 Consensus Algorithm

Average Consensus Algorithm. The average consensus algorithm is frequently used to obtain the mean value of all nodes in the network. Given the value of each node, the set of values of the network can be represented by $\left\{a_j\right\}_{j=1}^{N_{sn}}$.

By using the average consensus algorithm, the mean value $\frac{1}{N_{sn}}\sum_{j=1}^{N_{sn}} a_j$ can be obtained by multiple iterations. Each node initializes its estimate by $a_j(0) = a_j$ and iteratively exchanges information with its neighbors and updates its own estimate. We assume that each node has the value $a_j(i-1)$ before exchanging information at iteration step *i*, then the *j*th node exchanges its value with its neighbors $j' \in N_j$, which means the *j*th node sends its estimate to its neighbors and also receives its neighbors' estimates $a_{j'}(i-1)$. N_j denotes the set of nodes that connect with node *j*. The update stage of the *j*th node from (*i*-1)th iteration to *i*th iteration is given by

$$a_j(i) = a_j(i-1) + \varepsilon \sum_{j' \in N_j} \left(a_{j'}(i-1) - a_j(i-1)\right) \tag{6.22}$$

where ε is the rate parameter that has a value between 0 and $1/\Delta_{\max}$, and $\Delta_{\max}$ is the maximum degree of the node in the network, which is the number of connections it has to other nodes. Note that the convergence will be faster if a larger ε is chosen.

Example 6.1: For convenience, let's denote **A** as the adjacency matrix with the element $\mathbf{A}_{ij} = 1$ if the *i*th node is connected to the *j*th node. The Laplacian matrix is then given by $\boldsymbol{L}$

$$\boldsymbol{L}_{ij} = \begin{cases} -a_{ij} & i \neq j \\ \sum_{i \neq j} a_{ij} & i = j \end{cases} \tag{6.23}$$

Assume the matrix **A** is given by

$$\mathbf{A} = \begin{bmatrix} 0 & 1 & 0 & 0 & 1 \\ 1 & 0 & 1 & 0 & 0 \\ 0 & 1 & 0 & 1 & 0 \\ 0 & 0 & 1 & 0 & 1 \\ 1 & 0 & 0 & 1 & 0 \end{bmatrix} \tag{6.24}$$

The initial value of nodes in the network is given by $[1,2,3,4,5]^T$, which contains the value for all nodes in the network. $\varepsilon = 0.05$. The maximum number of iteration is set to 500. The trajectory of each node value is shown in Fig. 6.4. It can be seen all nodes converge to the same value when the iteration number is greater than 100.

In real applications, the average consensus algorithm may need to be revised for realistic information transmitting problems, such as packet drop, and quantification.

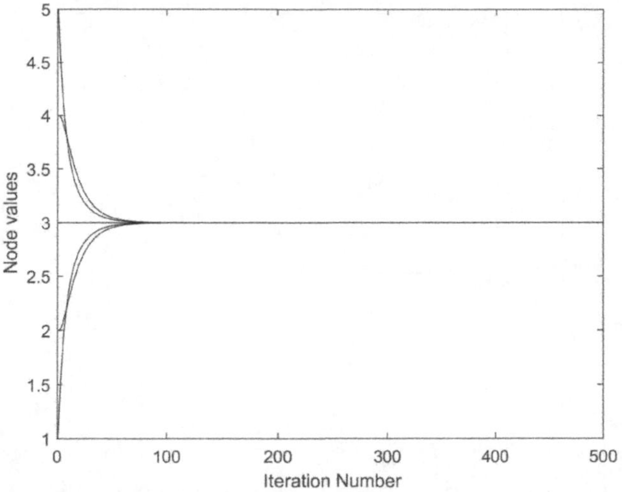

Fig. 6.4 Illustration of consensus using the average consensus algorithm.

Consensus with Packet Drop. The consensus step can also be described by the following matrix form

$$\mathbf{x}_i(k+1) = \sum_{j=1}^{N_{sn}} \boldsymbol{C}_{ij}\mathbf{x}_j(k) \tag{6.25}$$

where the subscript '*i*' denotes the *i*th sensor, *k* is the iteration step, $\boldsymbol{C}_{ij} = I - \varepsilon \cdot \boldsymbol{L}_{ij}$ and ε is a small number in [0,1]. Note that $\boldsymbol{C}$ satisfies the constraints, $\mathbf{1}^T\boldsymbol{C} = \mathbf{1}^T$ and $\boldsymbol{C}\mathbf{1} = \mathbf{1}$. $\mathbf{1}$ is the vector with all elements being 1.

In order to compensate the packet drop, a binary variable γ is introduced. Specifically, the consensus step can be revised as (Wang et al. 2015)

$$\begin{aligned}\mathbf{x}_i(k+1) &= \sum_{j=1}^{N_{sn}} \left(\gamma_{ij}(k)\boldsymbol{C}_{ij}\mathbf{x}_j(k) + \left(1-\gamma_{ij}(k)\right)\boldsymbol{C}_{ij}\mathbf{x}_i(k)\right) \\ &= \sum_{j=1, j\neq i}^{N_{sn}} \gamma_{ij}(k)\boldsymbol{C}_{ij}\mathbf{x}_j(k) + \left(\sum_{j=1, j\neq i}^{N_{sn}} \left(1-\gamma_{ij}(k)\right)\boldsymbol{C}_{ij} + \boldsymbol{C}_{ii}\right)\mathbf{x}_i(k)\end{aligned} \tag{6.26}$$

In Eq. (6.26), the local node value $\mathbf{x}_i(k)$ is actually used if the packet drop happened, $\gamma_{ij}(k) = 0, j \neq i$. If there is no packet drop, $\gamma_{ij}(k) = 1$.

Alternatively, the consensus step can be revised as (Wang et al. 2015)

$$\mathbf{x}_i(k+1) = \frac{\boldsymbol{C}_{ii}\mathbf{x}_i(k) + \sum_{j=1, j\neq i}^{N_{sn}} \gamma_{ij}(k)\boldsymbol{C}_{ij}\mathbf{x}_j(k)}{\boldsymbol{C}_{ii} + \sum_{j=1, j\neq i}^{N_{sn}} \gamma_{ij}(k)\boldsymbol{C}_{ij}} \tag{6.27}$$

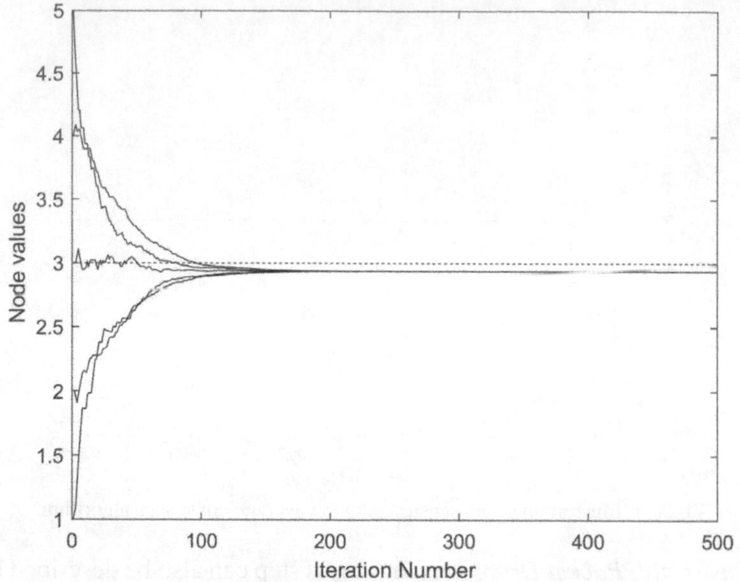

Fig. 6.5 Illustration of the consensus algorithm with package loss.

Note that both methods reduce to the conventional consensus algorithm if there is no packet drop.

Example 6.2: In this example, all initial conditions are the same as Example 6.1, In addition, it is assumed for each communication time, it is possible that the packet is lost with the probability $p = 0.5$. The packet loss rate p is the possibility when $\gamma_{ij}(k) = 0$. The trajectory of each node value is shown in Fig. 6.5. It can be seen that all nodes have a very close performance when the iteration number is greater than 150. However, the final value of each node is different from the mean value of the original node values due to the loss of packet. The dashed line represents the actual average value.

Average Consensus with Noise. It is possible that the transmitted value is corrupted by the random noise. In this case, the iteration step for node j can be revised by

$$a_j(i) = a_j(i-1) + \varepsilon \sum_{j' \in N_j} \left(a_{j'}(i-1) + n_{j'}(i-1) - a_j(i-1) \right) \quad (6.28)$$

where n denotes the noise term. The subscript is the sensor index.

A simple strategy to mitigate the effect of noise is by the Monte Carlo method. The consensus step is called multiple times, for example, p times. Then the value after consensus can be described by (Kar and Moura 2009)

$$\hat{\mathbf{x}} \approx \frac{1}{p}\sum_{i=1}^{p}\hat{\mathbf{x}}^{(i)} \tag{6.29}$$

where $\hat{\mathbf{x}}^{(i)}$ denotes the value of the network for the ith consensus. Specifically, for the ith consensus, Eq. (6.28) is used to achieve the consensus value $\hat{\mathbf{x}}^{(i)}$.

Gossip Algorithm. So far, we have presented the deterministic consensus algorithms or protocols. Now we introduce a class of randomized algorithms, named gossip algorithms, which can solve the average consensus problem by a sequence of pairwise averages. The gossip algorithm can be used to achieve consensus by randomly exchanging information between neighbors. One node is randomly selected first. The neighbor of this node is then randomly chosen. The average value of the chosen two nodes is then calculated and replace their own values. The consensus can be achieved by multiple iterations. The gossip algorithms, compared to deterministic protocols, are simple. In addition, the gossip algorithms achieve high stability under stress and disruptions (Kempe et al. 2003). To save communication energy, the geographic gossip algorithm was proposed (Dimakis et al. 2008). It assumes that each node knows its own geographic location. In the geographic gossip algorithm, the geographic information is used to create an overlay communication network in which the edge is assigned a cost equal to the number of hops in communication between different nodes. By using the overlay communication network, far-away neighbors can exchange information (Dimakis et al. 2008). The complete algorithm is summarized in (Dimakis et al. 2008). There are other extensions of the gossip algorithm, such as the gossip consensus with quantized communication (Carli et al. 2010).

In realistic applications, information sharing also requires consideration of many constraints, such as quantization and power consumption. Specific encoding and decoding strategies are used to achieve average consensus with quantization of the data. It has been proven that arbitrary bits can be used to achieve consensus if the encoding and decoding strategies are provided (Li et al. 2011).

Although different variants of the consensus algorithms are introduced separately, they can be used jointly in real applications. Next, we introduce consensus-based distributed nonlinear filters using the average algorithm as a basic component.

6.3.2 Consensus-based Filter

Giving the consensus algorithms, the consensus-based filter can be designed to achieve the estimation consensus in different ways. In the following, we introduce two typical consensus-based filters.

Kalman Consensus Filter. The Kalman Consensus Filter (KCF) is a widely used consensus-based distributed estimation algorithm (Olfati-Saber 2009). The prediction step of the KCF is the same as that of other filters. The update step for each sensor is given by (Kamal et al. 2011, Kamal et al. 2013)

$$\begin{aligned}\hat{\mathbf{x}}_{k|k,j} &= \hat{\mathbf{x}}_{k|k-1,j} + \left(\breve{\mathbf{Y}}_{k|k-1,j} + \sum_{j'\in N_j\cup j}\mathbf{I}_{k,j'}\right)^{-1}\left(\sum_{j'\in N_j\cup j}\mathbf{i}_{k,j'} - \sum_{j'\in N_j\cup j}\mathbf{I}_{k,j'}\hat{\mathbf{x}}_{k|k-1,j}\right) \\ &+ \frac{\varepsilon}{1+\mathrm{Tr}\left(\left(\breve{\mathbf{Y}}_{k|k-1,j}\right)^{-1}\right)}\left(\breve{\mathbf{Y}}_{k|k-1,j}\right)^{-1}\sum_{j'\in N_j}\left(\hat{\mathbf{x}}_{k|k-1,j'} - \hat{\mathbf{x}}_{k|k-1,j}\right)\end{aligned} \tag{6.30}$$

and

$$\breve{\mathbf{Y}}_{k|k,j} = \breve{\mathbf{Y}}_{k|k-1,j} + \sum_{j'\in N_j\cup j}\mathbf{I}_{k,j'} \tag{6.31}$$

where ε is a parameter which affects the rate of convergence. $\mathrm{Tr}(\cdot)$ denotes the trace operation.

Although KCF is popular, it cannot achieve the same performance as the centralized nonlinear filters. In addition, it assumes all sensors have measurements.

Information Weighted Consensus Filter. The information weighted consensus filter (IWCF) is derived from the centralized information filter. The IWCF aims to use the consensus algorithm to calculate the centralized state estimation.

The update step of the centralized information filter, Eqs. (6.9) and (6.10), can be written as

$$\begin{aligned}\hat{\mathbf{x}}_{k|k} &= \left(\breve{\mathbf{Y}}_{k|k}\right)^{-1}\left(\breve{\mathbf{y}}_{k|k}\right) = \left(\breve{\mathbf{Y}}_{k|k-1} + \sum_{j=1}^{N_{sn}}\mathbf{I}_{k,j}\right)^{-1}\left(\breve{\mathbf{y}}_{k|k-1} + \sum_{j=1}^{N_{sn}}\mathbf{i}_{k,j}\right) \\ &= \left(\sum_{j=1}^{N_{sn}}\left(\frac{\breve{\mathbf{Y}}_{k|k-1}}{N_{sn}} + \mathbf{I}_{k,j}\right)\right)^{-1}\left(\sum_{j=1}^{N_{sn}}\left(\frac{\breve{\mathbf{y}}_{k|k-1}}{N_{sn}} + \mathbf{i}_{k,j}\right)\right)\end{aligned} \tag{6.32}$$

and

$$\breve{\mathbf{Y}}_{k|k} = \sum_{j=1}^{N_{sn}}\left(\frac{\breve{\mathbf{Y}}_{k|k-1}}{N_{sn}} + \mathbf{I}_{k,j}\right). \tag{6.33}$$

For the IWCF, it is assumed that the *a priori* information for each node is the same. Under this assumption, we have $\hat{\mathbf{x}}_{k|k-1,j} = \hat{\mathbf{x}}_{k|k-1}$, $\mathbf{P}_{k|k-1,j} = \mathbf{P}_{k|k-1}$. Then, Eq. (6.32) can be rewritten as

$$\hat{\mathbf{x}}_{k|k} = \left(\sum_{j=1}^{N_{sn}}\left(\frac{\breve{\mathbf{Y}}_{k|k-1,j}}{N_{sn}} + \mathbf{I}_{k,j}\right)\right)^{-1}\left(\sum_{j=1}^{N_{sn}}\left(\frac{\breve{\mathbf{y}}_{k|k-1,j}}{N_{sn}} + \mathbf{i}_{k,j}\right)\right) \tag{6.34}$$

and
$$\breve{\mathbf{Y}}_{k|k} = \sum_{j=1}^{N_{sn}} \left(\frac{\breve{\mathbf{Y}}_{k|k-1,j}}{N_{sn}} + \mathbf{I}_{k,j} \right) \tag{6.35}$$

Define the terms $\mathbf{V} = \frac{\breve{\mathbf{Y}}}{N_{sn}} + \mathbf{I}$, $\mathbf{v} = \frac{\breve{\mathbf{Y}}}{N_{sn}}\mathbf{x} + \mathbf{i}$ and let $\mathbf{V}_j(0) = \frac{\breve{\mathbf{Y}}_j}{N_{sn}} + \mathbf{I}_j$, $\mathbf{v}_j(0) = \frac{\breve{\mathbf{Y}}_j}{N_{sn}}\mathbf{x}_j + \mathbf{i}_j$. Using the average consensus algorithm, when the iteration number goes infinity, we have

$$\lim_{i \to \infty} \mathbf{V}_j(i) = \frac{\sum_{j=1}^{N_{sn}} \mathbf{V}_j(0)}{N_{sn}} \tag{6.36}$$

and
$$\lim_{i \to \infty} \mathbf{v}_j(i) = \frac{\sum_{j=1}^{N_{sn}} \mathbf{v}_j(0)}{N_{sn}} \tag{6.37}$$

Hence, the update equations are given by

$$\begin{aligned} \hat{\mathbf{x}}_{k|k} &= \left(\sum_{j=1}^{N_{sn}} \left(\frac{\mathbf{Y}_{k|k-1,j}}{N_{sn}} + \mathbf{I}_{k,j} \right) \right) \left(\sum_{j=1}^{N_{sn}} \left(\frac{\mathbf{y}_{k|k-1,j}}{N_{sn}} + \mathbf{i}_{k,j} \right) \right) \\ &= \lim_{i \to \infty} \left(N_{sn} \mathbf{V}_j(i) \right)^{-1} \left(N_{sn} \mathbf{v}_j(i) \right) = \lim_{i \to \infty} \left(\mathbf{V}_j(i) \right)^{-1} \left(\mathbf{v}_j(i) \right) \end{aligned} \tag{6.38}$$

and
$$\breve{\mathbf{Y}}_{k|k} = \lim_{i \to \infty} N_{sn} \mathbf{V}_j(i) \tag{6.39}$$

It is evident that Eqs. (6.38) and (6.39) can obtain the same value as the centralized information filter but using the decentralized algorithm, i.e., average consensus.

The information weighted consensus filter uses the local prior information and the measurement information in the consensus step. There is no requirement that all sensors have measurements. It has a strong relationship with the information consensus filter and the measurement consensus filter (Battistelli et al. 2015). For the information consensus filter, the posterior information state and the information covariance of each sensor achieve agreement via the average consensus algorithm. Note that the information consensus filter can be realized using the covariance intersection algorithm (Battistelli et al. 2015) which will be described in Section 6.4. For the measurement consensus algorithm, the information contribution of each sensor, $\mathbf{i}_j$ and $\mathbf{I}_j$, achieves agreement via the average consensus algorithm while the prior information is not exchanged between different sensors. The main advantage of the information consensus filter is that the estimates are bounded as the covariance intersection can be used for any number of information exchanges. However, the novel information is underweighted (Battistelli et al. 2015). The measurement consensus algorithm can partially remedy the drawback of the information consensus filter by introducing weights $\bar{\omega}$ to adjust the importance of the novel information. The measurement consensus algorithm, however, needs a sufficient number of information exchanges to provide a bounded estimation error for each node (Battistelli et al. 2015). Assuming the information contribution after the average consensus step for sensor j is given as $\bar{\mathbf{i}}_j$ and $\bar{\mathbf{I}}_j$, the posterior local information state $\breve{\mathbf{y}}_{k|k,j}$ and information matrix $\breve{\mathbf{Y}}_{k|k,j}$ for sensor j are then given by

$$\breve{\mathbf{y}}_{k|k,j} = \breve{\mathbf{y}}_{k|k-1,j} + \bar{\omega}_{k,j}\bar{\mathbf{i}}_{k,j} \tag{6.40}$$

$$\breve{\mathbf{Y}}_{k|k,j} = \breve{\mathbf{Y}}_{k|k-1,j} + \bar{\omega}_{k,j}\bar{\mathbf{I}}_{k,j} \tag{6.41}$$

where $\bar{\omega}_{k,j}$ are the introduced weights.

To take advantage of both the information consensus filter and the measurement consensus filter, the hybrid consensus filter can be used. The main idea is to use the average consensus algorithm to achieve the agreement of the prior information state and information matrix, besides the consensus on the information contribution of each sensor. Equations (6.40) and (6.41) then can be rewritten as

$$\breve{\mathbf{y}}_{k|k,j} = \bar{\breve{\mathbf{y}}}_{k|k-1,j} + \bar{\omega}_{k,j}\bar{\mathbf{i}}_{k,j} \tag{6.42}$$

$$\breve{\mathbf{Y}}_{k|k,j} = \bar{\breve{\mathbf{Y}}}_{k|k-1,j} + \bar{\omega}_{k,j}\bar{\mathbf{I}}_{k,j} \tag{6.43}$$

where $\bar{\breve{\mathbf{y}}}_{k|k-1,j}$ and $\bar{\breve{\mathbf{Y}}}_{k|k-1,j}$ are the prior information state and prior information matrix after the consensus step.

Note that the information weighted consensus filter is equivalent to the hybrid consensus filter using Eqs. (6.42) and (6.43) when $\bar{\omega}_{k,j} = N_{sn}$.

Remark 6.2: For real applications, the iteration step number i cannot be infinite, as the maximum iteration number has to be given for a finite time.

For the distributed filter, it is necessary to consider the resilient consensus in the presence of misbehaving sensors. The resilient consensus aims to design a robust protocol so that the sensor network is robust to the effects of adversaries. The Byzantine agent is considered in (Pasqualetti et al. 2009). In (LeBlanc and Koutsoukos 2011), the authors proposed the so-called Adversarial Robust Consensus Protocol (ARC-P) to solve the consensus problem whenever there are more normal sensors than adversarial sensors. In (LeBlanc et al. 2012), the authors provided the necessary and sufficient conditions for normal sensors to reach consensus with consideration of the influence of malicious nodes under different assumptions. The consensus problem for discrete-time multi-agent systems in the presence of both adversaries and transmission delays is discussed in (Wu et al. 2014).

6.4 Covariance Intersection Strategy

Covariance intersection is an algorithm to combine two or more estimates when the correlation between them is unknown. It provides a method to update the local estimate using the estimates of other sensors. The covariance intersection gives a common upper bound of the actual estimation error variances. It is robustness to unknown correlations. In addition, the fused estimate is consistent and the accuracy outperforms the single estimate (Li et al. 2015).

6.4.1 Covariance Intersection

In the Kalman filter, the prior estimation error and the measurement error are assumed uncorrelated. This assumption, however, can be violated in the decentralized estimation problem. The sensor node may receive the information originally from itself. Using this information in the update equation of the Kalman filter can lead to serious problems, because in this case, the uncertainty is supposed to remain at the same level, but is mistakenly reduced by the conventional update equation of the Kalman filter. The covariance intersection is proposed to solve such a fusion problem when the information from multiple sensors is correlated (Chen et al. 2002).

The covariance intersection aims to provide a consistent estimate of the covariance matrix when two random variables are linearly combined. The consistency means that the estimated covariance is always an upper-bound of the true variance even if the correlation of the two variables are unknown (Chen et al. 2002).

Given two estimates $(\hat{\mathbf{x}}_1, \mathbf{P}_1)$ and $(\hat{\mathbf{x}}_2, \mathbf{P}_2)$, the fused estimate $\hat{\mathbf{x}}_f$ is rewritten as the linear combination of these two given by

$$\hat{\mathbf{x}}_f = \mathbf{p}_f \hat{\mathbf{x}}_1 + \mathbf{q}_f \hat{\mathbf{x}}_2 \tag{6.44}$$

where $\mathbf{p}_f$ and $\mathbf{q}_f$ are coefficient matrices. To guarantee the unbiased estimate, $\mathbf{p}_f + \mathbf{q}_f = I$, where I is the identity matrix. Using Eq. (6.44), the covariance matrix is given by (Noack et al. 2017)

$$\mathbf{P}_f = \mathbf{p}_f \mathbf{P}_1 \mathbf{p}_f^T + \mathbf{p}_f \mathbf{P}_{12} \mathbf{q}_f^T + \mathbf{q}_f \mathbf{P}_{21} \mathbf{p}_f^T + \mathbf{q}_f \mathbf{P}_2 \mathbf{q}_f^T \tag{6.45}$$

It can be seen in Eq. (6.45) that $\mathbf{P}_f$ depends on the cross-covariance terms $\mathbf{P}_{12}$ and $\mathbf{P}_{21}$. However, the cross-covariance matrix is often unknown. Hence,

$\mathbf{P}_f$ cannot be directly obtained using Eq. (6.45). The approximation of $\mathbf{P}_f$ is often required.

The covariance intersection proposed in (Julier and Uhlmann 1997) uses the following coefficient matrices.

$$\mathbf{p}_f = \omega \mathbf{P}_f \mathbf{P}_1^{-1}, \tag{6.46}$$

$$\mathbf{q}_f = (1-\omega)\mathbf{P}_f \mathbf{P}_2^{-1} \tag{6.47}$$

The covariance matrix is given by

$$\mathbf{P}_f = \left(\omega \mathbf{P}_1^{-1} + (1-\omega)\mathbf{P}_2^{-1}\right)^{-1} \tag{6.48}$$

Using Eqs. (6.44), (6.45), (6.46), and (6.47), we have

$$\mathbf{P}_f^{-1}\hat{\mathbf{x}}_f = \omega \mathbf{P}_1^{-1}\hat{\mathbf{x}}_1 + (1-\omega)\mathbf{P}_2^{-1}\hat{\mathbf{x}}_2 \tag{6.49}$$

Equation (6.48) can be rewritten as

$$\mathbf{P}_f^{-1} = \omega \mathbf{P}_1^{-1} + (1-\omega)\mathbf{P}_2^{-1} \tag{6.50}$$

Note that the information state and information matrix are represented by $\breve{\mathbf{y}} = \mathbf{P}^{-1}\mathbf{x}$ and $\breve{\mathbf{Y}} = \mathbf{P}^{-1}$. Hence, the covariance intersection can be understood as the weighted fusion of the information.

The covariance intersection algorithm can be extended to fuse multiple estimates and the parameter ω can be chosen according to the covariance matrix. The fused state for the distributed filter can be obtained by (Niehsen 2002)

$$\left(\hat{\mathbf{P}}_j\right)^{-1}\hat{\mathbf{x}}_j = \sum_{j'\in\tilde{N}_j} \omega_{j',j}\left(\mathbf{P}_{j'}\right)^{-1}\mathbf{x}_{j'} \tag{6.51}$$

where the covariance $\hat{\mathbf{P}}_j$ is given by

$$\left(\hat{\mathbf{P}}_j\right)^{-1} = \sum_{j'\in\tilde{N}_j} \omega_{j',j}\left(\mathbf{P}_{j'}\right)^{-1}. \tag{6.52}$$

$\tilde{N}_j$ is the set including all neighbors of the jth sensor. Note that $\mathbf{x}_{j'}$ and $\mathbf{P}_{j'}$ are the estimates given by the j'th sensor and the weights are given by

$$\omega_{j',j} = \frac{1/tr\left(\mathbf{P}_{j'}\right)}{\sum_{j'\in\tilde{N}_j} 1/tr\left(\mathbf{P}_{j'}\right)}. \tag{6.53}$$

where '$\mathrm{tr}(\cdot)$' is the trace operator.

Remark 6.3: There are different ways to calculate the weights in Eq. (6.53) (Reinhardt et al. 2012, Franken and Hupper 2005). The weights are critical to the fusion results (Mokhtarzadeh and Gebre-Egziabher 2016).

The distributed estimation algorithm can be obtained by using Eqs. (6.51)–(6.53) to update the state estimation of each sensor node.

Note that the covariance intersection is consistent, i.e., $\mathbf{P}_f \geq E\left[\left(\hat{\mathbf{x}}_f - \mathbf{x}\right)\left(\hat{\mathbf{x}}_f - \mathbf{x}\right)^T\right]$. However, it is usually too conservative. To provide less conservative but consistent result, many improved methods are proposed, such as the inverse covariance intersection (Noack et al. 2017), which computes the bound on the intersection of inverse covariance ellipsoids. The common information $(\hat{\mathbf{x}}_c, \mathbf{P}_c)$ of the two estimates $(\hat{\mathbf{x}}_1, \mathbf{P}_1)$ and $(\hat{\mathbf{x}}_2, \mathbf{P}_2)$ is considered. It can be proved that the following inequality exists (Noack et al. 2017).

$$\mathbf{P}_c^{-1} \leq \left(\omega \mathbf{P}_1 + (1-\omega)\mathbf{P}_2\right)^{-1} \tag{6.54}$$

The fused information matrix can be given by

$$\mathbf{P}_f^{-1} = \mathbf{P}_1^{-1} + \mathbf{P}_2^{-1} - \mathbf{P}_c^{-1} \tag{6.55}$$

Because the common information is counted in $\mathbf{P}_1^{-1} + \mathbf{P}_2^{-1}$. The inverse covariance intersection thus uses the following information fusion equation.

$$\mathbf{P}_f^{-1} = \mathbf{P}_1^{-1} + \mathbf{P}_2^{-1} - (\omega \mathbf{P}_1 + (1-\omega)\mathbf{P}_2)^{-1} \tag{6.56}$$

The fused estimate $\hat{\mathbf{x}}_f$ can be calculated by Eq. (6.44). The coefficient matrices in Eq. (6.44) are chosen by (Noack et al. 2017)

$$\mathbf{p}_f = \mathbf{P}_f\left(\mathbf{P}_1^{-1} - \omega\left(\omega \mathbf{P}_1 + (1-\omega)\mathbf{P}_2\right)^{-1}\right) \tag{6.57}$$

$$\mathbf{q}_f = \mathbf{P}_f\left(\mathbf{P}_2^{-1} - (1-\omega)\left(\omega \mathbf{P}_1 + (1-\omega)\mathbf{P}_2\right)^{-1}\right) \tag{6.58}$$

$\omega \in [0,1]$ is a parameter with

$$\omega^* = \arg\min\left(\left(\omega \mathbf{P}_1 + (1-\omega)\mathbf{P}_2\right)^{-1}\right) \tag{6.59}$$

Note that the unbiased estimate using Eqs. (6.57) and (6.58) can be obtained by using Eq. (6.44). In addition, the inverse covariance intersection algorithm is consistent for the chosen $\mathbf{p}_f$ and $\mathbf{q}_f$.

Example 6.4: Given two estimates $(\hat{\mathbf{x}}_1, \mathbf{P}_1)$ (estimate 1) and $(\hat{\mathbf{x}}_2, \mathbf{P}_2)$ (estimate 2) with $\hat{\mathbf{x}}_1 = \hat{\mathbf{x}}_2 = [0,0]^T$, $\mathbf{P}_1 = \text{diag}([3,3])$, and $\mathbf{P}_2 = \begin{bmatrix} 1 & 0.3 \\ 0.3 & 5 \end{bmatrix}$, the error ellipse of

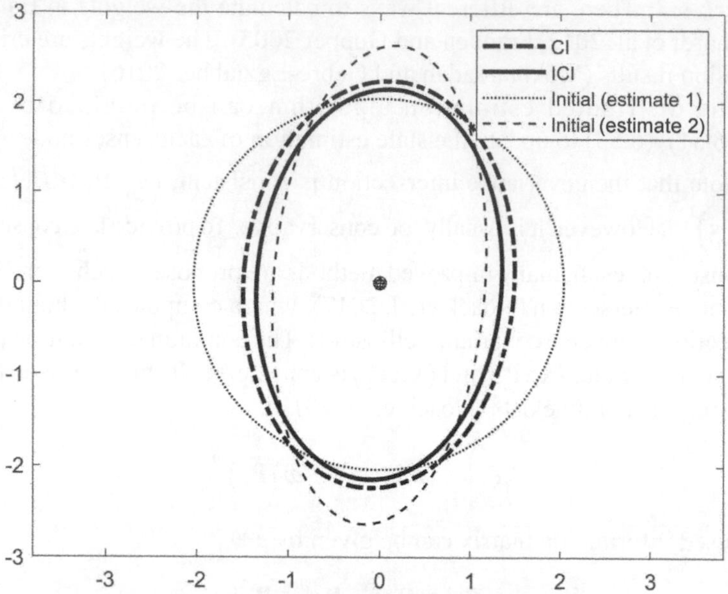

Fig. 6.6 Error ellipse of the fused result using CI and ICI.

the fused result with confidence level 0.5 using the covariance intersection (CI) and inverse covariance intersection (ICI) are shown in Fig. 6.6. It can be seen that the CI gives the reasonable fused solution. The ICI gives less conservative fused solution than that using the CI.

6.4.2 Iterative Covariance Intersection

The covariance intersection is often used once to fuse the information of different sensors. The iterative covariance intersection iterates the fusion results between different nodes several times. Observations from different sensors are used more sufficiently via multiple iterations. It is intuitively feasible for the distributed processing. The consensus can be achieved by using iterative covariance intersection, which will be shown in Section 6.4.4. The algorithm can be summarized in Table 6.1 (Hlinka et al. 2014).

Table 6.1 Iterative covariance intersection.

Given initial information state and information matrix from local state and covariance.
For $i = 1, \cdots, I_{\max}$ ($I_{\max}$ is the maximum iteration number)
Send/Receive information state and information covariance from neighboring sensor nodes.
Using the covariance intersection to update local information state and information matrix.
End For
Recover the local state and covariance from the information state and information matrix.

Example 6.5: Assume the matrix **A** is given by Eq. (6.24). The initial mean value of the network is given by $[1.1, 1.2, 1.3, 1.4, 1.5]^T$ and the covariance value of the network is given by $[2, 1, 2, 1, 2]^T$. After multiple iterations, we can obtain the fused state and variance for these five nodes given by 1.3 and 1.4114, respectively.

The trajectories of each node in terms of state and covariance are shown in Fig. 6.7 and Fig. 6.8, respectively. It can be seen that all sensor nodes have close values finally, which demonstrates that the iterative covariance intersection can be used to achieve the consensus of different nodes.

6.4.3 Distributed Batch Covariance Intersection

According to the covariance intersection, for the centralized fusion, the weights can be chosen as

$$\omega_j = \frac{1/tr(\mathbf{P}_j)}{\sum_{j \in N_{sn}} 1/tr(\mathbf{P}_j)} \tag{6.60}$$

The information matrix can be written as

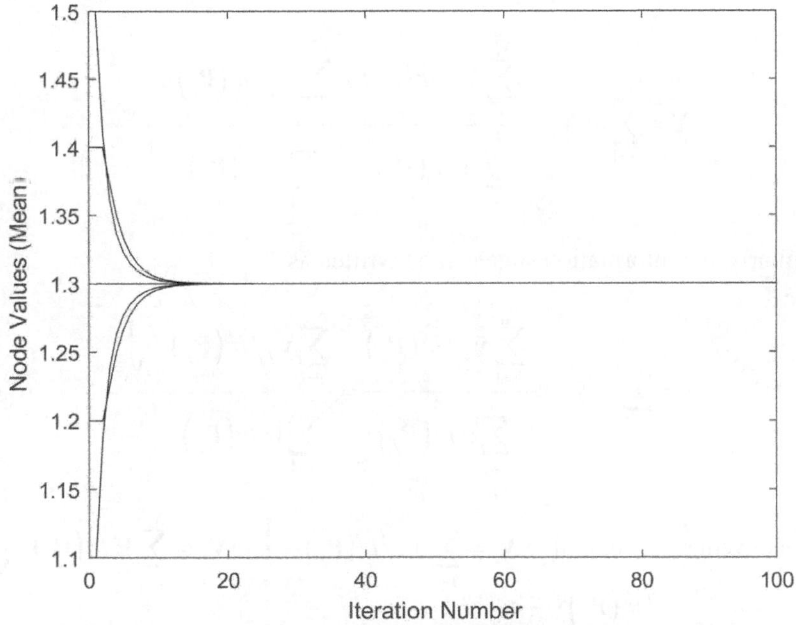

Fig. 6.7 Illustration of the state values using the iterative covariance intersection algorithm.

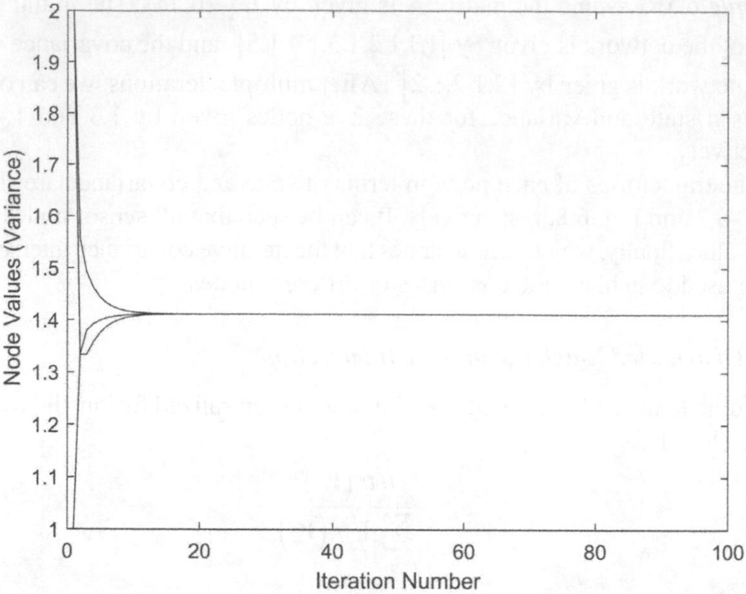

Fig. 6.8 Illustration of covariance value of nodes using the iterative covariance intersection algorithm.

$$\breve{\mathbf{Y}} = \sum_{j=1}^{N_{sn}} \omega_j \breve{\mathbf{Y}}_j = \frac{\sum_{j=1}^{N_{sn}} \breve{\mathbf{Y}}_j / tr\left(\mathbf{P}_j\right)}{\sum_{j=1}^{N_{sn}} 1/tr\left(\mathbf{P}_j\right)} = \frac{\sum_{j=1}^{N_{sn}} \breve{\mathbf{Y}}_j / tr\left(\mathbf{P}_j\right) \cdot \frac{1}{N_{sn}}}{\sum_{j=1}^{N_{sn}} 1 / tr\left(\mathbf{P}_j\right) \cdot \frac{1}{N_{sn}}} \tag{6.61}$$

Similarly, the information state can be written as

$$\breve{\mathbf{y}} = \sum_{j=1}^{N_{sn}} \omega_j \breve{\mathbf{y}}_j = \frac{\sum_{j=1}^{N_{sn}} \breve{\mathbf{y}}_j / tr\left(\mathbf{P}_j\right)}{\sum_{j=1}^{N_{sn}} 1/tr\left(\mathbf{P}_j\right)} = \frac{\sum_{j=1}^{N_{sn}} \breve{\mathbf{y}}_j / tr\left(\mathbf{P}_j\right) \cdot \frac{1}{N_{sn}}}{\sum_{j=1}^{N_{sn}} 1 / tr\left(\mathbf{P}_j\right) \cdot \frac{1}{N_{sn}}} \tag{6.62}$$

For convenience, we define $\mathbf{V}_1 = \sum_{j=1}^{N_{sn}} \breve{\mathbf{Y}}_j / tr\left(\mathbf{P}_j\right) \cdot \frac{1}{N_{sn}}$, $\mathbf{V}_2 = \sum_{j=1}^{N_{sn}} 1 / tr\left(\mathbf{P}_j\right) \cdot \frac{1}{N_{sn}}$, and $\mathbf{V}_3 = \sum_{j=1}^{N_{sn}} \breve{\mathbf{y}}_j / tr\left(\mathbf{P}_j\right) \cdot \frac{1}{N_{sn}}$.

It can be seen that $\mathbf{V}_1$, $\mathbf{V}_2$, and $\mathbf{V}_3$ can be obtained by the average consensus algorithm, which is called the distributed batch covariance intersection.

Note that, different from the iterative covariance intersection algorithm, the distributed fusion algorithm using the batch covariance iteration can achieve the result of the centralized covariance intersection fusion.

Example 6.6: All initial conditions are the same as in Example 6.5. The trajectories of each node in terms of state and covariance are shown in Fig. 6.9 and Fig. 6.10, respectively. Similar to the iterative covariance intersection algorithm, using the distributed batch covariance intersection algorithm, all nodes converge to very close values finally. But the distributed batch covariance intersection algorithm provides the consistent result with the centralized batch covariance intersection algorithm, which is different from the iterative covariance intersection algorithm as shown in Example 6.5.

6.4.4 Analysis

When the covariance intersection (CI) method is used for data fusion, a basic assumption is made that the estimate at each sensor node is consistent (Julier and Uhlmann 2009). If it is assumed that each node's local estimate is consistent, i.e., $\mathbf{P}_{k|k,j} \geq E\left[\left(\hat{\mathbf{x}}_{k|k,j} - \mathbf{x}_k\right)\left(\hat{\mathbf{x}}_{k|k,j} - \mathbf{x}_k\right)^T\right]$, then the final estimate is still consistent because the CI is applied. Without this assumption, consistency is not guaranteed through the CI technique.

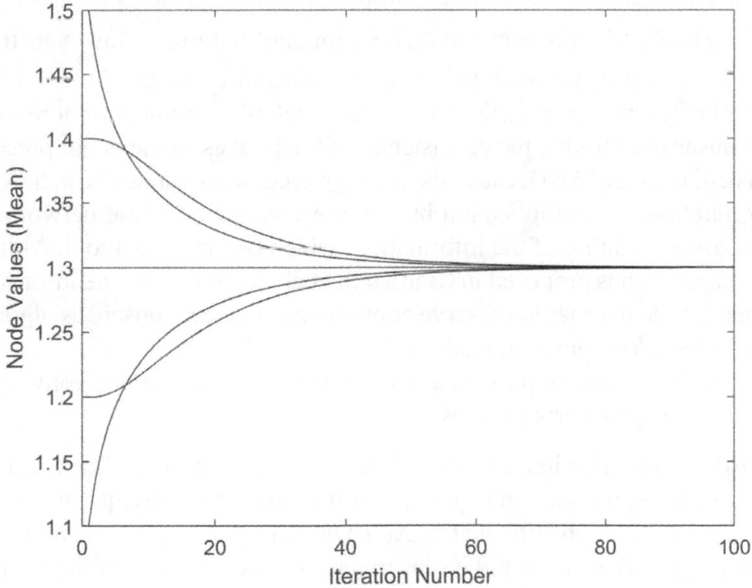

Fig. 6.9 Illustration of the state value of nodes using the distributed batch covariance intersection algorithm.

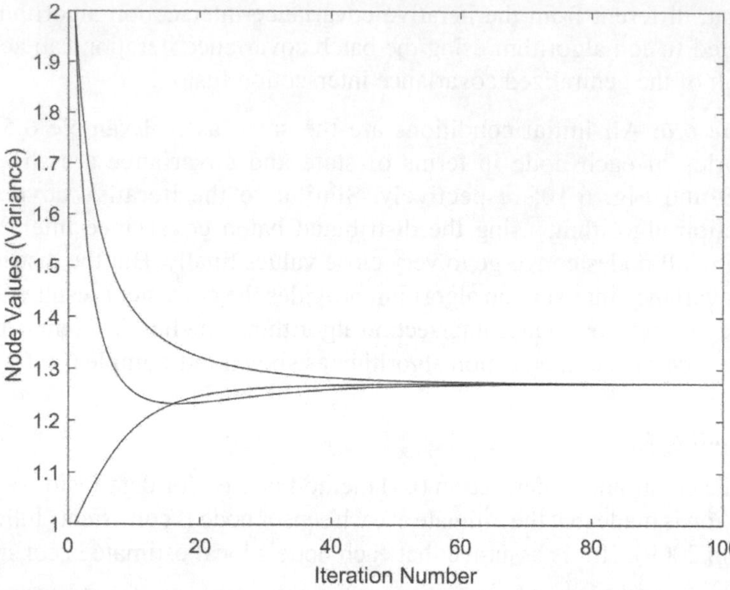

Fig. 6.10 Illustration of covariance value of nodes using the distributed batch covariance intersection algorithm.

It was shown in (Hlinka et al. 2014) that the final covariance and estimate from each node converge to a common value, i.e., $\lim_{t\to\infty} \mathbf{P}_{k,j}(t) = \mathbf{P}_k$ and $\lim_{t\to\infty} \hat{\mathbf{x}}_{k,j}(t) = \hat{\mathbf{x}}_k$. '$t$' represents the tth iteration, not the time. However, for the consensus-based distributed estimation (Battistelli et al. 2015), even if the local estimate obtained at each node is consistent, if the number of iterations of consensus is not infinite, the consistency of the final estimate is not preserved (Battistelli et al. 2015). Because the average consensus cannot be achieved in a few iterations, a multiplication by |N|, the cardinality of the network, will lead to overestimating of the information, which one tries to avoid. Although another approach is proposed in (Battistelli et al. 2015) to fuse the information from each node in order to preserve consistency, the new consensus algorithm results in more computation load.

In the following, we provide a more complete analysis of the convergence in the following two propositions.

Proposition 6.1: The iterative covariance intersection using Eq. (6.51) and (6.52) can be represented in a general form of $\boldsymbol{\eta}(t+1) = \mathbf{A}(t)\boldsymbol{\eta}(t)$ where each (j, j') entry of the transition matrix $\mathbf{A}(t)$ denoted by $a_{j,j'}(t)$ corresponds to the weight $\omega_{j',j}(t)$. Assume that the sensor network is connected. If there exists a positive constant $\alpha < 1$ and the following three conditions are satisfied

(a) $a_{j,j}(t) \geq \alpha$ for all j,t;

(b) $a_{j,j'}(t) \in \{0\} \cup [\alpha,1]$, $j \neq j'$;

(c) $\sum_{j'=1}^{N_{sn}} a_{j,j'}(t) = 1$ for all j,j',t;

the estimates using the iterative covariance intersection method reach a consensus value.

Proof: The proof uses the Theorem 2.4 in (Olshevsky and Tsitsiklis 2011). If the connected sensor network satisfies these three conditions, $\boldsymbol{\eta}(t)$ using the algorithm:

$$\boldsymbol{\eta}(t+1) = \mathbf{A}(t)\boldsymbol{\eta}(t) \tag{6.63}$$

converges to a consensus value. For the scalar case (the dimension of the state is one), $a_{j,j'}(t)$ corresponds to $\omega_{j',j}(t)$.

The jth element of $\boldsymbol{\eta}(t)$ corresponds to the information state $\breve{\mathbf{y}}_{k|k,j}(t)$. For the vector case, the transition matrix $\mathbf{A}(t) \otimes I_n$ should be applied where $\otimes$ denotes the Kronecker product and n is the dimension of the state. For the matrix case, each column of the matrix can be treated as the vector case.

As seen from Eq. (6.53), the weight $\omega_{j',j}(t)$ only depends on the covariance matrix. Here we assume that the covariance in the first iteration is upper bounded and for any t, there is no covariance matrix equal to 0 (no uncertainty). As long as the node j and node j' are connected, $\omega_{j',j}(t) \in (0,1)$. Thus, the condition (b) is satisfied. In addition, from Eq. (6.53), $\sum_{j'=1}^{N_{sn}} \omega_{j',i}(t) = 1$ always holds, i.e., the transition matrix $\mathbf{A}(t)$ is always row stochastic. Therefore, the condition (c) is satisfied.

For any arbitrary large t, say $t_{\max}$, the non-zero weight set $\{\omega_{j',j}(t), t = 1, \cdots, t_{\max}\}$ for all j, j', is a finite set since the number of nodes and the number of iterations are finite. There always exists a minimum value in this finite set. Thus, α can be chosen to be $0 < \alpha \leq \min\{\omega_{j',j}(t)\}$ such that the conditions (a) and (b) are satisfied.

According to the Theorem 2.4 in (Olshevsky and Tsitsiklis 2011) for the agreement algorithm (6.63), the estimate $\boldsymbol{\eta}(t)$ reaches a consensus value. ■

Proposition 6.2: If the assumption and conditions in Proposition 6.1 are satisfied, the consensus estimate using the iterative covariance intersection strategy is unique.

Proof: Let $U_{0,t} = \mathbf{A}(t)\mathbf{A}(t-1)\cdots\mathbf{A}(0)$ be the backward product of the transition matrices and $\lim_{t\to\infty} U_{0,t} = U^*$ according to the Proposition 6.1. On the other hand, when the consensus is achieved, the covariance matrix or the information matrix

$\breve{\mathbf{Y}}_{k|k,j'}$ associated with each node becomes the same. According to Eq. (6.53), the weights $\omega_{j',j}(t)$ converge to the same value. Thus, $\lim_{t\to\infty} \mathbf{A}(t) = \mathbf{A}^*$ and $\mathbf{A}^*\mathbf{1} = \mathbf{1}$ since $\mathbf{A}^*$ is a row stochastic matrix where $\mathbf{A}^* = [\mathbf{a}_1\mathbf{a}_2\cdots\mathbf{a}_n]^T$ with $\mathbf{a}_j$ being the row vector of the matrix $\mathbf{A}^*$. Furthermore, because $\breve{\mathbf{Y}}_{k|k,j'}$ converges to the same value, from Eq. (6.53), all the non-zero weights $\omega_{j',j}(t)$ or all non-zero entries of the row vector $\mathbf{a}_j$ are identical and equal to the reciprocal of the degree of the jth node, i.e., $\frac{1}{\delta_j}$ (where $\delta_j \triangleq$ the degree of the jth node $\triangleq$ cardinality of N_j). Hence, $\mathbf{A}^*$ is deterministic given the connected sensor network.

$\mathbf{A}^*$ is irreducible since the sensor network is connected. Moreover, the diagonal elements of $\mathbf{A}^*$ are all positive (equal to the reciprocal of the degree of each node). Hence, 1 is a unique maximum eigenvalue of $\mathbf{A}^*$ (Horn and Johnson 1990) and in fact $\mathbf{A}^*$ is a primitive matrix (Horn and Johnson 1990).

In the sense of consensus, $\lim_{t\to\infty} \boldsymbol{\eta}(t) = U^*\boldsymbol{\eta}(0)$, we have $\mathbf{A}^*U^* = U^*$ or $(\mathbf{A}^* - I)U^* = \mathbf{0}$ (Note, it is not possible for U^* to be $\mathbf{0}$ since it is the backward product of non-negative matrices). The column of U^* belongs to the null space of $\mathbf{A}^* - I$. Since 1 is the unique maximum eigenvalue of $\mathbf{A}^*$, 0 is the unique eigenvalue of $\mathbf{A}^* - I$ and the dimension of the null space of $\mathbf{A}^* - I$ is 1. Thus, $\mathbf{1}$ (or any scalar multiplication of $\mathbf{1}$) is the unique vector belonging to the null space of $\mathbf{A}^* - I$. Therefore, U^* is ergodic, i.e., $U^* = \mathbf{1}[\alpha_1, \alpha_2, \cdots, \alpha_n]$ where α_i is a scalar constant. According to the Theorem 4.20 in (Horn and Johnson 1990), $[\alpha_1, \alpha_2, \cdots, \alpha_n]$ and the consensus value of $\boldsymbol{\eta}(t)$ are unique. ■

The iterative covariance intersection update can be interpreted from the information theory as the process to minimize the Kullback-Leibler (KL) divergence (Battistelli and Chisci 2014). In the information theory, a measure of distance between different PDFs is given by the Kullback-Leibler (KL) divergence. Given the weight π_i associated with the PDF p_i, the fused PDF p_f is defined as

$$p_f = \arg\min_{p} \sum_{i=1}^{N_{sn}} \pi_i D(p \| p_i) \tag{6.64}$$

with $\sum_{i=1}^{N_{sn}} \pi_i = 1$ and $\pi_i \geq 0$. $D(p \| p_i)$ is the KL divergence defined as

$$D(p \| p_i) = \int p(x) \log \frac{p(x)}{p_i(x)} dx \tag{6.65}$$

The KL divergence is always non-negative and equal to zero only when $p(x) = p_i(x)$.

The solution to (6.64) turns out to be (Battistelli and Chisci 2014)

$$p_f(x) = \frac{\prod_{i=1}^{N_{sn}} [p_i(x)]^{\pi_i}}{\int \prod_{i=1}^{N_{sn}} [p_i(x)]^{\pi_i} dx} \tag{6.66}$$

The above equation is also the Chernoff fusion (Hurley 2002). Under the Gaussian assumption, the Chernoff fusion yields update equations identical to the covariance intersection (6.51)–(6.53).

Therefore, with the interpretation from the information theory, the iterative covariance intersection (6.51)–(6.53) is actually equivalent to solving an optimization problem repeatedly. For instance, at the tth iteration,

$$p_{f,j}(t+1) = \arg \min_{p_j(t+1)} \sum_{j' \in N_j} \omega_{j,j'} D(p_j(t+1) \| p_{j'}(t)) \text{ with } j = 1, \ldots, N_{sn} \tag{6.67}$$

When t approaches to $t_{\max}$, from the convergence property of the iterative covariance intersection discussed above, the cost for the optimization problem in Eq. (6.67) approaches 0 since $p_j(t_{\max}) = \overline{p}$ for all $j = 1, \ldots, N_{sn}$ and $D(\overline{p} \| \overline{p}) = 0$.

6.5 Diffusion-based Strategy

The consensus strategy has been used to design the distributed nonlinear estimation algorithm, such as space object tracking (Jia et al. 2016). One requirement to use the consensus strategy is sufficient communication between different sensors. However, communication resources may be limited in real applications. Hence, it is necessary to design a communication-efficient distributed estimation algorithm. In this section, a diffusion-based strategy is discussed for the distributed estimation. The diffusion strategy is essentially robust to node or link failures and has the ability to adapt to changes of network topology. Similar to the consensus strategy, in the diffusion strategy, information is processed locally and all nodes communicate with their neighbors to share the information (Arablouei et al. 2014). The information thus spreads throughout the network without using any central node or protocol. However, unlike the consensus strategy, the diffusion strategy does not require each node converge to the same estimate (Arablouei et al. 2014).

The diffusion strategy is usually applied to solve linear distributed estimation problems. Although the linearization technique can be used to extend the diffusion strategy to solve the nonlinear estimation problem, the accuracy of the linearization is limited. Hence, the grid-based numerical rules can be used to extend the traditional diffusion strategy. As shown in Fig. 6.11, there are two steps of the diffusion strategy based distributed nonlinear filters, the incremental update and diffusion update. For the incremental update, at

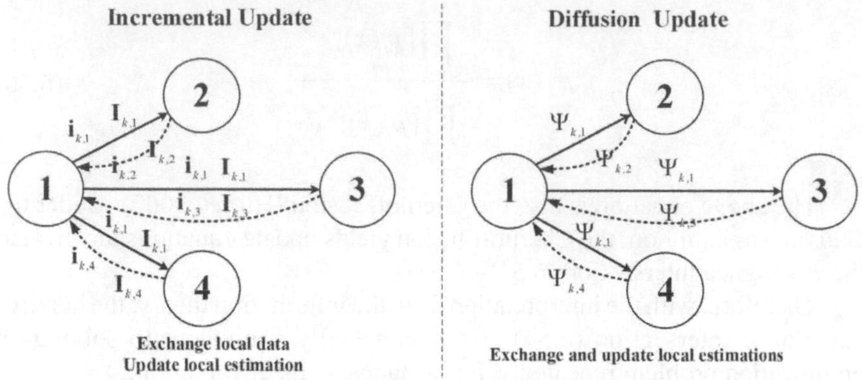

Fig. 6.11 Diffusion strategy based distributed nonlinear filter update at node 1.

each time step, each agent communicates with their neighbors the information state and information matrix to obtain an intermediate estimate of the system state. The diffusion update is then conducted based on the intermediate estimate given by the incremental update step (Hu et al. 2012). The symbol '$\Psi_{k,i}$' with $i = 1, \cdots, 4$ in Fig. 6.11 denotes the information shared by sensor i in the diffusion update step. Note that if the covariance intersection is used in the diffusion update step, $\Psi_{k,i}$ denotes $\hat{\mathbf{x}}_{k,i}$ and ${}_{k\ i}$. The main reason to use the diffusion strategy for some distributed estimation applications is that it only requires parameters to be transmitted between different sensors twice. Hence, it can save communication resources and afford efficient distributed and cooperative estimation.

The basic diffusion strategy is described next and the iterative diffusion strategy follows (Hu et al. 2012).

The diffusion Kalman filter using the diffusion strategy was first proposed in (Cattivelli and Sayed 2010). It includes two steps: the incremental update and the diffusion update. For each node, the incremental update uses the update equation of the filter to estimate the system state using measurement information from its neighbors. At this step, for the linear system, the local estimate has not been explored yet. For the nonlinear system, the local estimate is also used in the incremental update step. The diffusion update, thus, is used to fuse the local estimate to provide a better global estimate by exchanging local estimates' information across the entire network. Similar to the Kalman filter, the diffusion-based filtering algorithm can be presented in the normal form or the information form. In our opinion, the information form is simpler to use, which is given as follows.

There are different ways to choose the weighting coefficients $C_{j'j}$. The *covariance intersection* algorithm is often employed due to its simplicity.

Table 6.2 Diffusion-based distributed filter.

Incremental Update. In this step, each sensor broadcasts its information to its neighbors. For every sensor j, after receiving the information from its neighbors, the information state and information matrix are obtained by

$$\breve{\mathbf{y}}_{k|k,j} = \breve{\mathbf{y}}_{k|k,j} + \sum_{j' \in N_j} \mathbf{i}_{k,j'} \tag{6.68}$$

$$\breve{\mathbf{Y}}_{k|k,j} = \breve{\mathbf{Y}}_{k|k,j} + \sum_{j' \in N_j} \mathbf{I}_{k,j'} \tag{6.69}$$

where N_j is the neighboring set of sensor j, $\mathbf{i}_{k,j'}$ is the information state contribution and $\mathbf{I}_{k,j'}$ is the information matrix contribution.

Diffusion Update. In this step, the intermediate state estimation can be updated by

$$\hat{\mathbf{x}}_{k,j} = \sum_{j' \in \tilde{N}_j} C_{j',j} \hat{\mathbf{x}}_{k,j'} \tag{6.70}$$

Note that the weighting coefficients $C_{j',j}$ should satisfy the constraint $\sum_{j'=1}^{N_{sn}} C_{j'j} = 1$. $\tilde{N}_j$ is the neighboring set of sensor j, including itself.

The estimate state and covariance can be predicted using the same predict equations in the conventional nonlinear filter for each node.

Note that the covariance intersection algorithm fuses estimates under unknown correlation, which often gives conservative estimation results.

Similar to the extension of the covariance intersection, the iterative covariance intersection can also be used in the diffusion update step. Since the iterative diffusion update is identical to the iterative covariance intersection (ICI) algorithm (Hlinka et al. 2014), the properties of unbiasedness and consistency of the iterative diffusion-based estimation are the same as the ICI.

6.6 Distributed Particle Filter

In previous Sections 6.2–6.5, the Gaussian approximation filter is used. As in the single sensor nonlinear estimation, the particle filter can be used for the general distributed nonlinear estimation problem. There are many different distributed particle filters (DPF) in terms of consensus of the posterior distribution or the likelihood function.

Assume that identical sets of particles and weights, $\mathbf{x}_{k-1}^{(i)}, \overline{w}_{k-1}^{(i)}, i = 1, \cdots, N_s$, which represent the global posterior PDF at time k–1, are sampled and available at each agent. The propagation of the particle can be calculated locally, as shown in Chapter 3. The remaining part is how to update the particles and weights using the information of the measurements from multiple sensors. Specifically, the weights are calculated as follows (Hlinka et al. 2012)

$$w_k^{(i)} \propto \bar{w}_{k-1}^{(i)} \frac{p\left(\mathbf{y}_k \mid \mathbf{x}_k^{(i)}\right) p\left(\mathbf{x}_k^{(i)} \mid \mathbf{x}_{k-1}^{(i)}\right)}{q\left(\mathbf{x}_k^{(i)} \mid \mathbf{x}_{k-1}^{(i)}, \mathbf{y}_k\right)}, \quad \bar{w}_k^{(i)} = \frac{w_k^{(i)}}{\sum_{i=1}^{N_s} w_k^{(i)}}, \; i = 1, \cdots, N_s \quad (6.71)$$

where $\mathbf{y}_k \triangleq \left[\mathbf{y}_{k,1}, \cdots, \mathbf{y}_{k,j}, \cdots, \mathbf{y}_{k,N_{sn}}\right]$ is the global measurement vector including the measurements of all sensors. The superscript '(i)' denotes the index of the particles.
If we let $q\left(\mathbf{x}_k^{(i)} \mid \mathbf{x}_{k-1}^{(i)}, \mathbf{y}_k\right) = p\left(\mathbf{x}_k^{(i)} \mid \mathbf{x}_{k-1}^{(i)}\right)$, Eq. (6.71) can be rewritten as

$$\log\left(w_k^{(i)}\right) \propto \sum_{j=1}^{N_{sn}} \log\left(p\left(\mathbf{y}_{k,j} \mid \mathbf{x}_k^{(i)}\right)\right) \quad (6.72)$$

Note that for each sensor, $\mathbf{x}_{k-1}^{(i)}$, $\mathbf{x}_k^{(i)}$ are available. In addition, we assume the global likelihood function $p\left(\mathbf{y}_k \mid \mathbf{x}_k^{(i)}\right)$ can be obtained by the product of local likelihood functions, i.e., $p\left(\mathbf{y}_k \mid \mathbf{x}_k^{(i)}\right) = \prod_{j=1}^{N_{sn}} p\left(\mathbf{y}_{k,j} \mid \mathbf{x}_k^{(i)}\right)$.

Equation (6.72) can be approximated using the average consensus algorithm. The normalized $\bar{w}_k^{(i)}$ can then be updated and used in the next filtering cycle.

The consensus-based DPF described above can achieve the same estimation performance as the centralized particle filter. However, the communication demand is high. The average consensus algorithm is used for each particle by Eq. (6.72). To save communication resources, the parametric method was proposed. The particles are represented by the Gaussian distribution or Gaussian mixture model (GMM) first (Gu 2008). Then parameters of the Gaussian distribution or Gaussian mixture are transmitted over the network. Whether to use the GMM approximation of the posterior distribution is problem dependent. Communicating with GMM also needs to transmit the mean values and covariance matrices. The covariance matrices, however, require proportional information exchange of the state dimension. Hence, the GMM approximation can become costly if the state dimension is large and the number of the Gaussian components is large.

As shown in the particle filter (Chapter 3), it is not accurate to use $q\left(\mathbf{x}_k^{(i)} \mid \mathbf{x}_{k-1}^{(i)}, \mathbf{y}_k\right) = p\left(\mathbf{x}_k^{(i)} \mid \mathbf{x}_{k-1}^{(i)}\right)$. In (Farahmand et al. 2011), data adaptation for selecting the importance sampling distribution $q\left(\mathbf{x}_k^{(i)} \mid \mathbf{x}_{k-1}^{(i)}, \mathbf{y}_k\right)$ was proposed. The data-adapted important distribution accounts for the new measurements $\mathbf{y}_k$ in the particle generation process. The data-adaption process reduces the number of required particles and is helpful for the distributed processing.

To overcome the imperfect communication connectivity, the consensus/fusion filter (Mohammadi and Asif 2013) was proposed to address the problem that the convergence of the consensus algorithm cannot be achieved between two consecutive measurements. Other methods, such as the constrained sufficient statistic based DPF (Mohammadi and Asif 2015) and measurement dissemination-based distributed particle filter (Rosencrantz et al. 2003) are also available.

6.7 Multiple Sensor Estimation and Sensor Allocation

Sensor allocation is essential to the multiple sensor estimation problem since sensor resources are limited and the overall estimation performance of the system should be maintained. The multiple sensor estimation problem is often coupled with the sensor allocation problem as shown in Fig. 6.12. The former uses the results of the latter. The sensor allocation problem uses the information provided by the multiple sensor estimation, such as the covariance information, to give the best allocation results.

The sensor allocation problem can be formulated as an integer programming problem in general. There are many metrics to evaluate the performance of the sensor allocation, such as the trace of the covariance matrix, eigenvalue, L_2-norm, Frobenius norm, and determinant of the covariance (Yang et al. 2012). For convenience, we assume that there are N_t targets/objects and N_s sensors. The objective of the sensor management problem is to use the available sensors to estimate single or multiple targets/objects. The solution to the sensor management should be conflict-free since multiple sensors are used. The sensor management problem can be formulated as

$$\begin{aligned} \max_{x_{ij}} \quad & \sum_{j=1}^{N_t} J_j\left(x_{k,ij},k\right) \\ \text{s.t.} \quad & \sum_{j\in S_{k,j}} x_{k,ij} \le L_t, \quad \forall i=1,\cdots,N_s \\ & \sum_{i=1}^{N_s} x_{k,ij} \le 1 \quad \forall j=1,2,\cdots,N_t \\ & x_{k,ij} \in \{0,1\}, \quad \forall i=1,\cdots,N_s, \quad j=1,\cdots,N_t \end{aligned} \tag{6.73}$$

where $J_j\left(x_{k,ij},k\right)$ is the performance gain for the *j*th target at time *k*. $x_{k,ij}$ is the optimization variable, which equals to 1 if the *i*th sensor is used to observe the *j*th target. L_t is the capacity of the sensor. $S_{k,j}$ is a group of sensors that can track the *j*th target at time *k*. Note that there are some sensors that cannot observe some targets due to hardware constraints, such as the field of view or distance.

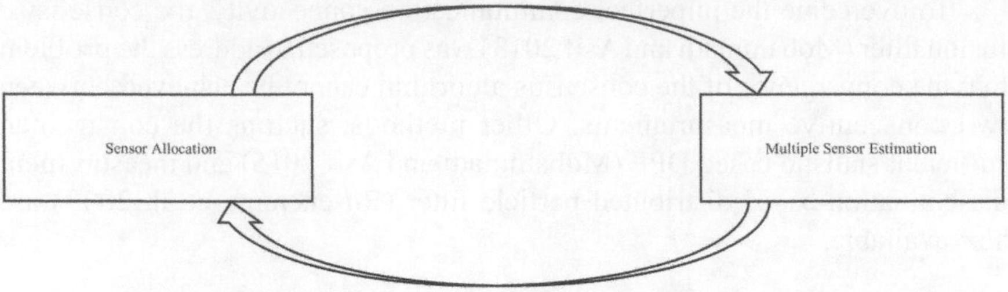

Fig. 6.12 Coupling of sensor allocation and multiple sensor estimation.

The 2nd constraint shown in Eq. (6.73) can be removed if multiple sensors can be used to observe the same object. In this case, the optimization problem can be rewritten as

$$\begin{aligned} &\max_{x_{k,ij}} \quad \sum_{j=1}^{N_t} J_j\left(x_{k,ij},k\right) \\ &\text{s.t.} \quad \sum_{j\in S_{k,j}} x_{k,ij} \le L_t, \quad \forall i=1,\cdots,N_s \\ &\qquad x_{k,ij}\in\{0,1\}, \quad \forall i=1,\cdots,N_s, \quad j=1,\cdots,N_t \end{aligned} \tag{6.74}$$

The objective function $J_j\left(x_{k,ij},k\right)$ is problem dependent. For example, it can be modeled to mitigate the risk of identification of targets or reduce the overall uncertainty of the sensor network. If the overall objective is to maximally reduce the target state uncertainty, the objective can be set to maximize the information gain. The information gain uses the entropy given by

$$J_j\left(x_{k,ij},k\right) = H\left(\hat{\mathbf{x}}_{k|k-1,j} \mid \mathbf{y}_{k-1,j}\right) - H\left(\hat{\mathbf{x}}_{k|k,j} \mid \mathbf{y}_{k,j}\right) \tag{6.75}$$

where $H(\cdot|\cdot)$ is the conditional entropy of the jth object given the observations of sensors up to time k. The specific form of $H\left(\hat{\mathbf{x}}_{k|k,j} \mid \mathbf{y}_{k,j}\right)$ is given by (Fu et al. 2012)

$$H\left(\hat{\mathbf{x}}_{k|k,j} \mid \mathbf{y}_{k,j}\right) = \gamma - \frac{1}{2}\log\left(\left|\breve{\mathbf{Y}}_{k|k,j}\right|\right) = \gamma - \frac{1}{2}\log\left(\left|\breve{\mathbf{Y}}_{k|k-1,j} + \sum_{i=1}^{N_s}\mathbf{I}_{k,j}x_{k,ij}\right|\right) \tag{6.76}$$

where γ is a constant. Intuitively, the determinant $|\cdot|$ defines the volume of the error ellipsoid (Yang et al. 2012). Because the information matrix $\breve{\mathbf{Y}}_{k|k,j}$ has the inverse proportional relation to the covariance matrix, $H(\cdot|\cdot)$ decreases with the decrease of the covariance, which defines the uncertainty. Hence, Eq. (6.75) defines the uncertainty reduction gain by using the sensor observations.

To solve the integer programming problem in Eq. (6.73) or Eq. (6.74), many commercial solvers, such as CPLEX and GUROBI, can be used.

Alternatively, the convex relaxation can be used for the integer constraints. The problem in Eq. (6.74) can then be rewritten as (Fu et al. 2012)

$$\begin{aligned} &\max_{k\,ij} \quad \sum J_j\left(x_{k\,ij}, k\right) \\ &\text{s.t.} \quad \sum_{j\in S_{k\,j}} x_{k\,ij} \le L_t, \quad \forall i = 1, \quad , N_s \\ &\qquad\quad 0 \le x_{k\,ij} \le 1, \quad \forall i = 1, \cdots, N_s, \quad j = 1, \cdots, N_t \end{aligned} \tag{6.77}$$

Then, the efficient solution for centralized or distributed sensor allocation can be derived based on the formulation of Eq. (6.77) (Fu et al. 2012).

Remark 6.4: There are myopic and non-myopic sensor management algorithms. The myopic sensor management algorithm uses the current state of the system at time k–1 to decide the next time action of sensors at time k. The non-myopic sensor management considers future actions of sensors from time k to time $k + W$, where W denotes the time window. The non-myopic sensor management algorithm is based on look-ahead optimization to determine the current action (Salvagnini et al. 2015). The non-myopic sensor management is more complicated than the myopic sensor management algorithm.

6.8 Summary

It can be foreseen that applications of the sensor network will become quickly prevalent due to the rapid development of the sensor technology and the inadequacy of a single sensor. Many conventional filtering algorithms have been integrated with distributed processing protocol, such as the consensus algorithm, to realize distributed multiple sensor estimation. The consensus-based strategies, covariance intersection-based strategies, and diffusion-based strategies are introduced in this chapter. The consensus can be achieved by using the iterative covariance intersection algorithm. The covariance intersection and the iterative covariance intersection algorithms can also be used in the diffusion-based algorithm. Specifically, they can be used as the diffusion update step in the diffusion-based filters. Compared to the consensus-based filters, the covariance intersection-based filters and the diffusion-based filters can flexibly balance the computational complexity and the communication cost. The consensus-based filter, however, requires consensus for each update step. We believe that practical multiple sensor fusion algorithms also need to balance their benefits with the computation and communication cost. In addition, how to allocate the sensor resource according to the objective in specific applications is also important. The centralized sensor allocation and estimation is easier than the distributed allocation algorithm. It is still an open research area, which is attracting extensive research efforts.

References

Amundson, I., M. Kushwaha, B. Kusy, P. Volgyesi, G. Simon, X. Koutsoukos and A. Ledeczi. 2007. Time synchronization for multi-modal target tracking in heterogeneous sensor networks. Workshop on Networked Distributed Systems for Intelligent Sensing and Control. Greece.

Arablouei, R., S. Werner, Y.F. Huang and K. Doğançay. 2014. Distributed least mean-square estimation with partial diffusion. IEEE Transactions on Signal Processing 62: 472—484.

Battistelli, G. and L. Chisci. 2014. Kullback–Leibler average, consensus on probability densities, and distributed state estimation with guaranteed stability. Automatica 50: 707–718.

Battistelli, G., L. Chisci, G. Mugnai, A. Farina and A. Graziano. 2015. Consensus-based linear and nonlinear filtering. IEEE Transactions on Automatic Control 60: 1410—1415.

Battistelli, G. and L. Chisci. 2014. Kullback–Leibler average, consensus on probability densities, and distributed state estimation with guaranteed stability. Automatica 50: 707–-718.

Carli, R., F. Fagnani, P. Frasca, and S. Zampieri. 2010. Gossip consensus algorithms via quantized communication. Automatica 46: 70—80.

Cattivelli, F. S. and A. H. Sayed. 2010. Diffusion strategies for distributed Kalman filtering and smoothing. IEEE Transactions on Automatic Control 55: 2069—2084.

Chen, L., P.O. Arambel and R.K. Mehra. 2002. Estimation under unknown correlation: covariance intersection revisited. IEEE Transactions on Automatic Control 47: 1879–-1882.

Dimakis, A. D. G., A. D. Sarwate and M. J. Wainwright. 2008. Geographic gossip: efficient averaging for sensor networks. IEEE Transactions on Signal Processing 56: 1205–-1216.

Durrant-Whyte, H. and T. C. Henderson. 2008. Multisensor data fusion. pp. 585–-610. In: B. Siciliano, B. and O. Khatib. [(eds.]). 2008. Springer Handbook of Robotics. Springer, Berlin.

Farahmand, S., S. I. Roumeliotis and G. B. Giannakis. 2011. Set-membership constrained particle filter: distributed adaptation for sensor networks. IEEE Transactions on Signal Processing 59: 4122–-4138.

Franken, D. and A. Hupper. 2005. Improved fast covariance intersection for distributed data fusion. 2005 7th International Conference on Information Fusion. USA. , 154–-160.

Fu, Y., Q. Ling and Z. Tian. 2012. Distributed sensor allocation for multi-target tracking in wireless sensor networks. IEEE Transactions on Aerospace and Electronic Systems 48: 3538–-3553.

Gu, D. 2008. Distributed EM algorithm for Gaussian mixtures in sensor networks. IEEE Transactions on Neural Networks 19: 1154—1166.

Hlinka, O., F. Hlawatsch and P.M. Djuric. 2012. Distributed particle filtering in agent networks: a survey, classification, and comparison. IEEE Signal Processing Magazine 30: 61–-81.

Hlinka, O., O. Slučiak, F. Hlawatsch and M. Rupp. 2014. Distributed data fusion using iterative covariance intersection. 2014 IEEE International Conference on Acoustics, Speech and Signal Processing (ICASSP). Italy. , 1861–-1865.

Horn, R.A. and C.R. Johnson. 1990. Matrix Analysis. Cambridge Uuniversity Ppress, New York.

Hu, J., L. Xie and C. Zhang. 2012. Diffusion Kalman filtering based on covariance intersection. IEEE Transactions on Signal Processing 60: 891–-902.

Hurley, M.B. 2002. An information theoretic justification for covariance intersection and its generalization. Proceedings of the Fifth International Conference on Information Fusion. USA. , 505–-511.

Jia, B., K.D. Pham, E. Blasch, D. Shen, Z. Wang and G. Chen. 2016. Cooperative space object tracking using space-based optical sensors via consensus-based filters. IEEE Transactions on Aerospace and Electronic Systems 52: 1908–-1936.

Julier, S.J. and J.K. Uhlmann. 1997. A non-divergent estimation algorithm in the presence of unknown correlations. Proceedings of the 1997 American Control Conference. USA. , 2369–-2373.

Julier, S.J. and J.K. Uhlmann. 2009. General decentralized data fusion with covariance intersection. pp. 319–-344. In: D. Hall, D. and J. Llinas (eds.). Handbook of Multisensor Data Fusion: Theory and Practice. CRC Press, Boca Raton.

Kamal, A.T., C. Ding, B. Song, J.A. Farrell and A.K. Roy-Cchowdhury. 2011. A generalized Kalman consensus filter for wide-area video network. Proceedings of IEEE Conference on Decision and Control and European Control Conference, . USA, 7863–-7869.

Kamal, A.T., J.A. Farrell and A.K. Roy-Chowdhury. 2013. Information weighted consensus filters and their application in distributed camera networks. IEEE Transactions on Automatic Control 58: 3112–-3125.

Kar, S. and J.M. Moura. 2009. Distributed consensus algorithms in sensor networks with imperfect communication: Link failures and channel noise. IEEE Transactions on Signal Processing 57: 355–-369.

Kempe, D., A. Dobra and J. Gehrke. 2003. Gossip-based computation of aggregate information. 44th Annual IEEE Symposium on Foundations of Computer Science. USA. , 1–-10.

Klein, L.A. 2004. Sensor and Data Fusion: A Tool for Information Assessment and Decision Making. SPIE Press, Bellingham .

LeBlanc, H.J. and X.D. Koutsoukos. 2011. Consensus in networked multi-agent systems with adversaries. Proceedings of the 14th International Conference on Hybrid Systems: Computation and Control. USA. , 281–-290.

LeBlanc, H.J., H. Zhang, S. Sundaram and X. Koutsoukos. 2012. Consensus of multi-agent networks in the presence of adversaries using only local information. Proceedings of the 1st International Conference on High Confidence Networked Systems. China. 1: 10.

Lee, D. 2008. Nonlinear estimation and multiple sensor fusion using unscented information filtering. IEEE Signal Processing Letters 15: 861–-864.

Li, T., M. Fu, L. Xie and J. F. Zhang. 2011. Distributed consensus with limited communication data rate. IEEE Transactions on Automatic Control 56: 279–-292.

Li, W., Z. Wang, G. Wei, L. Ma, J. Hu and D. Ding. 2015. A Survey survey on multisensor fusion and consensus filtering for sensor networks. Discrete Dynamics in Nature and Society 2015: 683071.

Liggins, I.I.M., D. Hall and J. Llinas. 2008. Handbook of Multisensor Data Fusion: Theory and Practice. CRC Press, Boca Raton.

Mahmoud, M. S. and H. M. Khalid. 2013. Distributed Kalman filtering: a bibliographic review. IET Control Theory & Applications 7: 483–-501.

Maskell, S.R., R.G. Everitt, R. Wright and M. Briers. 2006. Multi-target out-of-sequence data association: tracking using graphical models. Information Fusion 7: 434–-447.

Mohammadi, A. and A. Asif. 2013. Distributed particle filter implementation with intermittent/irregular consensus convergence. IEEE Transactions on Signal Processing 61: 2572–-2587.

Mohammadi, A. and A. Asif. 2015. Distributed consensus + innovation particle filtering for bearing/range tracking with communication constraints. IEEE Transactions on Signal Processing 63: 620–-635.

Mokhtarzadeh, H. and D. Gebre-Egziabher. 2016. Fusing data with unknown or uncertain level of correlation. Journal of Guidance, Control, and Dynamics 39: 1163–-1167.

Niehsen, W. 2002. Information fusion based on fast covariance intersection filtering. Proceedings of the Fifth International Conference on Information Fusion. USA. , 901–-904.

Noack, B., J. Sijs, M. Reinhardt and U. D. Hanebeck. 2017. Decentralized data fusion with inverse covariance intersection. Automatica 79: 35–-41.

Olfati-Saber, R. 2009. Kalman-Consensus consensus filter: optimality, stability, and performance. Joint 48th IEEE Conference on Decision and Control and 28th Chinese Control Conference. China. , 7036–-7042.

Olshevsky, A. and J.N. Tsitsiklis. 2011. Convergence speed in distributed consensus and averaging. SIAM Review 53: 747–-772.

Pasqualetti, F., A. Bicchi and F. Bullo. 2009. On the security of linear consensus networks. Joint 48th IEEE Conference on Decision and Control and 28th Chinese Control Conference. China. , 4894–-4901.

Reinhardt, M., B. Noack and U. D. Hanebeck. 2012. Closed-form optimization of covariance intersection for low-dimensional matrices. 2012 15th International Conference on Information Fusion. Singapore. , 1891–-1896.

Rosencrantz, M., G. Gordon and S. Thrun. 2003. Decentralized sensor fusion with distributed particle filters. Proceedings of the 19th conference Conference on Uncertainty in Artificial Intelligence. Mexico. , 493–-500.

Salvagnini, P., F. Pernici, M. Cristani, G. Lisanti, A.D. Bimbo and V. Murino. 2015. Non-myopic information theoretic sensor management of a single pan–tilt–zoom camera for multiple object detection and tracking. Computer Vision and Image Understanding 134: 74–-88.

Taghavi, E., R. Tharmarasa, T. Kirubarajan and M. Mcdonald. 2016. Multisensor-multitarget bearing-only sensor registration. IEEE Transactions on Aerospace and Electronic Systems 52: 1654–-1666.

Terejanu, G.A. 2013. Discrete Kalman Filter Tutorial. Retrieved from cse.sc.edu.

Wang, Y., C. Qian and X. Liu. 2015. Compensation strategy for distributed tracking in wireless sensor networks with packet losses. Wireless Networks 21: 1925–-1934.

Wu, Y., X. He, S. Liu and L. Xie. 2014. Consensus of discrete-time multi-agent systems with adversaries and time delays. International Journal of General Systems 43: 402–-411.

Yang, C., L. Kaplan and E. Blasch. 2012. Performance measures of covariance and information matrices in resource management for target state estimation. IEEE Transactions on Aerospace and Electronic Systems 48: 2594–-2613.

Application Uncertainty Propagation

7

Uncertainty propagation through dynamic systems widely exists and is often an essential need in many science and engineering disciplines, such as biology (Sankaran and Marsden 2011), computational physics (Xiu and Karniadakis 2003), and aerospace engineering (Prabhakar et al. 2010, Witteveen and Iaccarino 2010). From the perspective of nonlinear filtering, the uncertainty propagation can be viewed as the prediction step. In this chapter, the uncertainty propagation mainly refers to the long-term prediction. Due to the longtime propagation through nonlinear dynamics, the Gaussian assumption of uncertainty may not be valid anymore. It needs an alternative way to represent the uncertainty, such as point cloud or the Gaussian mixture. One advantage of grid-based methods is that all grid points can be processed independently. Hence, the grid-based method can use modern computational techniques, such as the graphics processing unit and cloud computing techniques. We will introduce typical uncertainty propagation methods in which the grid-based methods are involved, and integration of the grid-based methods with the parallel computing techniques.

Besides the grid/sample-based uncertainty propagation methods, other approaches, such as the semi-analytic method, are available. The state transition tensors (STTs) (Park and Scheeres 2006) use high order Taylor series terms to describe the localized nonlinear motion and employ an analytical way to map the initial uncertainties. The main difficulty is to derive the high order STTs of the dynamics. Alternatively, the uncertainty can be directly propagated via solving the Fokker-Planck equation (Sun and Kumar 2016). However, it is generally hard to do with limited computational resources. Hence, in this chapter, we focus on various grid/sample-based methods, and the application to the orbital uncertainty propagation problem.

Orbital uncertainty propagation is one of important applications of uncertainty propagation in space situational awareness (SSA), which concerns space surveillance and tracking. With a rapidly growing population of resident space objects, space surveillance becomes more and more challenging due to limited tracking resources. The lack of measurement or track updates requires precise long-term propagation of the state probability density function of the space object, often on the order of several orbital periods, between sparse measurements (Horwood et al. 2011).

Due to scarce measurement updates, the long-term propagation results in high uncertainty and inevitable non-Gaussianity, even though the uncertainty may be initialized as a Gaussian distribution. In order to characterize the non-Gaussian uncertainty, Monte Carlo samples and Gaussian mixtures (GM) are commonly used. The Monte Carlo methods demand excessive computational resources, which may become infeasible for high-dimensional dynamic systems. The GM is based on the result originally given by Alspach and Sorenson (1972, 1971) that any PDF of practical concern can be approximated as closely as desired with a sufficient number of Gaussian components. The GM and its improved variants have been widely used in the orbit propagation problem (Horwood and Poore 2011, Terejanu et al. 2008, DeMars et al. 2013) because the GM can better represent non-Gaussian PDFs. Additionally, it has the computation advantage of being parallelizable since many Gaussian predictors can run independently on each Gaussian component. The number of Gaussian components, the weight, mean, and variance of each component are usually the parameters to design the GM. In (Terejanu et al. 2008), a weight update rule of the GM was proposed for uncertainty propagation by minimizing an L_2 difference between the true predicted PDF and its GM approximation. Horwood and Poore (2011) proposed an adaptive GM for orbit uncertainty propagation by refining or coarsening the GM components based on a moment matching criterion. However, as the authors mentioned in (Horwood and Poore 2011), this scheme may suffer the curse of dimensionality since the refinement is performed along all dimensions of the distribution. In addition, the method tends to place too many Gaussian components on the tail of the PDF. Horwood and Poore (2011) proposed a more efficient adaptive GM scheme for orbit propagation by adopting the equinoctial orbital element space instead of the Cartesian earth-centered inertial frame. It is shown that the refinement of GM only needs to be performed along the semi-major axis coordinate to avoid an expensive full six-dimensional mixture. In addition, it can provide improved uncertainty consistency. DeMars et al. (2013) proposed an adaptive strategy to split the Gaussian component based on an entropy measure of nonlinearity during the orbital uncertainty propagation. A differential entropy for a linearized system is compared with the one for the nonlinear dynamics and the deviation is used to determine if the GM needs to be refined. Integrating the scalar entropy equation brings a computational advantage over the fully linearized implementation of a predictor.

In this chapter, we first introduce the grid-based method. Then different propagators are briefly reviewed. Because the grid-based method cannot directly represent the non-Gaussian distribution, we discuss the Gaussian mixture based uncertainty propagation and stochastic expansion based uncertainty propagation, which can use grid-based numerical integration techniques. Then parallel computation for the grid-based uncertainty propagation is introduced, including the GPU-based and the MapReduce based computational architectures.

7.1 Gaussian Quadrature-based Uncertainty Propagation

For either single Gaussian model or a GM model, propagation of the first two statistical moments, i.e., mean and covariance, is necessary. For linear Gaussian systems, the Gaussianity of the state is preserved through the time update of the Kalman filter (KF). The Gaussian PDF is fully determined by the mean and covariance and higher moments of the state can be computed from the mean and covariance. The nonlinear dynamic systems require more accurate estimation techniques to propagate the mean and covariance. Many methods have been developed for statistical moment propagation of nonlinear systems, such as the Monte Carlo (MC) method, the EKF (Jazwinski 1970) based on the first-order linearization of the dynamics, the unscented transformation (UT) (Julier and Uhlmann 2004), the Gauss-Hermite quadrature (GHQ) (Ito and Xiong 2000), the sparse grid quadrature (SGQ) (Jia et al. 2012, Jia et al. 2011), the spherical-radial cubature rules (CR) (Jia et al. 2013, Arasaratnam and Haykin 2009), and the polynomial chaos (PC) (Li and Xiu 2009). The EKF is the most widely used but its performance may degrade for long-term uncertainty propagation when the measurement is scarce and the nonlinear system contains large initial estimation errors, non-Gaussian uncertainty, large process noise, or high nonlinearities. In the recent two decades, the UT has attracted a great deal of attention and has been shown to outperform the EKF in many estimation problems. The UT can be regarded as a quadrature numerical integration rule for computing the mean and covariance of nonlinear functions of Gaussian random vectors (Jia et al. 2011). The CR is of the same class of quadrature rules. With slightly increased computational complexity, the UT and CR have been proved more accurate than the EKF and can thus be used to better propagate the uncertainty. The orbital uncertainty propagation addressed in (Horwood and Poore 2011, DeMars et al. 2013) used the UT or GHQ to propagate the mean and covariance of each Gaussian component in the GM model. However, the UT in (Julier and Uhlmann 2004) and CR in (Arasaratnam and Haykin 2009) are both the third-degree rules, which may not be sufficient for computing the moments of highly nonlinear systems, for instance, the magnetometer-based attitude estimation (Jia et al. 2011) and the problems with non-Gaussian random variables. It is especially challenging for the UT or CR to perform long-term orbital propagation with very sparse measurement updates. The GHQ rule (Ito and Xiong 2000) is accurate to arbitrary degrees and all weights of the quadrature points are positive. However, it is based on the tensor product of univariate GHQ rule and thus suffers the curse of dimensionality problem. The SGQ rules (Jia et al. 2012) and the high-degree (> 3) CR in (Jia et al. 2013) have been shown in Chapter 4 to be more accurate than the UT. Both rules have the computation complexity on the order of polynomials of the dimension while maintaining the estimation accuracy close to the GHQ rule. However, the UT, the high-degree CR, and the SGQ may generate negative quadrature weights, which can result in negative definite covariance matrix and degrade the propagation performance.

In this section, we compare the uncertainty propagation results using different quadrature rules based on the single Gaussian model. Consider the following deterministic dynamic system described by

$$\dot{\mathbf{x}} = f(\mathbf{x}) \tag{7.1}$$

where the initial state is denoted by the random variable $\mathbf{x}_0$ with the uncertainty described by the PDF $p_0(\mathbf{x})$, which is assumed a Gaussian distribution. It is also assumed that the uncertainty can be represented by the Gaussian distribution at any time. Hence, the key is to estimate the mean and covariance since the Gaussian distribution can be fully described by the first two moments. Specifically, the first and second moments at time t are calculated by

$$\hat{\mathbf{x}}_t = E[\mathbf{x}_t] = \int \mathbf{x}_t \cdot p_t(\mathbf{x}_t) d\mathbf{x}_t \tag{7.2}$$

and

$$M_t = E\left[\mathbf{x}_t \mathbf{x}_t^T\right] = \int \mathbf{x}_t \mathbf{x}_t^T \cdot p_t(\mathbf{x}_t) d\mathbf{x}_t \tag{7.3}$$

where $p_t(\mathbf{x}_t)$ denotes the PDF at time t. The covariance matrix is

$$\mathbf{P}_t = E\left[(\mathbf{x}_t - \hat{\mathbf{x}}_t)(\mathbf{x}_t - \hat{\mathbf{x}}_t)^T\right] \tag{7.4}$$

For the grid-based propagation algorithms, the initial points are generated via the linear transformation

$$\mathbf{x}_{0,i} = \mathbf{S}_0 \xi_i + \hat{\mathbf{x}}_0, \quad i = 1, \cdots, N \tag{7.5}$$

where $\mathbf{S}_0$ satisfies $\mathbf{S}_0 \mathbf{S}_0^T = \mathbf{P}_0$. ξ_i is the quadrature points corresponding to the standard normal distribution as shown in Chapter 4.

Given a set of N initial points $\{\mathbf{x}_{0,i}, W_i\}$, $i = 1, \cdots, N$, where $\mathbf{x}_{0,i}$ denotes the ith point and W_i denotes its associated weight. The moments are given by

$$\hat{\mathbf{x}}_t \approx \sum_{i=1}^{N} W_i \mathbf{x}_{t,i} \tag{7.6}$$

$$\mathbf{P}_t \approx \sum_{i=1}^{N} W_i \left(\mathbf{x}_{t,i} - \hat{\mathbf{x}}_t\right)\left(\mathbf{x}_{t,i} - \hat{\mathbf{x}}_t\right)^T \tag{7.7}$$

where $\mathbf{x}_{t,i}$ are computed by Eq. (7.1).

Remark 7.1: For the Monte Carlo method or the quasi-Monte Carlo method, the weights $W_i = 1/N$. In addition, many grid-based numerical integration techniques can be used in this framework. The main advantage of the grid-based uncertainty propagation is that the form is simple. The main disadvantage is that the propagation accuracy is highly constrained by the assumed Gaussian distribution.

Example 7.1: Consider a space object in a high elliptical orbit (HEO) confined to the equatorial plane. Define the state vector $\mathbf{x} = [x, y, \dot{x}, \dot{y}]^T$ with $[x, y]^T$ and $[\dot{x}, \dot{y}]^T$ being the position and velocity, respectively. Selecting the orbit with the position and velocity described in a two-dimensional coordinate allows clear visualization of the PDF.

The orbital dynamics under the influence of gravity only can be described by

$$\dot{\mathbf{x}} = \begin{bmatrix} \dot{x} \\ \dot{y} \\ -\mu x/r^3 \\ -\mu y/r^3 \end{bmatrix} \tag{7.8}$$

where $r = \sqrt{x^2 + y^2}$; μ is the gravitational constant.

The initial state distribution is assumed Gaussian with the initial mean and covariance given by

$$\hat{\mathbf{x}}_0 = [28000\text{km}, 0\text{km}, 0\,\text{km/s}, 4.1331436\,\text{km/s}]^T \tag{7.9a}$$

$$\mathbf{P}_0 = \text{diag}\left(\left[1km^2, 1km^2, 10^{-6}\,km^2/s^2, 10^{-6}\,km^2/s^2\right]\right) \tag{7.9b}$$

10,000 Monte Carlo samples are propagated through the nonlinear orbital dynamics (7.8) and are used to represent the true uncertainty distribution. The compact quadrature rule (CQR) given in Chapter 4.9 in the single Gaussian model is compared with the third-degree CR and the GHQ. The 3-point univariate GHQ and 4-point univariate GHQ are used to construct the multivariate GHQ with the 5th-degree and 7th-degree of accuracy, respectively. These rules are selected for comparison because they always generate positive weights and are numerically advantageous over the widely used UT. It is worth noting that the third-degree CR is the same as the UT when the UT parameter κ is set to zero although they are derived from two different mathematical principles. Moreover, the GHQ can achieve higher than the third-degree accuracy that is the best the UT can obtain.

To evaluate the propagation performance, we utilize the measure of the Kullback-Leibler (KL) divergence between the PDF generated from the Monte Carlo result and the one computed from the quadrature based methods. In particular, the PDF $\left(\mathbf{x};\hat{\mathbf{x}}_{MC},\mathbf{P}_{MC}\right)$ represents the single Gaussian PDF in which the mean $\hat{\mathbf{x}}_{MC}$ and covariance $\mathbf{P}_{MC}$ are computed from the Monte Carlo samples, $N\left(\mathbf{x};\hat{\mathbf{x}}_{QR},\mathbf{P}_{QR}\right)$ represents the single Gaussian PDF in which the mean $\hat{\mathbf{x}}_{QR}$ and covariance $\mathbf{P}_{QR}$ are computed by the quadrature rule-based methods. The KL distance between these two PDFs is defined by (Kullback 1997)

$$D_{KL}=\frac{1}{2}\left\{\ln\left(\frac{\left|\mathbf{P}_{QR}\right|}{\left|\mathbf{P}_{MC}\right|}\right)+\operatorname{tr}\left(\mathbf{P}_{QR}^{-1}\mathbf{P}_{MC}\right)+\left(\hat{\mathbf{x}}_{QR}-\hat{\mathbf{x}}_{MC}\right)^{T}\mathbf{P}_{QR}^{-1}\left(\hat{\mathbf{x}}_{QR}-\hat{\mathbf{x}}_{MC}\right)-n\right\} \quad (7.10)$$

where n = 4 is the dimension.

The performance of the different methods is presented in Fig. 7.1, which shows the KL divergence with respect to the number of orbit propagation periods. It can be seen in 10 orbit periods that the (5th-degree and 7th-degree) CQRs and the GHQs outperform the CR consistently in the single Gaussian model based orbital uncertainty propagation. In addition, the 5th-degree CQR and the 5th-degree GHQ behave very closely and are not distinguishable. The performance of both the CR and the GHQ rule degrades as the number of propagating orbits increases. The 7th-degree CQR, however, maintains satisfying results throughout the 10 orbit periods. The 7th-degree GHQ rule provides the close performance to the 7th-degree CQR but it degrades in the later portion of the propagation.

The computation complexity of these rules is directly related to the number of quadrature points, which is compared in Table 7.1.

As can be seen, the CQR achieves much better performance than the CR with a moderately increased number of points. The number of the 5th-degree CQR requires only 1/5 of the number of the 5th-degree GHQ points while generates nearly the same result as the 5th-degree GHQ. The number of the 7th-degree CQR points is only half of the number of the 5th-degree GHQ points but it exhibits much better performance than the 5th-degree GHQ. The number of the 7th-degree GHQ points is nearly six times of the number of the 7th-degree CQR points. However, its performance is not as good as the 7th-degree CQR.

Figure 7.2 shows the position and velocity PDF contours obtained from the 7th-degree CQR as well as the Monte Carlo samples in one period of the nominal orbit. As can be seen, although the 7th-degree CQR outperforms the CR

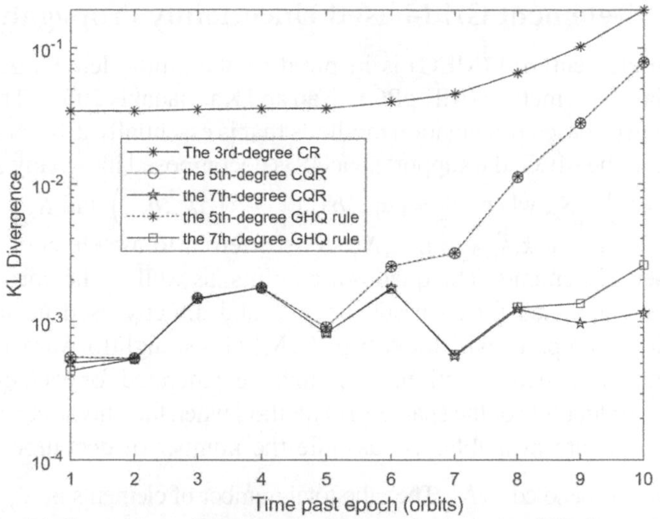

Fig. 7.1 Comparison of different quadrature rules in terms of KL divergence.

Table 7.1 Number of quadrature points of different numerical rules.

Quadrature rules	Number of points
CR	8
5th-degree CQR	16
7th-degree CQR	41
5th-degree GHQ Rule	81
7th-degree GHQ Rule	256

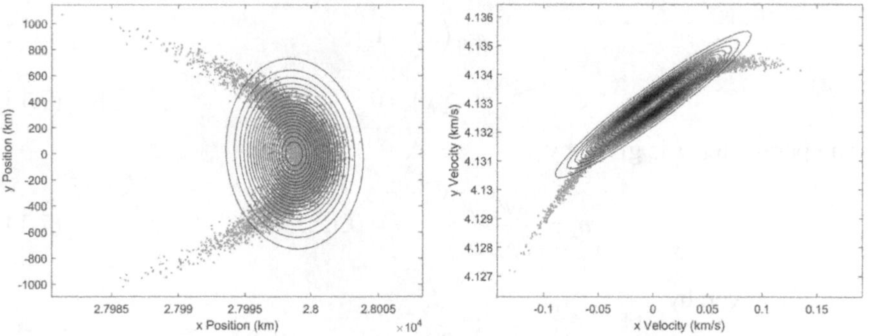

(a) position (a) velocity

Fig. 7.2 PDF contours with the 7th-degree CQR and the Monte Carlo samples.

and the GHQ rule, it still cannot represent the true PDF very well. It is because the single Gaussian model cannot adequately capture the non-Gaussianity when the Gaussian PDF propagates through the nonlinear orbital dynamics.

7.2 Multi-element Grid-based Uncertainty Propagation

The multi-element grid (MEG) is inspired by the multi-element generalized polynomial chaos method (ME-gPC) (Wan and Karniadakis 2006). The MEG is one of the grid-based propagation methods that is essentially different from the ME-gPC. In the MEG, the support space S is decomposed into many subspaces S_k, i.e., $S=\bigcup_{k=1}^{N_d} S_k$, where $S_k=\left(a_{k,1},b_{k,1}\right)\times\cdots\times\left(a_{k,n},b_{k,n}\right)$ and $S_{k_1}\bigcap S_{k_2}=\varnothing$, if $k_1\neq k_2$. Note that $k,k_1,k_2=1,\cdots,N_d$. S_k is referred to as 'element' and N_d is the number of elements. The quadrature points, as well as the corresponding weights, are obtained for each subspace S_k and directly used in uncertainty propagation. Compared with the ME-gPC, MEG is straightforward and easy to implement. After decomposition, the points are generated for each element by the tensor product rule or the sparse-grid method when the univariate quadrature points/weights are available. We assume the number of decompositions for dimension m is denoted as N_m. Then the total number of elements is $N_d=\prod_{m=1}^{n} N_m$, if the tensor product rule is used.

The univariate quadrature points are chosen as the roots of the orthogonal interpolation polynomials in each element. The procedure to obtain the orthogonal interpolation polynomials recursively for each element is introduced as follows.

The classical Stieltjes procedure and modified Chebyshev algorithm can be used to obtain the orthogonal polynomials. In this section, we introduce the Stieltjes procedure only.

The orthogonal polynomials $\psi(x)$ can be iteratively obtained by

$$\psi_{i+1}(x)=(x-a_i)\psi_i(x)-b_i\psi_{i-1}(x),\quad i=0,1,2,\cdots \tag{7.11}$$

$$\psi_0(x)=1 \tag{7.12}$$

$$\psi_{-1}(x)=0 \tag{7.13}$$

The parameter a_i is given by

$$a_i=\frac{\langle x\psi_i,\psi_i\rangle}{\langle \psi_i,\psi_i\rangle},\quad i=0,1,2,\cdots \tag{7.14}$$

and b_i is given by

$$b_i=\frac{\langle \psi_i,\psi_i\rangle}{\langle \psi_{i-1},\psi_{i-1}\rangle},\quad i=1,2,\cdots \tag{7.15}$$

with $b_0=\langle \psi_0,\psi_0\rangle$.

For example, if the ith dimension support space $(-\infty,\infty)$ is decomposed into two parts: $(-\infty,0)$ and $[0,\infty)$. For the first and the second part, the subspaces are

$(-\infty, 0)$ and $[0, \infty)$, respectively. The $\langle \psi_j(x), \psi_j(x) \rangle$ in Eqs. (7.14) and (7.15) for different subspaces can be obtained by $\langle \psi_j(x), \psi_j(x) \rangle = \int_{\Gamma^i} \psi_j(x)\psi_j(x) \rho^i(x)dx$ using different Γ^i, where Γ^i is the support space of the *i*th dimensional PDF $\rho^i(x)$.

In the next, the computation of points and weights is given. Denote q_j^i $(j = 1, \cdots, r)$ as the *i*th dimension points that are the roots of the *r*th order polynomial ψ_r^i, where the superscript '*i*' denotes the *i*th dimension and *r* is the number of the collocation points. By definition, the quadrature rule is described by

$$\int_{\Gamma^i} \eta^i(x) g(x) dx \cong \sum_{j=1}^{r} w_j^i g(q_j^i) \tag{7.16}$$

where $\eta^i(x) = \dfrac{\rho^i(x)}{\int_{\Gamma^i} \rho^i(x) dx}$. Given the collocation points, the interpolation polynomial can be constructed by

$$L^i(x) = \sum_{j=1}^{r} L_j^i(x) g(q_j^i) \tag{7.17}$$

where $L_j^i(x)$ is the Lagrangian basis polynomial of the *j*th point.
Now let $L^i(x)$ replace $g(x)$, we have

$$\begin{aligned} &\int_{\Gamma^i} \eta^i(x) g(x) dx \cong \sum_{j=1}^{r} w_j^i g(q_j^i) \\ &\Rightarrow \sum_{j=1}^{r} \int_{\Gamma^i} \eta^i(x) L_j^i(x) dx g(q_j^i) \cong \sum_{j=1}^{r} w_j^i g(q_j^i) \\ &\Rightarrow \int_{\Gamma^i} \eta^i(x) L_j^i(x) dx = w_j^i \end{aligned} \tag{7.18}$$

The multi-element grids $\mathbf{q}_j^k$ $(j = 1, \cdots, r_s)$ in each set S_k are obtained by the tensor product rule and one-dimensional quadrature points q_j^i $(j = 1, \cdots, r)$, with weights W_j^k $(j = 1, \cdots, r_s)$, where $r_s = r^n$ is the number of points used.

The point set of the multi-element grid for each element can also be represented as $\bigcup_{i=1}^{n} \bigcup_{j=1}^{r} \{q_j^i\}$ and the *k*th weight of the multi-dimensional collocation point can be obtained by

$$W_j^k = \prod_{i=1}^{n} w_j^i \tag{7.19}$$

where w_j^i can be obtained by Eq. (7.18).

The moment's value can be recovered from the points and weights of the multi-element grid method. The approximated mean can be obtained by

$$E[\mathbf{x}_t] = \sum_{k=1}^{N_d} \hat{\mathbf{x}}_t^k \Pr(\mathbf{x}_t \in S_k) \tag{7.20}$$

where $\hat{\mathbf{x}}_t^k$ is the mean value corresponding to the kth element at time t and $\Pr(\mathbf{x}_t \in S_k)$ is the probability when $\mathbf{x}_t \in S_k$.

The approximated variance can be obtained by

$$\sigma_i^2 = \sum_{k=1}^{N_d} \left[\sigma_{k,i}^2 + \left(\hat{\mathbf{x}}_{t,i}^k - \left(E[\mathbf{x}_t] \right)_i \right)^2 \right] \Pr(\mathbf{x}_t \in S_k) \tag{7.21}$$

where $\sigma_{k,i}^2$ is the variance corresponding to the kth element and the subscript 'i' denotes the ith component of the state.

For convenience, the procedure of using the MEG method in the uncertainty propagation problem is summarized in Table 7.2.

It should be pointed out that the Gauss-Hermite quadrature (GHQ) points and weights are identical to the points and weights of MEG with respect to the Gaussian distribution when only one element is used with support space $S = \times_{i=1}^{n} (-\infty, \infty)$ and the univariate quadrature points and weights are extended to multi-dimensional by the tensor product rule. In addition, the sparse-grid points and weights can be obtained when only one element is used and the univariate quadrature points and weights are extended by the sparse-grid method.

When the number of elements is large, there is no need to use high accuracy level quadrature rules for each element. In fact, the level-1 or level-2 sparse-grid quadrature can be used. If level-1 sparse-grid quadrature is used, only 1 point is used for each element. In this case, the final grid set is similar to the points of the Monte Carlo method. Both of them use some points with positive weights in the uncertainty propagation. However, these two methods are essentially different. For MEG, the points may contain different weights while the weights of the points of the Monte Carlo method are the same. The MEG is also different from the GHQ in this case. The location of the point of MEG depends on the selection of the elements that is different from the GHQ. In addition, the point of MEG can be obtained from an arbitrary distribution directly while GHQ cannot be used for non-Gaussian distribution.

Although MEG works for arbitrary multivariate PDF, we only show the MEG for univariate Gaussian distribution with different domain decomposition strategy. Figure 7.3 shows the MEG for univariate Gaussian distribution in the domain [–5,0] and (0,5]. Different markers represent the point belonging to different domains. Similarly, Fig. 7.4 shows the MEG for univariate Gaussian distribution with the three domain intervals, [–5, –2], [–2,2], and [2,5]. Note

Table 7.2 The multi-element grid for uncertainty propagation.

Initialization: S
Step 1: Divide the support space S into subspaces S_k.
Step 2: Generate multi-element points $\mathbf{q}_j^k$ $(j = 1, \cdots, r_s)$ and weights W_j^k $(j = 1, \cdots, r_s)$ for each subspace S_k
Step 3: Calculate the moments of each subspace S_k.
Step 4: Recover mean and covariance for each dimension by Eqs. (7.20) and (7.21).

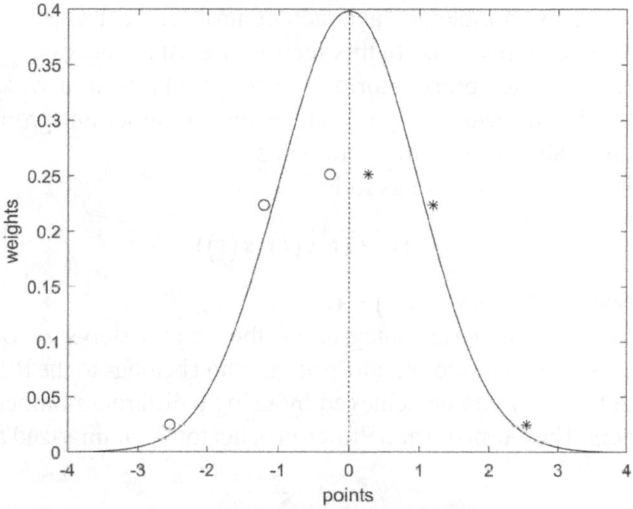

Fig. 7.3 MEG for univariate Gaussian distribution with two domain decompositions.

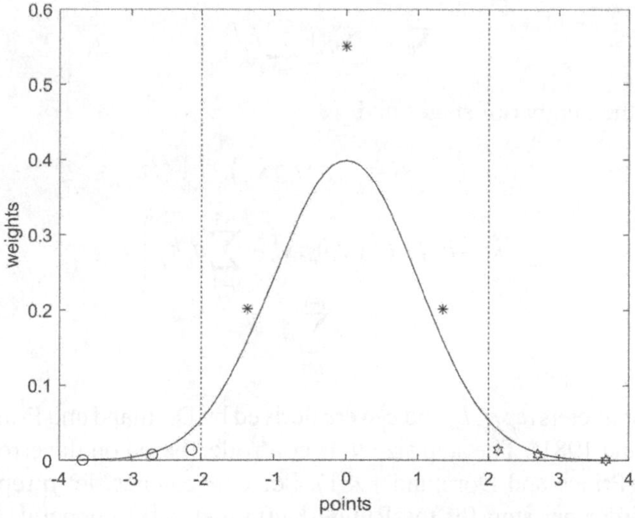

Fig. 7.4 MEG for univariate Gaussian distribution with three domain decompositions.

that MEG depends on the choice of domain decomposition. The main advantage of the MEG is that it is very flexible to use.

7.3 Uncertainty Propagator

In order to predict the uncertainty of the dynamic system accurately, the high-precision numerical integrator for solving the ordinary differential equation is required. The reasons why different propagators are designed and used are many. Different propagators have different advantages and disadvantages in terms of accuracy, efficiency, complexity, and step control. The complete comparison is out of the scope of this book. In this section, we list the necessary equations of Dormand and Prince propagator due to its popularity in a wide range of problems. In addition, we briefly introduce several important propagators in the orbital uncertainty propagation problems.
We consider a dynamic system as follows

$$\ddot{\mathbf{x}}(t) = \boldsymbol{f}\left(t, \mathbf{x}(t), \dot{\mathbf{x}}(t)\right) \tag{7.22}$$

with initial values $\mathbf{x}(t_0) = \mathbf{x}_0$, $\dot{\mathbf{x}}(t_0) = \dot{\mathbf{x}}_0$.

The widely used numerical integrator is the variable step-size fourth-order Runge-Kutta method. The Dormand-Prince method belongs to the Runge-Kutta family. High accuracy can be achieved by using a different number of stages and parameters. The Dormand and Prince propagator is summarized as follows.

$$\hat{\mathbf{x}}_{n+1} = \hat{\mathbf{x}}_n + \sum_{i=1}^{s} \hat{b}_i k_i \tag{7.23}$$

$$\overline{\mathbf{x}}_{n+1} = \hat{\mathbf{x}}_n + \sum_{i=1}^{s} b_i k_i \tag{7.24}$$

where s is the number of stages and

$$k_1 = h_n \boldsymbol{f}\left(t_n, \hat{\mathbf{x}}_n\right) \tag{7.25}$$

$$k_i = h_n \boldsymbol{f}\left(t_n + c_i h_n, \hat{\mathbf{x}}_n + \sum_{j=1}^{i-1} a_{ij} k_j\right) \tag{7.26}$$

$$c_i = \sum_{j=1}^{i-1} a_{ij} \tag{7.27}$$

The parameters a_{ij}, b_i, $\hat{b}_i$, and c_i were derived by Dormand and Prince (Prince and Dormand 1981). The step size h_n is controlled based on the error estimate $\overline{\mathbf{x}}_{n+1} - \hat{\mathbf{x}}_{n+1}$ (Prince and Dormand 1981). For convenience, let p represent the accuracy order per step for the Runge-Kutta method. In general, the higher order formula requires more stages than the lower order formula (Dormand and Prince 1980).

The Runge-Kutta method can be explicit or implicit. The explicit Runge-Kutta method uses results at the previous time steps to directly deduce the current value. It is generally unsuitable for the stiff equations and less stable than the implicit Runge-Kutta method. In (Aristoff et al. 2014), the authors introduce the variable-step Gauss-Legendre implicit-Runge-Kutta (VGL-IRK) method. It is shown that VGL-IRK can be much faster than the Dormand-Prince 8(7) algorithm and the modified Chebyshev-Picard iteration.

Besides the above-mentioned propagators, there are many others, such as Runge-Kutta Nystrom method (Dormand et al. 1987), Gauss-Jackon propagator, and various methods proposed by Shampine and Gordon (Shampine and Gordon 1975). Roughly speaking, the fourth order Runge-Kutta method can be used if the propagation time is short. It is simple and fast. To achieve a better performance with a small increase of the computational time, the VGL-IRK can be used. If the accuracy is the only concern, the Runge-Kutta Nystrom method is a strong candidate (Elgohary et al. 2015).

Example 7.2: Dormand and Prince predictor with order $p = 7$ is compared to the same predictor with the order $p = 4$. In this example, the initial condition of the space object is given in Eq. (7.9) and the propagation time is one orbit period. The propagation difference of 10000 random points for different predictors is shown in Fig. 7.5. It can be seen that the propagator with the different order has more than 0.2 km absolute error. In the figure, 'x' and 'y' denote the first and

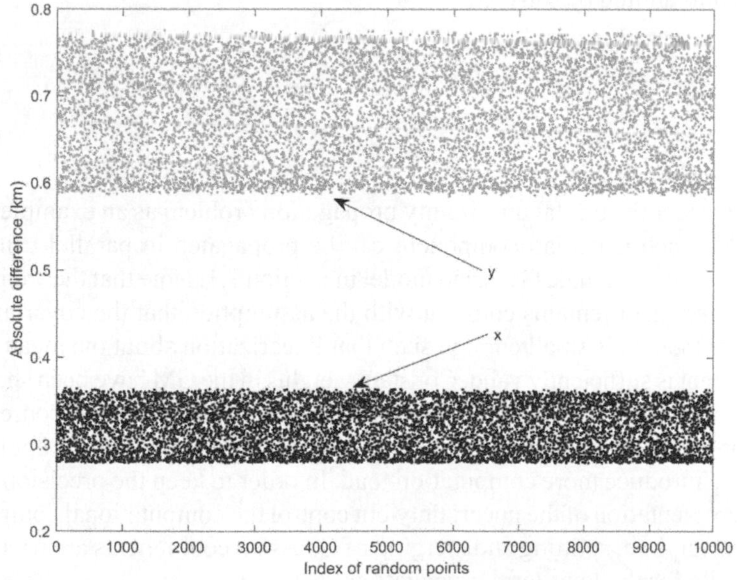

Fig. 7.5 The absolute difference of Dormand and Prince predictor with order 4 and 7.

the second value of the system state $\mathbf{x}$, respectively. Hence, it is recommended to use high order predictors if more accurate results are desired.

7.4 Gaussian Mixture based Uncertainty Propagation

In this section, the Gaussian mixture (GM) model is used to better represent the uncertainty after propagation. The initial state is assumed to be random with a Gaussian PDF q_0, which can be represented by

$$q_0 = N\left(\mathbf{x}_0; \hat{\mathbf{x}}_0, \mathbf{P}_0\right) = \frac{1}{\left|2\pi\mathbf{P}_0\right|^{1/2}} \exp\left\{-\frac{1}{2}\left(\mathbf{x}_0 - \hat{\mathbf{x}}_0\right)^T \mathbf{P}_0^{-1}\left(\mathbf{x}_0 - \hat{\mathbf{x}}_0\right)\right\} \quad (7.28)$$

where $N\left(\mathbf{x}_0; \hat{\mathbf{x}}_0, \mathbf{P}_0\right)$ denotes the normal distribution with $\hat{\mathbf{x}}_0$ and $\mathbf{P}_0$ being the initial mean of $\mathbf{x}(t)$ and the initial covariance, respectively. '$|\cdot|$' denotes the determinant operation. The uncertainty propagation problem is to predict the evolution of the PDF $q(\mathbf{x}(t),t)$ of the state $\mathbf{x}(t)$ at any time t.

A single Gaussian PDF is generally not sufficient to characterize the uncertainty after long-term propagation. One of the common solutions is to use the GM model. A GM representation of a PDF using N_l components is given by

$$q = \sum_{l=1}^{N_l} \alpha_l N\left(\mathbf{x}; \hat{\mathbf{x}}_l, \mathbf{P}_l\right) \quad (7.29)$$

where $\hat{\mathbf{x}}_l$ and $\mathbf{P}_l$ are the mean and covariance of the lth Gaussian component, respectively. α_l is the weight of the lth component. Note that the following constraints should be satisfied.

$$\alpha_l > 0, \quad 1 \le l \le N_l \quad (7.30)$$

and

$$\sum_{l=1}^{N_l} \alpha_l = 1 \quad (7.31)$$

Consider the orbital uncertainty propagation problem as an example using the GM. Each Gaussian component can be propagated in parallel using the algorithm for the single Gaussian model in Section 7.1. Note that the weight for each component remains constant with the assumption that the covariance of each component is small enough such that linearization about the mean of the component is sufficiently valid. Constant weights in the GM have been shown to be good enough for the orbital propagation problem (Horwood and Poore 2011, DeMars et al. 2013). Adaptive weights (Terejanu et al. 2008) may be applicable but will introduce more computation load. In order to keep the precision of the GM representation of the uncertainty but control the computational complexity in the meantime, splitting and merging of Gaussian components are necessary, especially for the long-term propagation.

When the nonlinearity causes more significant non-Gaussianity, a component of the GM can be split into smaller Gaussian components. A finer GM will make the uncertainty propagation more precise because Gaussians with smaller covariances remain more Gaussian than those with larger covariances under the nonlinear mapping (Horwood and Poore 2011, Horwood et al. 2011). There are many Gaussian component splitting algorithms (Horwood et al. 2011, Horwood and Poore 2011, DeMars et al. 2013) in the literature. We adopt the method in (DeMars et al. 2013). For the completeness of presentation, the splitting method is summarized in the following two steps.

(1) Split a univariate Gaussian PDF by minimizing an L_2 distance (divergence) between the Gaussian PDF and a GM. The parameters of the GM determined from this optimization process are the mean $\tilde{m}_i$, covariance $\tilde{\sigma}_i$, and weight $\tilde{\alpha}_i$ for each Gaussian component in the GM. Table 7.3 gives a splitting library (DeMars et al. 2013) with three Gaussian components.
(2) Split a multivariate Gaussian PDF. The idea is to apply a univariate splitting library in a specified direction, i.e., the principal (eigenvector) direction of the covariance matrix. Then, in the eigenvector coordinate system, the multivariate Gaussian PDF can be represented by a product of the univariate Gaussian PDFs. Suppose a Gaussian component $N\left(\mathbf{x};\hat{\mathbf{x}}_{l_a},\mathbf{P}_{l_a}\right)$ with weight α_{la} is split into smaller Gaussian components, i.e.,

$$\alpha_{l_a} N\left(\mathbf{x};\hat{\mathbf{x}}_{l_a},\mathbf{P}_{l_a}\right) \approx \sum_{i=1}^{N_{sp}} \bar{\alpha}_i N\left(\mathbf{x};\bar{\mathbf{x}}_i,\bar{\mathbf{P}}_i\right) \tag{7.32}$$

where $\bar{\mathbf{x}}_i$ and $\bar{\mathbf{P}}_i$ are the mean and covariance of the split Gaussian components, respectively; N_{sp} is the number of split components; $\bar{\alpha}_1$, $\bar{\alpha}_2$, and $\bar{\alpha}_3$ are the weights of each component.
The parameters are given by

$$\bar{\mathbf{x}}_i = \hat{\mathbf{x}}_{l_a} + \sqrt{\lambda_j}\,\tilde{m}_i \mathbf{v}_j \tag{7.33}$$

$$\bar{\alpha}_i = \alpha_{l_a} \cdot \tilde{\alpha}_i \tag{7.34}$$

$$\bar{\mathbf{P}}_i = \mathbf{V}\mathbf{\Lambda}_i\mathbf{V}^T \tag{7.35}$$

Table 7.3 Parameters of three Gaussian components after splitting a univariate Gaussian PDF.

	$i=1$	$i=2$	$i=3$
$\tilde{\alpha}_i$	0.2252246249	0.5495507502	0.2252246249
$\tilde{m}_i$	−1.0575154615	0	1.0575154615
$\tilde{\sigma}_i$	0.6715662887	0.6715662887	0.6715662887

$$\boldsymbol{\Lambda}_i = \boldsymbol{\Lambda} \odot \boldsymbol{\Xi} \tag{7.36}$$

where $\boldsymbol{\Xi} = \text{diag}\left(\left[1,\cdots,\tilde{\sigma}_j,\cdots,1\right]\right)$, $\boldsymbol{\Lambda} = \text{diag}\left(\left[\lambda_1,\cdots,\lambda_j,\cdots,\lambda_n\right]\right)$ with $\lambda_j, 1 \le j \le n$ being the *j*th largest eigenvalue of the covariance matrix $\mathbf{P}_{la}$. '$\odot$' denotes the element-wise product. $\mathbf{v}_j$ is the *j*th eigenvector of $\mathbf{P}_{la}$ corresponding to λ_j, i.e., the *j*th column of the eigenvector matrix $\mathbf{V}$. '$\mathbf{V}$' and '$\boldsymbol{\Lambda}$' can be obtained by factorization of the covariance, i.e., $\mathbf{P}_{la} = \mathbf{V}\boldsymbol{\Lambda}\mathbf{V}^T$ using Cholesky decomposition or singular value decomposition. $\tilde{\alpha}_i, \tilde{m}_i$, and $\tilde{\sigma}_j = \tilde{\sigma}_i$ are given in Table 7.3 (DeMars et al. 2013).

Remark 7.2: We split the Gaussian component along the eigen-direction with the largest eigenvalue of the covariance matrix, which represents the direction of the largest uncertainty. One Gaussian component is split into three finer Gaussian components. If more than three components are required, one can follow the same procedure described in (DeMars et al. 2013).

The split can be performed on the Gaussian component with large non-Gaussianity. Many measures can be used to assess the non-Gaussianity (DeMars et al. 2013, Jia and Xin 2017). For example, the third moment can be utilized as the measure, which is defined as

$$\mathbf{M}_a = \sum_{i=1}^{n} \left| E\left[\left(x_i - \hat{x}_i\right)^3\right] \right| \tag{7.37}$$

where x_i and $\hat{x}_i$ are the *i*th component of the state vector $\mathbf{x}$ and its mean $\hat{\mathbf{x}}$, respectively. n is the dimension of $\mathbf{x}$, and '$|\cdot|$' denotes the absolute value. A tolerance or threshold ε_s is set to assess the non-Gaussianity. If $\mathbf{M}_a \ge \varepsilon_s$, this Gaussian component will be split into finer Gaussian components. Otherwise, it is unchanged. Note that ε_s is an adjustable parameter and is problem dependent.

When the GM is propagated through the nonlinear dynamics with further splitting, the number of Gaussian components may grow quickly and the computation complexity increases as a result. This issue can be resolved by GM reduction methods (Runnalls 2007, Williams 2003), such as pruning Gaussian components with negligible weights, joining near Gaussian components, and regeneration of GM via the Kullback-Leibler approach. As one of the strategies, near-Gaussian components are joined to reduce the number of Gaussian components. Assume that there are two Gaussian components $N\left(\mathbf{x};\bar{\mathbf{x}}_1,\bar{\mathbf{P}}_1\right)$ and $N\left(\mathbf{x};\bar{\mathbf{x}}_2,\bar{\mathbf{P}}_2\right)$ with weights α_1 and α_2, respectively. The Gaussian component after joining operation is denoted by $N\left(\mathbf{x};\bar{\mathbf{x}}_0,\bar{\mathbf{P}}_0\right)$ with weight α_0, where the parameters can be obtained by

$$\alpha_0 = \alpha_1 + \alpha_2 \tag{7.38}$$

$$\overline{\mathbf{x}}_0 = \frac{1}{\alpha_0}(\alpha_1 \overline{\mathbf{x}}_1 + \alpha_2 \overline{\mathbf{x}}_2) \tag{7.39}$$

$$\overline{\mathbf{P}}_0 = \frac{1}{\alpha_0}\left(\alpha_1 \overline{\mathbf{P}}_1 + \alpha_2 \overline{\mathbf{P}}_2 + \frac{\alpha_1 \alpha_2}{\alpha_0}(\overline{\mathbf{x}}_1 - \overline{\mathbf{x}}_2)(\overline{\mathbf{x}}_1 - \overline{\mathbf{x}}_2)^T\right) \tag{7.40}$$

The GM components with the smallest distance are merged and the distance is calculated by the Mahalanobis distance d_{12} (Williams 2003) defined by

$$d_{12}^2 = \frac{\alpha_1 \alpha_2}{\alpha_0}(\overline{\mathbf{x}}_1 - \overline{\mathbf{x}}_2)^T \mathbf{P}_d^{-1}(\overline{\mathbf{x}}_1 - \overline{\mathbf{x}}_2) \tag{7.41}$$

and $\mathbf{P}_d$ is the combined covariance of N_l Gaussian components (assuming the lth component with mean $\overline{\mathbf{x}}_l$ and covariance $\overline{\mathbf{P}}_l$) at time k with $\mathbf{P}_d = \sum_l \alpha_l \left[\overline{\mathbf{P}}_l + (\overline{\mathbf{x}}_l - \tilde{\mathbf{x}})(\overline{\mathbf{x}}_l - \tilde{\mathbf{x}})^T\right]$ and $\tilde{\mathbf{x}} = \sum_l \alpha_l \overline{\mathbf{x}}_l$.

To balance the propagation accuracy and computational complexity, a simple criterion using a preset tolerance or threshold ε_m is set to perform the GM reduction. If $d_{12}^2 \le \varepsilon_m$, the GM reduction is performed.

Otherwise, the GM is unchanged. Note that ε_m is an adjustable parameter and is problem dependent.

The algorithm is summarized in Table 7.4. Note that the compact quadrature rule (CQR) given in Chapter 4.9 is used because positive weights can be ensured.

Example 7.3: GM Model based Orbital Uncertainty Propagation.
The initial conditions are the same as in Example 7.1. The propagation time is five orbit periods, which is 325823.9853 seconds. 10,000 Monte Carlo samples are used to represent the true uncertainty distribution. The initial uncertainty distribution is represented by a GM with five Gaussian distributions. Figure 7.6 and Fig. 7.7 show the position and velocity marginal pdf contours using the GM model with the 7th-degree CQR, as well as Monte Carlo samples in gray dots to represent the true distribution. It can be seen that the results based on the CQR is much better than those in Section 7.1, Example 7.1, Fig. 2, but still do not match the real distribution very well.

The performance of the fixed GM model will not be improved if we increase the number of Gaussian components. It is because with the fixed GM it is hard to capture the dynamic uncertainty distribution when it is propagated over a long time. Once the GM structure is fixed, the capability of this GM is constrained since the number of Gaussian components and the weights are fixed, and the predicted mean and covariance for each Gaussian component are deterministic during propagation. As a result, with these limited deterministic parameters, it is hard to capture a growing uncertainty as the time evolves. Hence, it is necessary to perform the GM splitting and reduction adaptively using the proposed algorithm in Table 7.4 if more precise propagation is desired.

Table 7.4. Orbital uncertainty propagation based on GM and the CQR.

Initialization: $\mathbf{q}_{k-1} = \sum_{l=1}^{N_l} \alpha_l N\left(\mathbf{x}_{k-1}; \hat{\mathbf{x}}_{l,k-1}, \mathbf{P}_{l,k-1}\right)$ with $k = 1$

Given the propagation time T, divide T into N_T propagation intervals denoted by $[t_{k-1}\ t_k]$, $k = 1, \cdots, N_T$

Step 1: Generate the compact quadrature points and weights from each Gaussian component using $\hat{\mathbf{x}}_{l,k-1}$ and $\mathbf{P}_{l,k-1}$.

Step 2: For each Gaussian component

Propagate the compact quadrature points until t_k

Compute the mean $\hat{\mathbf{x}}_{l,k}$ and the third moment $\mathbf{M}_a$.

If $\mathbf{M}_a \geq \varepsilon_s$ for the Gaussian component denoted by $N\left(\mathbf{x}; \hat{\mathbf{x}}_{l_a,k}, \mathbf{P}_{l_a,k}\right)$ with a weight α_{l_a}

Split this component into finer Gaussian components $\alpha_{l_a} N\left(\mathbf{x}; \hat{\mathbf{x}}_{l_a,k}, \mathbf{P}_{l_a,k}\right) \approx \sum_{i=1}^{N_{sp}} \bar{\alpha}_i N\left(\mathbf{x}; \bar{\mathbf{x}}_i, \bar{\mathbf{P}}_i\right)$

Generate new compact quadrature points for the new Gaussian components using $\bar{\mathbf{x}}_i, \mathbf{P}_i$.

Update the number of Gaussian components: $N_l = N_l + N_{sp} - 1$

End If

End For

Step 3: Sort the Gaussian components according to the weights in an ascending order.

For each Gaussian component i

Compare to other Gaussian components j that have not been paired to calculate the Mahalanobis distance d_{12}^2 using (7.41).

For each Gaussian component pair (i,j),

If $d_{12}^2 \leq \varepsilon_m$

Merge these two Gaussian components using (7.38)–(7.40).

Generate new compact quadrature points for the new merged Gaussian component using $\bar{\mathbf{x}}_0$, $\mathbf{P}_0$ (7.39), (7.40).

Update the number of Gaussian components: $N_l = N_l - 1$.

End IF

End For

End For

If $t_k = T$, go to Step 4

End IF

$k = k + 1$

Return to Step 2

Step 4: Compute the uncertainty distribution $\mathbf{q}_T = \sum_{l=1}^{N_{tot}} \alpha_{l,T} N\left(\mathbf{x}_T; \hat{\mathbf{x}}_{l,T}, \mathbf{P}_{l,T}\right)$ at the end of the propagation time T where $\hat{\mathbf{x}}_{l,T}$ and $\mathbf{P}_{l,T}$ are obtained using quadrature rules, N_{tot} is the total number of Gaussian components after splitting and reduction.

In the following, we name the propagation Algorithm in Table 7.4 adaptive GM based on the CQR.

Figure 7.8 and Fig. 7.9 show results using the adaptive GM based on the 7th-degree CQR. It can be clearly seen that it achieves significantly better performance than the previous fixed GM model. The number of Gaussian components in different propagation intervals is shown in Fig. 7.10 and it varies with the increase of the propagation time due to the splitting and reduction operations.

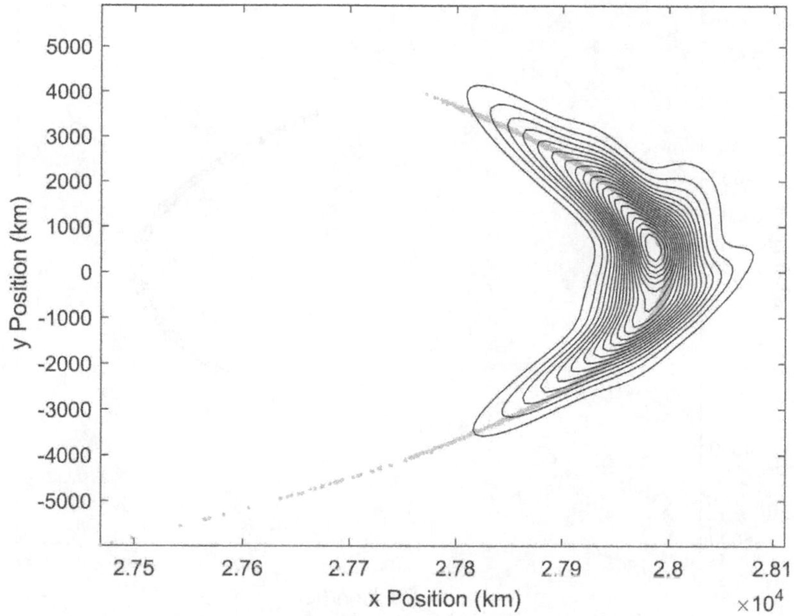

Fig. 7.6 PDF contours of the GM with the 7th-degree CQR and Monte Carlo samples (position).

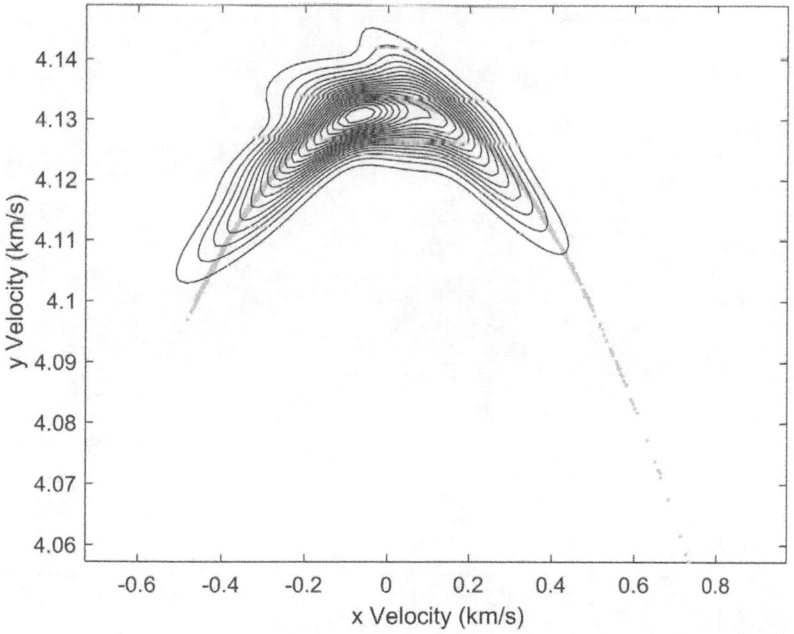

Fig. 7.7 PDF contours of the GM with the 7th-degree CQR and Monte Carlo samples (velocity).

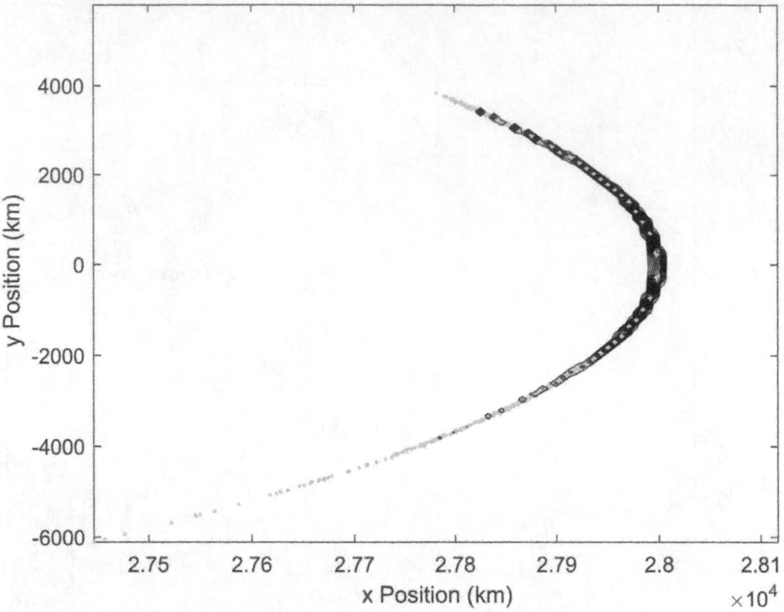

Fig. 7.8 PDF contours of the adaptive GM with the 7th-degree CQR and Monte Carlo samples (position).

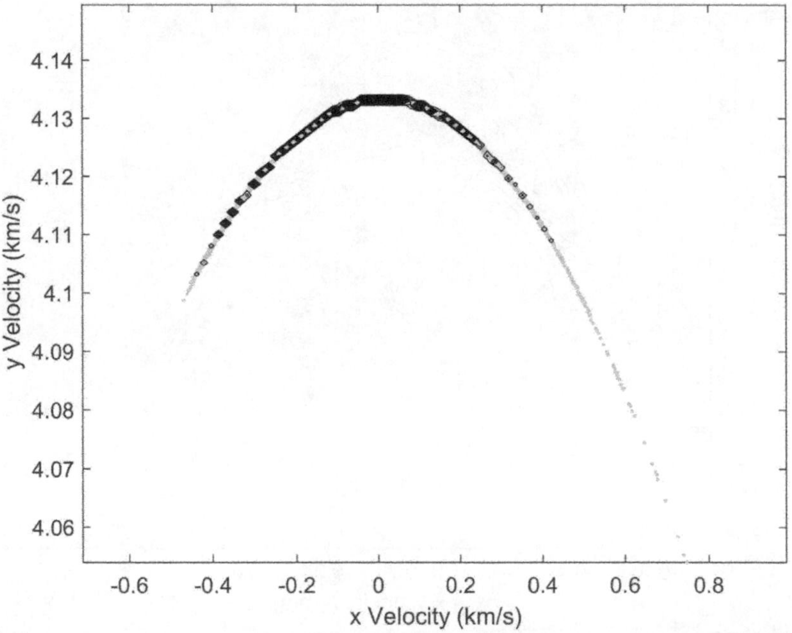

Fig. 7.9 PDF contours of the adaptive GM with the 7th-degree CQR and Monte Carlo samples (velocity).

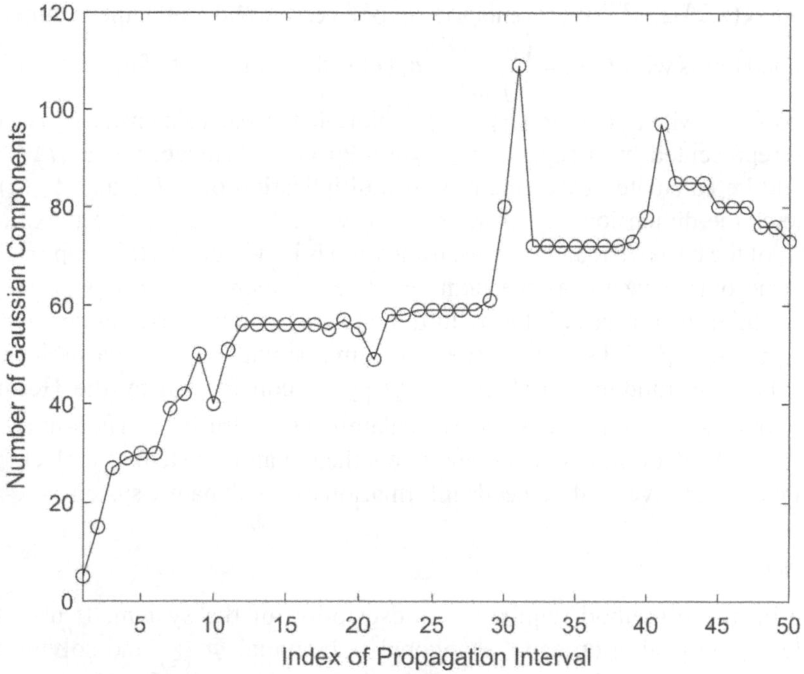

Fig. 7.10 Number of Gaussian components in different propagation intervals.

7.5 Stochastic Expansion based Uncertainty Propagation

Two stochastic expansion based uncertainty propagation methods, the stochastic Galerkin and the stochastic collocation (SC), are introduced. Both methods can use the grid-based techniques to compute the corresponding coefficients of the stochastic expansion.

7.5.1 Generalized Polynomial Chaos

Different from the Gaussian quadrature based uncertainty propagation method, the polynomial chaos (PC) has the ability to represent the high order moments information of the PDF. The PC was first introduced by Wiener using the Hermite polynomials and the normal random variables. The generalized polynomial chaos (gPC) generalized the result to various continuous and discrete distributions using the Askey-scheme (Xiu and Karniadakis 2002). For gPC, the input and the output of the dynamic system are modeled as random variables. The evolution of the uncertainty via the dynamic system can be represented as

$$\hat{\mathbf{x}}(t,\xi) \approx \sum_{j=0}^{P} \mathbf{c}_j(t)\boldsymbol{\Psi}_j(\xi) \tag{7.42}$$

where $\hat{\mathbf{x}}(t,\xi)\in\mathbb{R}^{m\times 1}$ represents the output vector. P denotes the number of retained terms with $P+1=\dfrac{(p+n)!}{p!n!}$. $c_j(t)$ is the coefficient of the polynomial chaos $\boldsymbol{\Psi}_j(\xi)$ with $\boldsymbol{c}_j(t),\boldsymbol{\Psi}_j(\xi)\in\mathbb{R}^{m\times 1}$. Note that for each dimension of $\hat{\mathbf{x}}(t,\xi)$, it is represented by a separated polynomial chaos. Hence, the $\boldsymbol{c}_j(t)\boldsymbol{\Psi}_j(\xi)$ should be calculated by element-wise multiplication of $c_j(t)$ and $\boldsymbol{\Psi}_j(\xi)$. n denotes the dimensionality of the random variable ξ and p is the maximum order of the basis. In real problems, truncation is inevitable. $\hat{\mathbf{x}}(t,\xi)$ approaches the true output vector as the number of terms increases. $\boldsymbol{\Psi}_j(\xi)$ is the jth polynomial chaos basis. ξ is the multi-dimensional random variables denoted by $\xi=[\xi_1,\cdots,\xi_n]^T$. $\xi_i, 1\le i\le n$ is the one-dimensional random variable. For the Gaussian random variables, the $\boldsymbol{\Psi}_j(\xi)$ is constructed by the Hermite polynomials. There are two ways to calculate $c_j(t)$: intrusive and non-intrusive methods. The non-intrusive method treats the dynamic system as a black box while the intrusive method needs information of the dynamic system.

Stochastic Galerkin

The intrusive method requires the description of the system. It uses the Galerkin projection on each orthogonal polynomial $\boldsymbol{\Psi}_j(\xi)$ and solving the resulting deterministic ordinary differential equations (ODEs) to obtain the coefficients $\boldsymbol{c}_j(t)$. Assume the dynamic equation is given in Eq. (7.1), and it can be rewritten as

$$\sum_{j=0}^{P}\dot{\boldsymbol{c}}_j(t)\boldsymbol{\Psi}_j(\xi)=\boldsymbol{f}\left(t,\sum_{j=0}^{P}\boldsymbol{c}_j(t)\boldsymbol{\Psi}_j(\xi)\right) \tag{7.43}$$

Using the Galerkin projection and orthogonal properties of $\boldsymbol{\Psi}_j(\xi)$, we have

$$\dot{c}_{i,k}(t)\left\langle\boldsymbol{\Psi}_k(\xi),\boldsymbol{\Psi}_k(\xi)\right\rangle=\left\langle\boldsymbol{f}\left(t,\sum_{j=0}^{P}c_{i,j}(t)\boldsymbol{\Psi}_j(\xi)\right),\boldsymbol{\Psi}_k(\xi)\right\rangle, \tag{7.44}$$

$$i=1,\cdots,n;\quad k=0,1,\cdots,P$$

where $\boldsymbol{c}_{i,k}(t)$ denotes the coefficients corresponding to the ith component of $\boldsymbol{c}_k(t)$. $\langle\cdot,\cdot\rangle$ denotes the inner product with respect to the PDF of ξ. Note that the inner product in Eq. (7.44) can be calculated using the grid-based numerical rules.

Stochastic Collocation

The coefficients of the gPC can also be obtained by (Xiu 2007)

$$c_j(t) = \sum_{m=1}^{N_p} \mathbf{x}(t, \gamma_m) \psi_j(\gamma_m) W_m, \quad j = 0, \cdots, P \tag{7.45}$$

where $\{\gamma_m, W_m\}_{m=1}^{N_p}$ denotes the set of points and weights. For the Gaussian distribution, they can be easily computed via the Gaussian quadrature introduced in Chapter 4. The $\mathbf{x}(t, \gamma_m)\psi_j(\gamma_m)$ should be calculated by element-wise multiplication of $\mathbf{x}(t, \gamma_m)$ and $\psi_j(\gamma_m)$. The multi-dimensional points γ_m are typically obtained from univariate quadrature points via the tensor product rule. The weight W_m is obtained from the univariate quadrature weights by the product rule. To reduce the number of points required, the sparse-grid or the anisotropic sparse grid rules in Chapter 4 can be used. Note that it is not necessary for the random variable to be Gaussian, and it can have an arbitrary PDF.

There are many toolboxes available to implement gPC, such as UQLab (Marelli and Sudret 2015) and Dakota (Wojtkiewicz et al. 2001).

***Example* 7.4:** Assume the initial uncertainty (variable x) follows a standard normal distribution. We want to find the uncertainty of the output $f(x)$ with

$$f(x) = \frac{x}{2} + 25\frac{x}{1+x^2} + 8\cos(1.2). \tag{7.46}$$

The uncertainty quantification results using gPC and the Monte Carlo method are compared. For gPC with order p, we use $p + 1$ Gauss-Hermite points to calculate the coefficients. The samples via the Monte Carlo method and gPC with $p = 15$ are shown in Fig. 7.11.

The root mean square error (RMSE) is used to evaluate the difference between the results of gPC and Monte Carlo. It is defined as follows.

$$RMSE = \sqrt{\frac{\sum_{j=1}^{N_{mc}} \left(f(x_j)^{mc} - f(x_j)^{gPC} \right)^2}{N_{mc}}} \tag{7.47}$$

where $N_{mc} = 1000$. The RMSEs for different p are shown in Fig. 7.12. It can be seen that the RMSE decreases when the accuracy order of the gPC, p, increases. It means with more terms in Eq. (7.42), the uncertainty can be more accurately described.

***Example* 7.5:** The orbital uncertainty propagation problem is used again to show the performance of gPC. For convenience, we use the software 'DAKOTA' (Wojtkiewicz et al. 2001). The sparse grid is used to calculate the coefficients of the polynomial expansion. Three different levels (level 1, 2, and 3) of the sparse-grid are used. They use 9, 57, and 233 points, respectively. To evaluate the performance of gPC, we compare the propagated values of random points

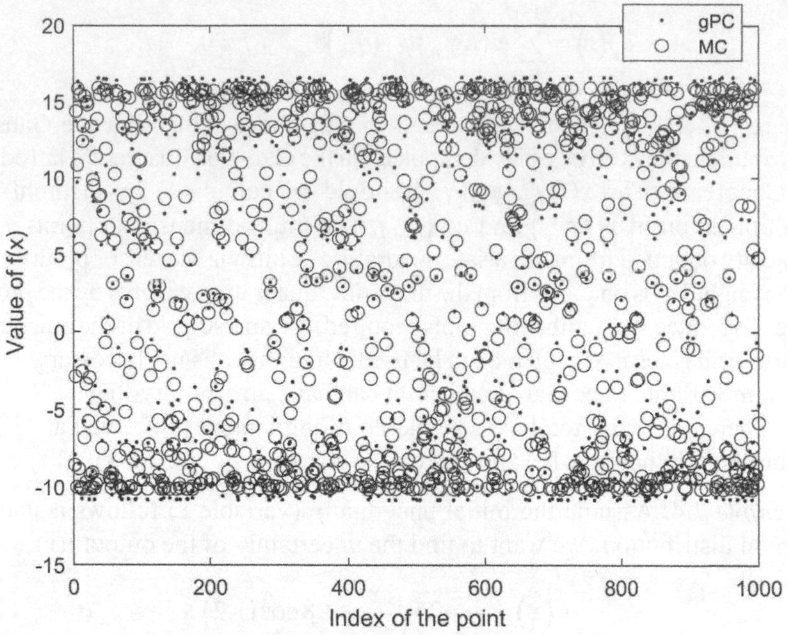

Fig. 7.11 Samples of Monte Carlo and gPC ($p = 15$).

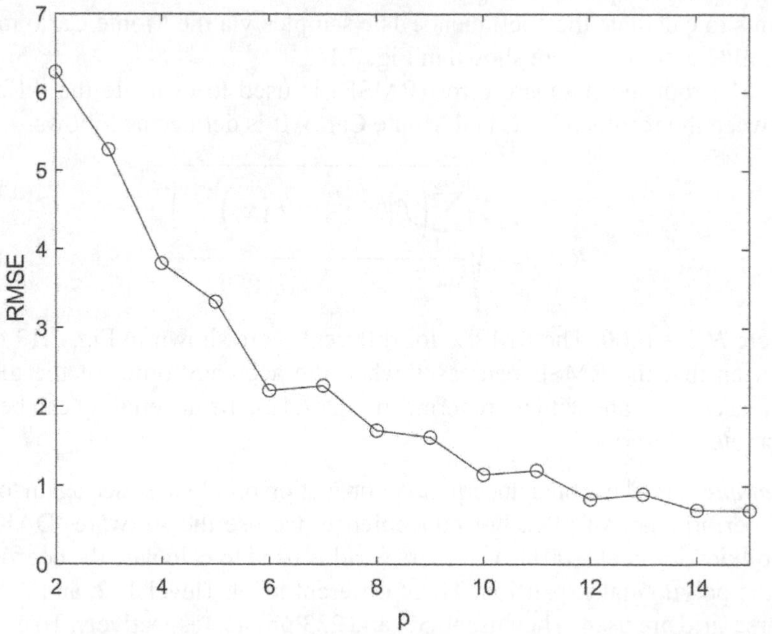

Fig. 7.12 RMSEs versus p.

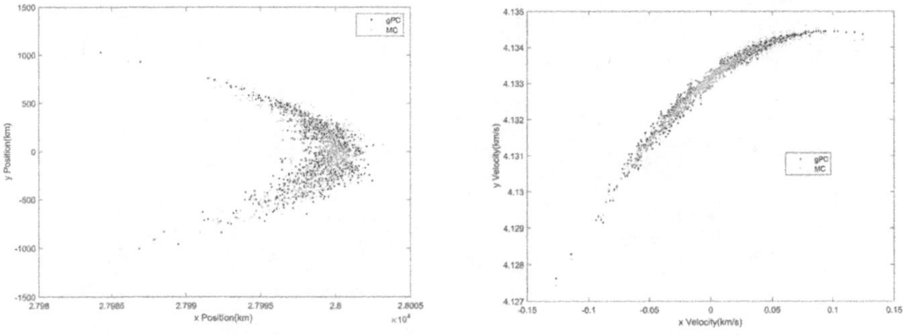

(a) Position (b) Velocity

Fig. 7.13 Samples obtained via Monte Carlo and gPC.

by the Monte Carlo (MC) method and values calculated via gPC using the same seed of random points. The propagated points using MC and points calculated via gPC with a level-2 sparse grid (DAKOTA) are shown in Fig. 7.13. It can be seen that their trends are close. Note that the level used in DAKOTA is slightly different from the level used in Chapter 4. The level-L sparse-grid in DAKOTA is a level-L+1 sparse grid in Chapter 4.

To further evaluate the performance of the gPC using different sparse grid levels, the RMSE is defined as

$$RMSE_i = \sqrt{\frac{\sum_{j=1}^{N_{mc}} \left(\mathbf{x}_{i,j}^{mc} - \mathbf{x}_{i,j}^{gPC}\right)^2}{N_{mc}}}. \tag{7.48}$$

where N_{mc} is the number of points, the subscript 'i' denotes the ith dimension and the subscript 'j' denotes the jth point. The RMSEs with gPC using different sparse grid levels are shown in Fig. 7.14. It can be seen that the RMSE decreases for all state variables with an increase of the level of the sparse grid. Hence, by using more points and higher order gPC, the approximation accuracy is improved. In real problems, the order of the gPC and number of sparse-grid points are problem dependent. In this example, the sparse-grid level 2 (DOKOTA) is sufficient to provide acceptable results describing the rough uncertainty of the space object.

7.5.2 Arbitrary Generalized Polynomial Chaos

The conventional gPC requires the given description of the initial uncertainty distribution. Hence, to use the stochastic collocation algorithm, the quadrature points and weights, as well as the orthogonal polynomial basis for the initial uncertainty distribution, should be provided. For some applications, however, the initial uncertainty distribution is hard to obtain (Jia and Xin 2018). For

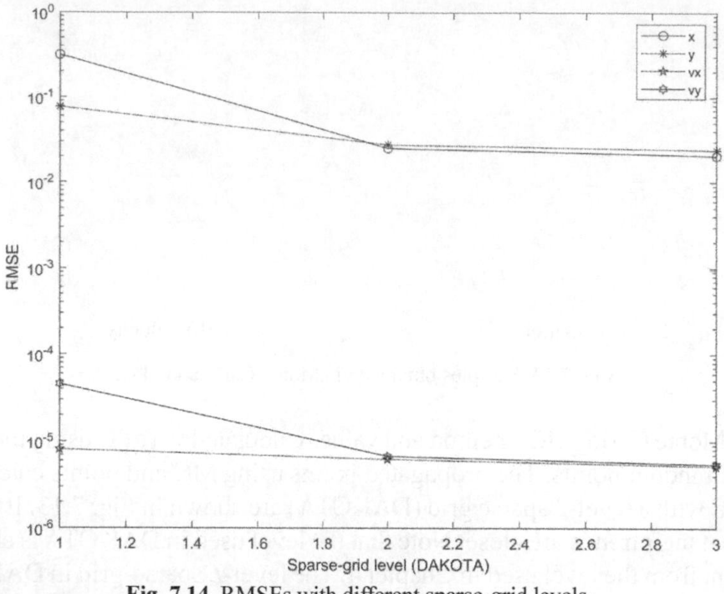

Fig. 7.14 RMSEs with different sparse-grid levels.

example, the initial orbit uncertainty distribution using the admissible area initialization method is hard to describe. Nonetheless, in this case, samples can be easily obtained via the Monte Carlo method. The moments' information of the initial uncertainty distribution, hence, can be calculated using Monte Carlo samples. Fortunately, the orthogonal polynomials used in PC can be generated from moments (Oladyshkin and Nowak 2012, Ahlfeld et al. 2016). It has been proved that the PC can be formed if the Hankel matrix of moments is given (Ahlfeld et al. 2016),

$$\mathbf{M} = \begin{bmatrix} m_0 & m_1 & \cdots & m_p \\ m_1 & m_2 & \cdots & m_{p+1} \\ \vdots & \vdots & \ddots & \vdots \\ m_p & m_{p+1} & \cdots & m_{2p} \end{bmatrix} \tag{7.49}$$

with m_p is the pth order moment.

The orthogonal polynomial ψ_j for PC is then given by

$$\psi_j = s_{1j}\,\xi^0 + s_{2j}\,\xi^1 + \cdots + s_{jj}\,\xi^{j-1},\ j = 1, \ldots, p+1 \tag{7.50}$$

where s_{ij} is the element in the following matrix

$$\mathbf{R}^{-1} = \begin{bmatrix} s_{11} & s_{12} & \cdots & s_{1,p+1} \\ 0 & s_{22} & \cdots & s_{2,p+1} \\ 0 & 0 & \ddots & \vdots \\ 0 & 0 & 0 & s_{p+1,p+1} \end{bmatrix} \tag{7.51}$$

with $\mathbf{R}$ obtained by Cholesky decomposition $\mathbf{R}^T\mathbf{R} = \mathbf{M}$.

Note that the orthogonal polynomial ψ_j can be recursively obtained as follows.

$$b_j\psi_j(\xi) = (\xi - a_j)\psi_{j-1}(\xi) - b_{j-1}\psi_{j-2}(\xi) \tag{7.52}$$

where the coefficients a_j and b_j are given by

$$a_j = \frac{r_{j,j+1}}{r_{j,j}} - \frac{r_{j-1,j}}{r_{j-1,j-1}}, j = 1,2,\cdots,p \tag{7.53}$$

$$b_j = \frac{r_{j+1,j+1}}{r_{j,j}}, j = 1,2,\cdots,p-1 \tag{7.54}$$

$r_{0,0} = 1$, $r_{0,1} = 0$, and $r_{i,j}$ is the entry of $\mathbf{R}$ on the ith row and jth column.

To use the collocation method, points γ_j and weights w_j are given by eigenvalues and eigenvectors of the matrix $\mathbf{J}$, respectively.

$$\mathbf{J} = \begin{bmatrix} a_1 & b_1 & & & & \\ b_1 & a_2 & b_2 & & & \\ & b_2 & a_3 & b_3 & & \\ & & & & & \\ & & & b_{p-2} & a_{p-1} & b_{p-1} \\ & & & & b_{p-1} & a_p \end{bmatrix} \tag{7.55}$$

Note $w_j = v_{1,j}^2$ with $v_{1,j}$ being the first component of the normalized jth eigenvector (Ahlfed et al. 2016). The above-described uncertainty quantification method is called the arbitrary polynomial chaos (aPC).

Remark 7.3: When the orthogonal polynomials with different orders are given, the arbitrary polynomial chaos can be obtained via the same method as shown in the conventional polynomial chaos.

The uncertainty increases with the time. For the long-term uncertainty propagation, the given order of aPC may not be adequate to retain the accuracy. Thus, higher-order aPCs are required. However, it demands higher order quadrature points and weights. Furthermore, higher order moments are required for the higher order aPCs. The high order moments are numerically hard to calculate accurately. Alternatively, the multi-element approach can be used to attain the better accuracy but using relatively lower-degree polynomials. We name this method as multiple element aPC (ME-aPC) (Jia and Xin 2018).

Remark 7.4: If the output variable is multi-dimensional, each dimension has different aPC representations.

In order to show the result of the aPC, the orbital propagation using the admissible region method is given as an example. Conventionally, initial orbit determination (IOD) can be solved using multiple observations at different times (Vallado 2013). For instance, the Herrick-Gibbs algorithm can be used to initialize an orbit from three radar observations while the Gaussian algorithm can be used to initialize the orbit from three optical observations. In recent years, admissible region based orbit initialization has become popular (Farnocchia et al. 2010, Murphy et al. 2015, Hussein et al. 2018, DeMars et al. 2012, Fujimoto and Scheeres 2012). Based on physical and geometric constraints, the admissible region can be obtained. Samples can be generated in the admissible region using the Monte Carlo method and then recover the initial uncertainty of the space object.

The geometric relation of the space object and the observer is illustrated in Fig. 7.15.

Based on the geometric relation, we have the following equation

$$\mathbf{r} = \mathbf{q} + \boldsymbol{\rho} \tag{7.56}$$

and

$$\dot{\mathbf{r}} = \dot{\mathbf{q}} + \dot{\boldsymbol{\rho}} \tag{7.57}$$

where $\mathbf{r}$ is the position of the space object in the inertial coordinate system, $\mathbf{q}$ is the position of the ground-based or space-based observer (SBO) in the inertial coordinate system, $\boldsymbol{\rho}$ is the position of the space object with respect to the observer. $\boldsymbol{\rho}$ and $\dot{\boldsymbol{\rho}}$ can be rewritten as

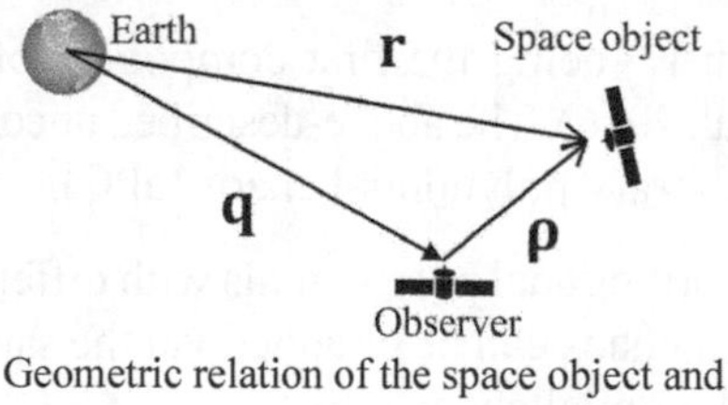

Fig. 7.15 Geometric relation of the space object and the observer.

$$\boldsymbol{\rho} = \rho \cdot \mathbf{u}_\rho \tag{7.58}$$

and

$$\dot{\boldsymbol{\rho}} = \dot{\rho}\mathbf{u}_\rho + \rho\dot{\alpha}\mathbf{u}_\alpha + \rho\dot{\delta}\mathbf{u}_\delta \tag{7.59}$$

where

$$\mathbf{u}_\rho = [\cos\alpha\cos\delta, \sin\alpha\cos\delta, \sin\delta]^T \tag{7.60}$$

$$\mathbf{u}_\alpha = [-\sin\alpha\cos\delta, \cos\alpha\cos\delta, 0]^T \tag{7.61}$$

$$\mathbf{u}_\delta = [-\cos\alpha\sin\delta, -\sin\alpha\sin\delta, \cos\delta]^T \tag{7.62}$$

α and δ are defined by Eqs. (7.63) and (7.64) in the topocentric reference system that can be arbitrarily selected (Farnocchia et al. 2010).

$$\alpha = \tan^{-1}\left(\mathbf{u}_\rho(2)/\mathbf{u}_\rho(1)\right) \tag{7.63}$$

$$\delta = \sin^{-1}\left(\mathbf{u}_\rho(3)\right) \tag{7.64}$$

The geometric relation is shown in Fig. 7.16.

There are many constraints in real applications. Hence, the final admissible region (*AR*) is given by the intersection of different admissible regions AR_i described by different physical constraints (Farnocchia et al. 2010, Murphy et al. 2015, Hussein et al. 2018).

$$AR = \bigcap_i AR_i \tag{7.65}$$

Constraint 1

First, the energy should satisfy (Farnocchia et al. 2010)

$$\varepsilon(\rho, \dot{\rho}) = \frac{1}{2}\left\|\dot{\mathbf{r}}(\rho, \dot{\rho})\right\|^2 - \frac{\mu}{\left\|\mathbf{r}(\rho)\right\|} \le 0 \tag{7.66}$$

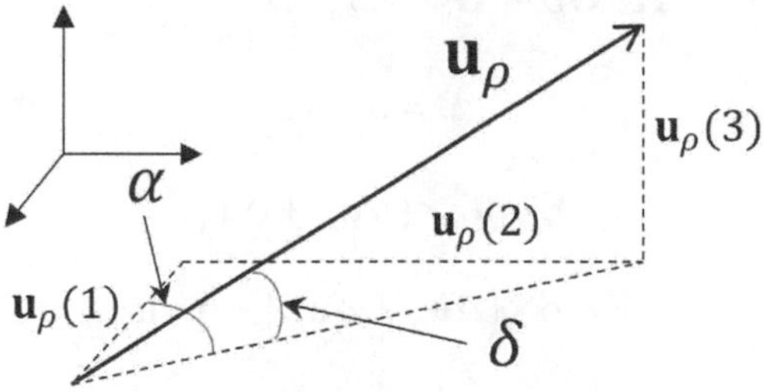

Fig. 7.16 Illustration of angle α and δ.

Equation (7.66) can be rewritten as

$$2\varepsilon(\rho,\dot{\rho}) = \dot{\rho}^2 + c_1\dot{\rho} + T(\rho) - \frac{2\mu}{\sqrt{S(\rho)}} \le 0 \tag{7.67}$$

where
$$T(\rho) = c_2\rho^2 + c_3\rho + c_4 \tag{7.68}$$

$$S(\rho) = \rho^2 + c_5\rho + c_0 \tag{7.69}$$

$$c_0 = \|\mathbf{q}\|^2 \tag{7.70a}$$

$$c_1 = 2\dot{\mathbf{q}} \cdot \mathbf{u}_\rho \tag{7.70b}$$

$$c_2 = \dot{\alpha}^2 \cos^2\delta + \dot{\delta}^2 \tag{7.70c}$$

$$c_3 = 2\left(\dot{\alpha}\dot{\mathbf{q}} \cdot \mathbf{u}_\alpha + \dot{\delta}\dot{\mathbf{q}} \cdot \mathbf{u}_\delta\right) \tag{7.70d}$$

$$c_4 = \|\dot{\mathbf{q}}\|^2 \tag{7.70e}$$

$$c_5 = 2\mathbf{q} \cdot \mathbf{u}_\rho \tag{7.70f}$$

Constraint 2
Another constraint is given by (Farnocchia et al. 2010)

$$\left(r_{\min}^2 - \|\mathbf{D}\|^2\right)\dot{\rho}^2 - P(\rho)\dot{\rho} - U(\rho) + r_{\min}^2 T(\rho) - \frac{2r_{\min}^2\mu}{\sqrt{S(\rho)}} \le 0 \tag{7.71}$$

where
$$P(\rho) = 2\mathbf{D}\cdot\mathbf{E}\rho^2 + 2\mathbf{D}\cdot\mathbf{F}\rho + 2\mathbf{D}\cdot\mathbf{G} - r_{\min}^2 c_1 \tag{7.72}$$

$$\begin{aligned} U(\rho) = {} & \|\mathbf{E}\|^2\rho^4 + 2\mathbf{E}\cdot\mathbf{F}\rho^3 + \left(2\mathbf{E}\cdot\mathbf{G} + \|\mathbf{F}\|^2\right)\rho^2 + \\ & 2\mathbf{F}\cdot\mathbf{G}\rho + \|\mathbf{G}\|^2 - 2r_{\min}\mu \end{aligned} \tag{7.73}$$

$$\mathbf{D} = \mathbf{q} \times \mathbf{u}_\rho \tag{7.74}$$

$$\mathbf{E} = \mathbf{u}_\rho \times \left(\dot{\alpha}\mathbf{u}_\alpha + \dot{\delta}\mathbf{u}_\delta\right) \tag{7.75}$$

$$\mathbf{F} = \mathbf{q} \times \left(\dot{\alpha}\mathbf{u}_\alpha + \dot{\delta}\mathbf{u}_\delta\right) + \dot{\mathbf{q}} \times \mathbf{u}_\rho \tag{7.76}$$

$$\mathbf{G} = \mathbf{q} \times \dot{\mathbf{q}} \tag{7.77}$$

$r_{\min}$ is the *a priori* knowledge of the semi-major axis.

Constraint 3

Besides Eq. (7.71), we can use the following constraint,

$$\varepsilon(\rho,\dot{\rho}) \leq -\frac{\mu}{2r_{\max}} \tag{7.78}$$

$$r_{\max} = R_{\oplus} + h \tag{7.79}$$

where $R_{\oplus}$ is the earth radius and h is the maximum altitude of the space object.

Constraint 4

If *a priori* knowledge of the eccentricity is known, we can use the following constraint. The specific angular momentum is given by (Hussein et al. 2018)

$$\mathbf{h} = \mathbf{h}_1\dot{\rho} + \mathbf{h}_2\rho^2 + \mathbf{h}_3\rho + \mathbf{h}_4 \tag{7.80}$$

where

$$\mathbf{h}_1 = \mathbf{q} \times \mathbf{u}_{\rho}, \tag{7.81a}$$

$$\mathbf{h}_2 = \mathbf{u}_{\rho} \times \left(\dot{\alpha}\mathbf{u}_{\alpha} + \dot{\delta}\mathbf{u}_{\delta}\right), \tag{7.81b}$$

$$\mathbf{h}_3 = \mathbf{u}_{\rho} \times \dot{\mathbf{q}} + \mathbf{q} \times \left(\dot{\alpha}\mathbf{u}_{\alpha} + \dot{\delta}\mathbf{u}_{\delta}\right), \tag{7.81c}$$

$$\mathbf{h}_4 = \mathbf{q} \times \dot{\mathbf{q}} \tag{7.81d}$$

Giving a specific eccentricity, we have,

$$a_4\dot{\rho}^4 + a_3\dot{\rho}^3 + a_2\dot{\rho}^2 + a_1\dot{\rho} + a_0 = 0 \tag{7.82}$$

where

$$a_4 = \left\|\mathbf{h}_1\right\|^2, \tag{7.83a}$$

$$a_3 = \tilde{P}(\rho) + \left\|\mathbf{h}_1\right\|^2 c_1, \tag{7.83b}$$

$$a_2 = \tilde{U}(\rho) + \left\|\mathbf{h}_1\right\|^2 F(\rho) + c_1\tilde{P}(\rho), \tag{7.83c}$$

$$a_1 = F(\rho)\tilde{P}(\rho) + c_1\tilde{U}(\rho) \tag{7.83d}$$

$$a_0 = F(\rho)\tilde{U}(\rho) + \mu^2\left(1 - e^2\right) \tag{7.83e}$$

$$\tilde{P}(\rho) = (2\mathbf{h}_1 \cdot \mathbf{h}_2)\rho^2 + (2\mathbf{h}_1 \cdot \mathbf{h}_3)\rho + (2\mathbf{h}_1 \cdot \mathbf{h}_4) \tag{7.84}$$

$$\tilde{U}(\rho)=\|\mathbf{h}_2\|^2\rho^4+2(\mathbf{h}_2\cdot\mathbf{h}_3)\rho^3+\left(2(\mathbf{h}_2\cdot\mathbf{h}_4)+\|\mathbf{h}_3\|^2\right)\rho^2 \qquad (7.85)$$
$$+2(\mathbf{h}_3\cdot\mathbf{h}_4)\rho+\|\mathbf{h}_4\|^2$$

$$F(\rho)=c_2\rho^2+c_3\rho+c_4-\frac{2\mu}{\sqrt{\rho^2+c_5\rho+c_0}} \qquad (7.86)$$

Note that it is possible to initialize the orbit of the space object if the *AR* is a small area. The *AR* describes various constraints on ρ and $\dot{\rho}$. However, if the constrained area is too large, it is too vague to initialize the space object. Hence, *a priori* knowledge of the orbit is very helpful to initialize the orbit.

As a demonstration, we show a typical *AR* obtained from an SBO observation. The state of the space object in Earth-centered inertial (ECI) coordinate system is given in Eq. (7.87). Similarly, the state of SBO is given by Eqs. (7.90) and (7.91).

$$\mathbf{x}^{so}=\left[\left(\mathbf{x}^{so}_{pos}\right)^T,\left(\mathbf{x}^{so}_{vel}\right)^T\right]^T \qquad (7.87)$$

$$\mathbf{x}^{so}_{pos}=10^4\times[3.9583,-1.4667,0.1035]^T \text{ km} \qquad (7.88)$$

$$\mathbf{x}^{so}_{vel}=[1.0583,2.8815,0.0842]^T \text{ km/s} \qquad (7.89)$$

$$\mathbf{x}^{sbo}_{pos}=10^3\times[2.0034,-3.7516,5.5598]^T \text{ km} \qquad (7.90)$$

$$\mathbf{x}^{sbo}_{vel}=[-3.9263,4.5965,4.5164]^T \text{ km/s} \qquad (7.91)$$

Based on the constraints described in this section, Fig. 7.17 illustrates the *AR*. Note that the dotted line corresponds to the equality constraint in Eq. (7.66). The black solid line and the dashed ellipse correspond to the equality of Eq. (7.78) (i.e., $\varepsilon(\rho,\dot{\rho})=-\frac{\mu}{2r_{\max}}$) and Eq. (7.82), respectively with the *a priori* information $e\leq 0.1$ and $3.5\times10^4\,\text{km}\leq r_{\max}\leq 4.5\times10^4\,\text{km}$, the *AR* is colored in grey. The random points are generated in this *AR* and they fulfill all constraints. The true value is denoted as a star. It can be seen the random points (in grey color) that satisfy all the constraints are close to the star (Jia and Xin 2018).

To initialize the propagator, for each random point $[\rho,\dot{\rho}]^T$ that satisfies all the constraints, by Eqs. (7.56)–(7.59), the vector $\boldsymbol{\rho}$ and $\dot{\boldsymbol{\rho}}$ can be calculated. The corresponding $\mathbf{r}$ and $\dot{\mathbf{r}}$ then can be obtained.

We use aPC with order $p = 3$ and the propagation time is 24 hours. Two-dimensional samples generated in the AR are used to calculate the value of the Hankel matrix of moments $\mathbf{M}$. In this example, $\mathbf{M}\in\mathbb{R}^{4\times4}$. The number of

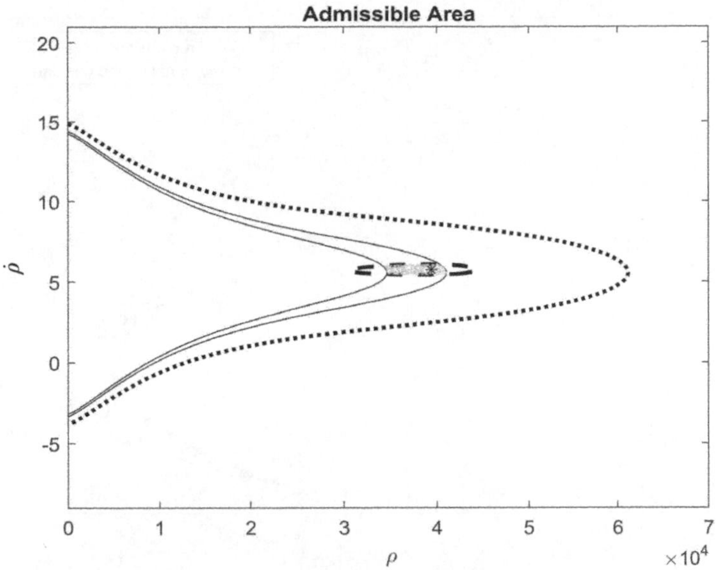

Fig. 7.17 A typical admissible region by SBO.

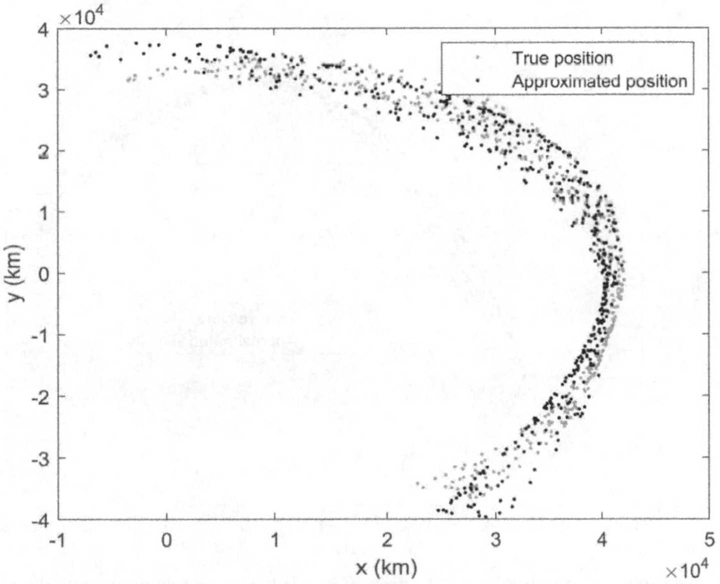

Fig. 7.18 The positional points of directly propagated samples and samples calculated by aPC along x and y axes (24 hours).

collocation points is 9. The results are shown in Fig. 7.18–Fig. 7.21. It can be seen that the directly propagated samples and the samples calculated via aPC has a small discrepancy.

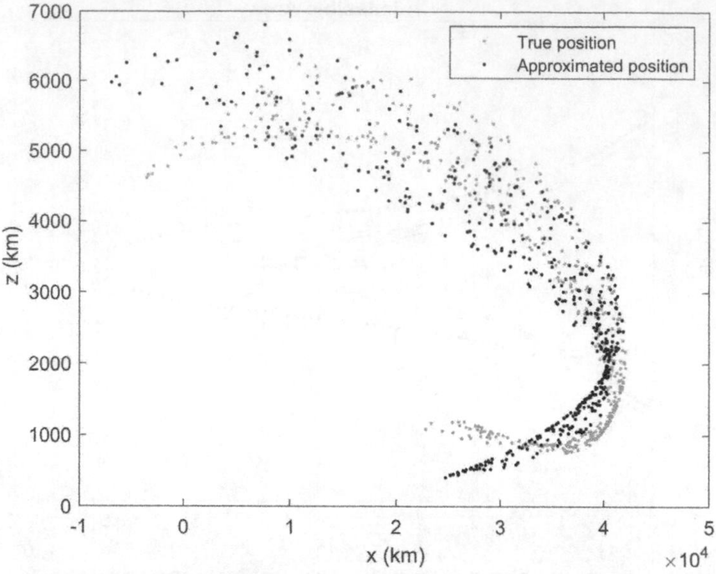

Fig. 7.19 The positional points of directly propagated samples and samples calculated by aPC along x and z axes (24 hours).

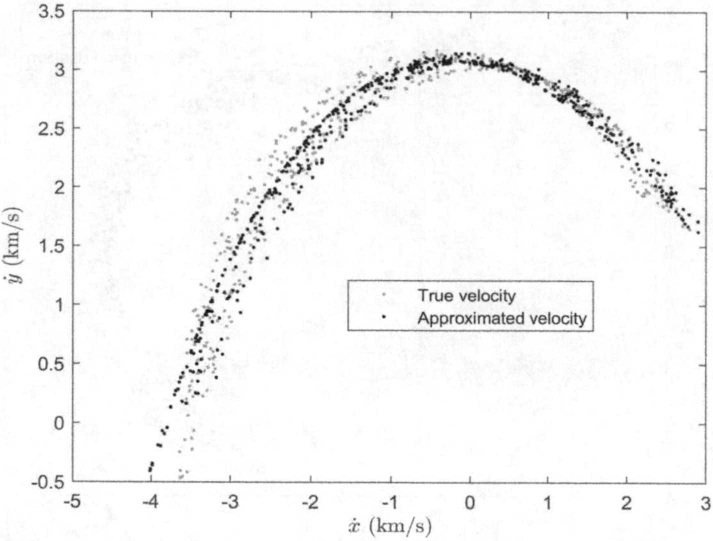

Fig. 7.20 The velocity points of directly propagated samples and samples calculated by aPC along x and y axes (24 hours).

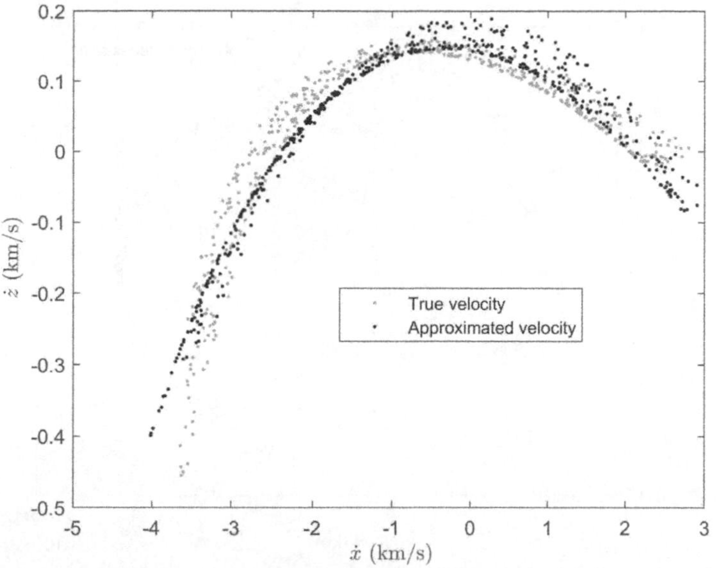

Fig. 7.21 The velocity points of directly propagated samples and samples calculated by aPC along x and z axes (24 hours).

To further improve the performance, we use the multi-element aPC (ME-aPC). It is worth noting that the *AR* is not a regular shape in general. Instead of decomposing the *AR* directly, we use random samples to describe the *AR* and then organize these samples into non-overlapping subsets. For example, given a threshold of $\bar{\rho}$, the set of random points $\{(\rho,\dot{\rho})\}$ can be decomposed into two subsets, $\{(\rho,\dot{\rho}) \mid \rho \geq \bar{\rho}\}$ and $\{(\rho,\dot{\rho}) \mid \rho < \bar{\rho}\}$. For each of these two subsets, aPC is used. The uncertainty is then described by the combination of these two aPCs.

For this specific orbital propagation problem, the *AR* is split into four areas according to the random variable ρ and the aPC (p = 3) is employed for each one. Specifically, three threshold values of $\bar{\rho}$ with $\rho_{\min} + \frac{\rho_{\max} - \rho_{\min}}{4}$, $\rho_{\min} + \frac{\rho_{\max} - \rho_{\min}}{2}$, and $\rho_{\min} + \frac{3(\rho_{\max} - \rho_{\min})}{4}$ are used to decompose the *AR*. Note that $\rho_{\max}$ and $\rho_{\min}$ are the maximum and minimum value of ρ, respectively. The final results are shown in Fig. 7.22–Fig. 7.25. Compared to Fig. 7.18–Fig. 7.21, it can be seen that the discrepancy of the directly propagated samples and samples calculated via aPC is reduced, which clearly demonstrates the effectiveness of the ME-aPC (Jia and Xin 2018).

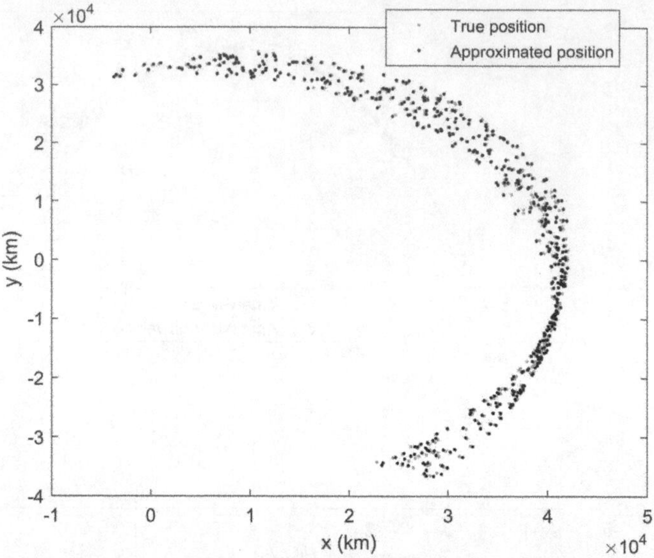

Fig. 7.22 The positional points of directly propagated samples and samples calculated by ME-aPC along x and y axes (ME-aPC, 24 hours).

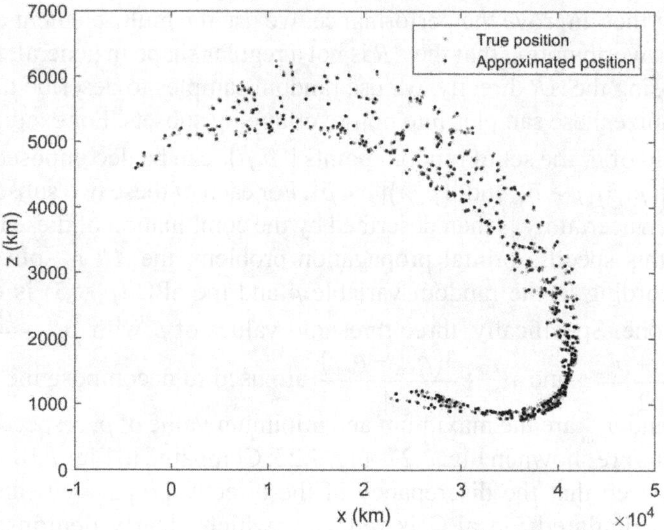

Fig. 7.23 The positional points of directly propagated samples and samples calculated by ME-aPC along x and z axes (ME-aPC, 24 hours).

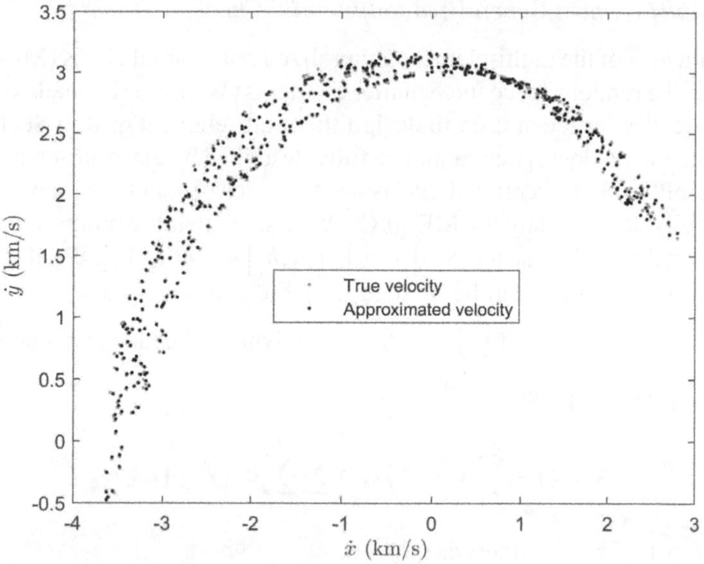

Fig. 7.24 The velocity points of directly propagated samples and samples calculated by ME-aPC along x and y axes (ME-aPC, 24 hours).

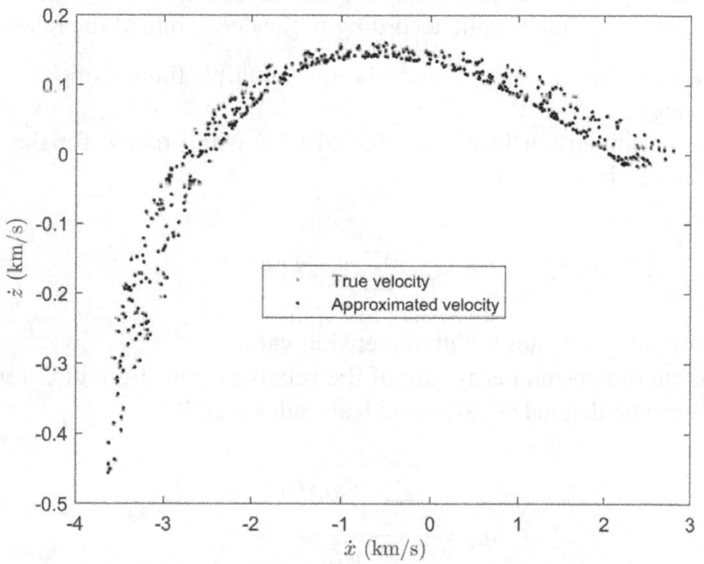

Fig. 7.25 The velocity points of directly propagated samples and samples calculated by ME-aPC along x and z axes (ME-aPC, 24 hours).

7.5.3 Multi-element Generalized Polynomial Chaos

The main idea of the multi-element generalized polynomial chaos (ME-gPC) is to divide the random space into smaller elements (Wan and Karniadakis 2006). The same idea has been used to design the multi-element grid in Section 7.2. However, the random space cannot be fully described by the multi-element grid. For ME-gPC, the conventional gPC is used for each element. The random space can be fully described by the ME-gPC. We assume that the random variable is defined on the random space $S=[a_1,b_1]\times[a_2,b_2]\times\cdots\times[a_n,b_n]$. We also assume that the random space can be divided into K elements, S_k, $k=1,\cdots,K$ where $S_i\bigcap S_j=\varnothing$, if $i\neq j$, and $\bigcup_{i=1}^{K}S_i=S$. The polynomial chaos expansion on the random space S is given by

$$\hat{\mathbf{x}}(t,\xi)=\sum_{k=1}^{K}\hat{\mathbf{x}}_k(t,\xi)I_k\approx\sum_{k=1}^{K}\sum_{j=0}^{P}\mathbf{c}_{j,k}\boldsymbol{\Psi}_{j,k}(t,\xi)I_k \tag{7.92}$$

where $I_k=1$, if $\xi\in S_k$, otherwise, $I_k=0$.

Note that orthogonal polynomials for an arbitrary piece of probability density functions can be constructed, as shown in Section 7.2. In addition, one can use a pre-defined random space decomposition strategy. Alternatively, the random space can be split according to the decay rate of the relative error.

Remark 7.5: The random space can be split multiple times using the adaptive split strategy.

The ith dimension local variance of gPC with order p for the element $\xi\in S_k$ is given by

$$\sigma_{i,k,p}^2=\sum_{j=0}^{P}c_{i,j,k}^2E[\psi_{j,k}^2] \tag{7.93}$$

The subscript 'i' denotes the ith dimension value.

The ith dimension decay rate of the relative error of the gPC for the kth element can be defined as (Wan and Karniadakis 2006)

$$\eta_{i,k}=\frac{\sum_{j=N_{p-1}+1}^{P}c_{i,j,k}^2E[\psi_{j,k}^2]}{\sigma_{i,k,p}^2} \tag{7.94}$$

Note that η_k indicates the decay rate of the kth element. N_{p-1} is the number of gPC basis with the maximum order p–1. The decomposition also depends on how important the kth element is, which can be described by the probability of the kth element, i.e., $\Pr(I_k = 1)$. Hence, the criterion for decomposition can be defined as

$$\eta_{i,k}^{\gamma} \Pr(I_k = 1) \geq \theta, \quad 0 < \gamma < 1 \tag{7.95}$$

where θ is a threshold.

In addition, the most sensitive dimension is often used as the dimension to split the kth element. The sensitivity for the ith dimension is given by (Wan and Karniadakis 2006)

$$r_{i,k} = \frac{c_{i,k,p}^2 E[\psi_{i,k,p}^2]}{\sum_{j=N_{p-1}+1}^{P} c_{j,k}^2 E[\psi_{j,k}^2]} \tag{7.96}$$

The subscript 'i,k,p' denotes the ith dimensional terms with the polynomial order p for the kth element.

Example 7.6: Assume the initial uncertainty (variable $\mathbf{x}$) is a 2-dimensional standard normal distribution. We study the uncertainty of the output $f(\mathbf{x})$ with

$$f(\mathbf{x}) = 2x_1^2 x_2^2 \tag{7.97}$$

The samples obtained via the Monte Carlo method and gPC with $p = 5$ are shown in Fig. 7.26. The absolute error of samples between the Monte Carlo method and gPC is shown in Fig. 7.27. To improve the accuracy, the ME-gPC algorithm is used and the results are shown in Fig. 7.28 and Fig. 7.29. By comparing the results shown in Fig. 7.27 and Fig. 7.29, it can be seen that the accuracy is improved by using ME-gPC. For comparison, the random space of x_1 $(-\infty, \infty)$ is decomposed into two parts $(-\infty, 0)$ and $[0, \infty)$. Note that in order to keep the numerical convenience, $-\infty$ is substituted by a numerical value, for example, −10 or −5 for the standard normal distribution. The polynomials corresponding to the standard normal distribution in $[0,10]$ and $[-10,0)$ with different orders are shown in Fig. 7.30 and Fig. 7.31, respectively. The polynomials corresponding to the standard normal distribution in $[-10,10]$ with different orders are shown in Fig. 7.32. It can be seen that these polynomials are different.

To show the effectiveness of ME-gPC clearly, the adaptive split strategy is used and multiple decompositions are processed. The RMSE of the samples is shown in Fig. 7.33. It can be seen that RMSE decreases with the increase of the number of splits, which verifies the effectiveness of ME-gPC.

There has been much promising progress in the research of the polynomial chaos to propagate uncertainty, such as the sparse polynomial chaos expansion

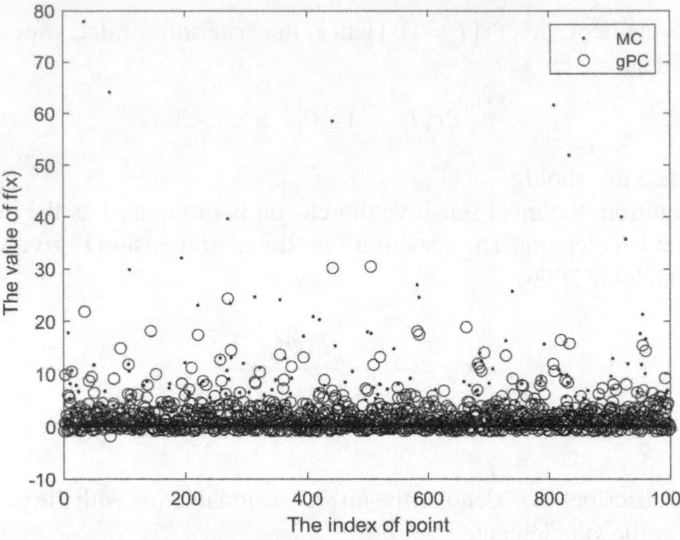

Fig. 7.26 Samples obtained via the Monte Carlo method and gPC with $p = 5$.

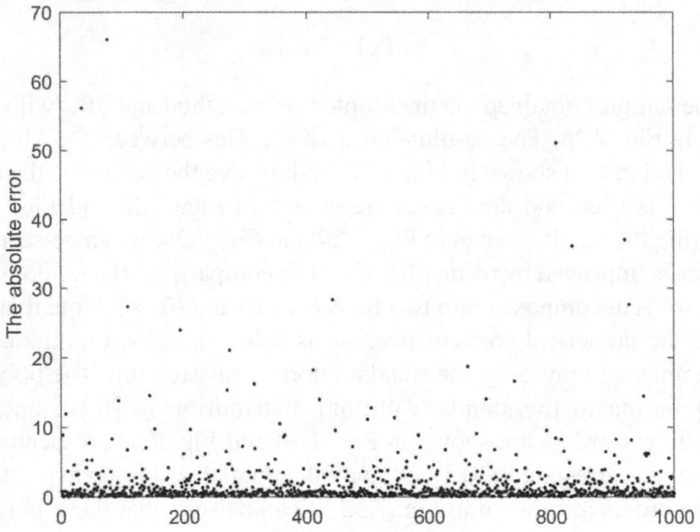

Fig. 7.27 Absolute error of samples between the Monte Carlo method and gPC with $p = 5$.

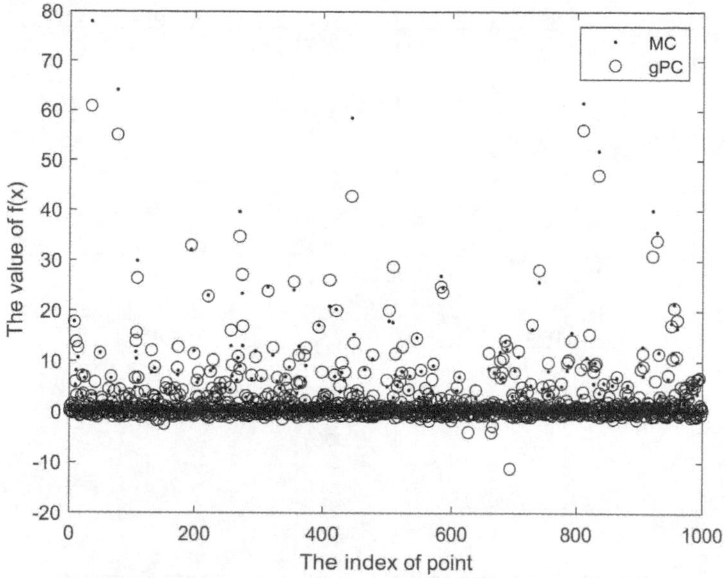

Fig. 7.28 Samples obtained via the Monte Carlo method and ME-gPC with $p = 5$.

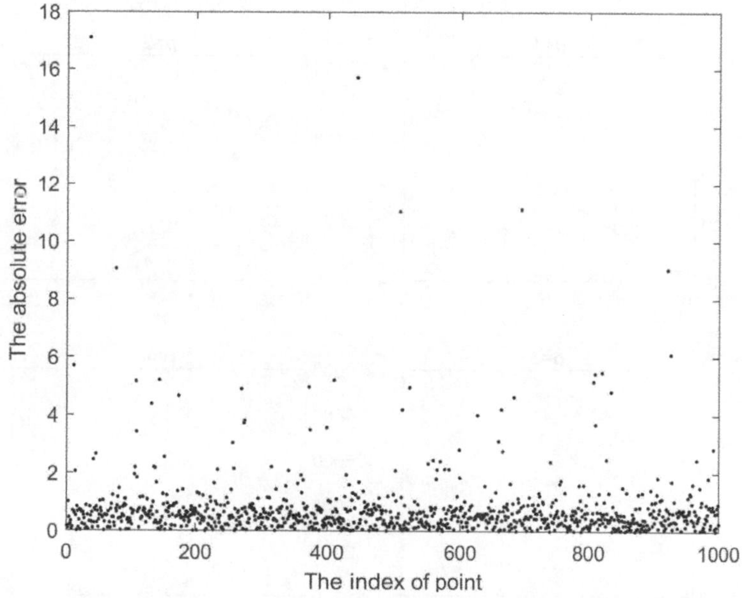

Fig. 7.29 The absolute error of samples between the Monte Carlo method and ME-gPC with $p = 5$.

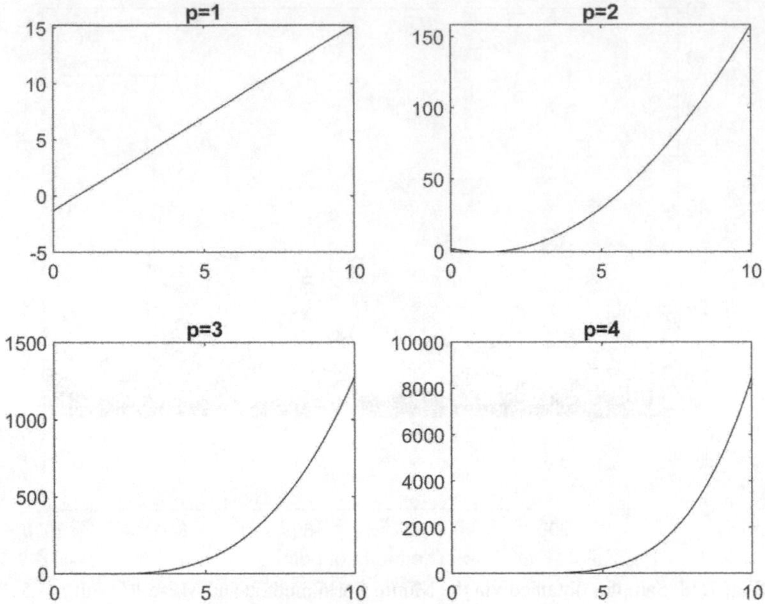

Fig. 7.30 Polynomials corresponding to the split Gaussian distribution ((0,10]).

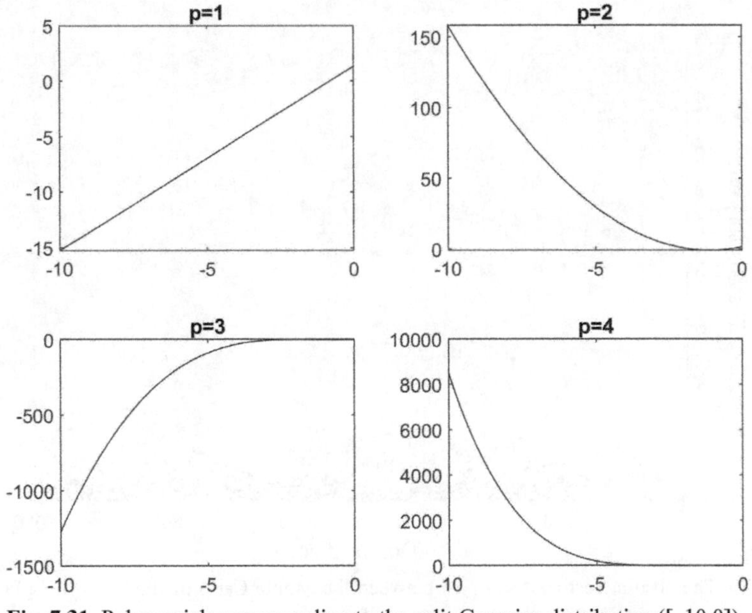

Fig. 7.31 Polynomials corresponding to the split Gaussian distribution ([–10,0]).

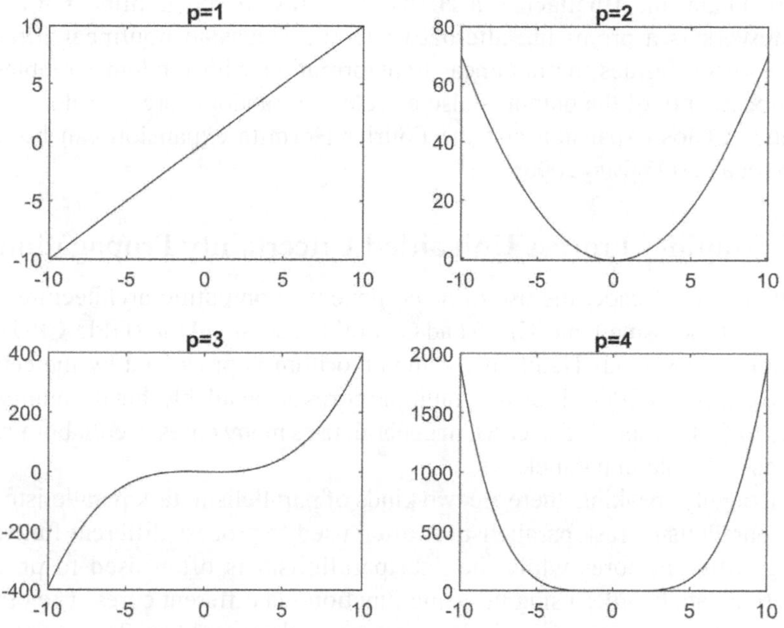

Fig. 7.32 Polynomials corresponding to the Gaussian distribution ([–10,10]).

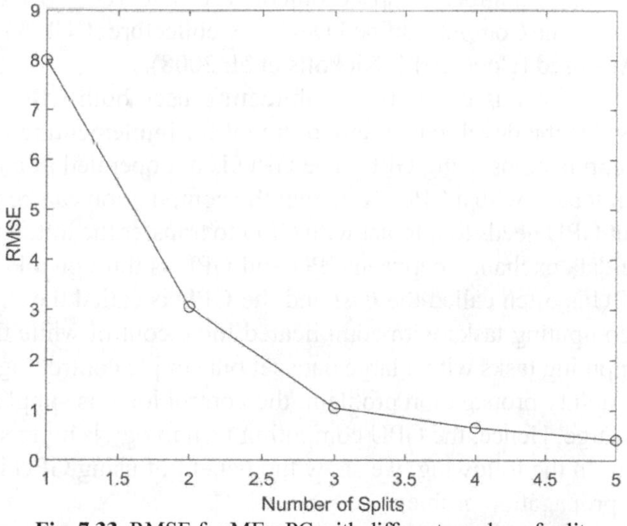

Fig. 7.33 RMSE for ME-gPC with different number of splits.

based on least angle regression (Blatman and Sudret 2011). For arbitrary nonlinear transformation, the polynomial chaos can be used to quantize the output uncertainty. Hence, the gPC can be used for nonlinear filters (Yu et al.

2016, Dutta and Bhattacharya 2010). The gPC based nonlinear filtering framework is a promising alternative to the Gaussian nonlinear filtering framework. Besides the nonlinear transformation with random variables, if the uncertainty of the output is also affected by random forces or noises, the Wiener Chaos expansion and the Fourier-Hermite expansion can be used (Hou et al. 2006, Luo 2006).

7.6 Graphics Process Unit aided Uncertainty Propagation

Over the last decade, the use of heterogeneous computing architecture with Graphic Processing Unit (GPU) and Central Processing Unit (GPU-CPU) has become widespread. Traditionally, the algorithm is processed by the central processing unit (CPU). Typically, multiple cores are available, but the number of cores of CPU is small. The GPU, in general, uses many cores to collaboratively process the data in parallel.

Roughly speaking, there are two kinds of parallelism: task parallelism and data parallelism. Task parallelism is often used to process different functions using different cores while the data parallelism is often used to process different data bundles using the same function via different cores. The orbital uncertainty propagation is typically a data parallelism problem. The uncertainty can be represented by weighted particles. The number of particles can be very large since a large number of space objects are involved. To facilitate the implementation, the Compute Unified Device Architecture (CUDA) developed by NVIDIA is used (Cook 2012, Nickolls et al. 2008).

The heterogeneous computing architecture uses both CPU and GPU. CUDA provides the development environment for implementing the parallel computing applications using GPU. The GPU is not operated as a standalone system but interacts with a CPU. Note that the computation can be performed by GPU. But GPU needs to interact with CPU to transfer the initial values and results. The data exchange between CPU and GPU is through a PCI-Express bus. The CPU is often called the *host* and the GPU is called the *device*. CPU is good at computing tasks with complicated logic control while the GPU is good at computing tasks with a large data set but simple control logic. For the orbital uncertainty propagation problem, the control logic is simple while the data can be large. Hence, the GPU computing technology is more suitable for this problem. In the following, we show the benefit of using GPU in the orbit uncertainty propagation problem.

Example 7.7: Assume the initial state is given by

$$\mathbf{x}_0 = [42164\text{km}, 0\text{km}, 0\text{km}, 0\,\text{km/s}, 3.074549\,\text{km/s}, 0.02683\,\text{km/s}]^T \tag{7.98}$$

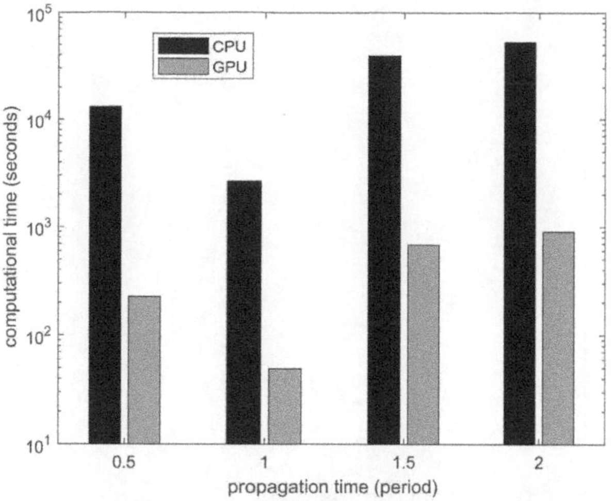

Fig. 7.34 Computation time of CPU and GPU versus propagation time.

$$\mathbf{P}_0 = \text{diag}\left(\left[1\text{km}^2, 1\text{km}^2, 10^{-2}\text{km}^2, 10^{-6}\,\text{km}^2/\text{s}^2, 10^{-6}\,\text{km}^2/\text{s}^2, 10^{-8}\,\text{km}^2/\text{s}^2\right]\right) \tag{7.99}$$

The Monte Carlo method with 10,000 samples is used. The difference between the CPU and GPU propagation result is calculated by

$$\text{RMSE}=\sqrt{\frac{1}{10^4}\sum_{i=1}^{10^4}\sum_{j=1}^{6}\left(\mathbf{x}_{j,\text{cpu}}^{(i)} - \mathbf{x}_{j,\text{gpu}}^{(i)}\right)} \tag{7.100}$$

where $\mathbf{x}_{j,\text{cpu}}^{(i)}$ and $\mathbf{x}_{j,\text{gpu}}^{(i)}$ are the jth propagated state corresponding to the ith point using CPU and GPU, respectively.

Figure 7.34 shows the computation time of the orbit propagation using CPU and GPU on a laptop with Intel Core i7-3720QM CPU and NVIDIA Quadro K1000M GPU. It can be seen that the computational time using GPU is consistently less than that using CPU for various propagation times. Besides the computational efficiency, we show the difference in terms of RMSE using CPU and GPU in Fig. 7.35. It can be seen that the difference of RMSEs using CPU and GPU increases with the propagation times. However, the difference is negligible. Similarly, Fig. 7.36 shows that the GPU consistently uses much less time than CPU when various numbers of samples are used.

7.7 MapReduce aided Uncertainty Propagation

MapReduce is a programming model, and widely used in processing big data. It has been implemented in various cloud computing framework. Cloud computing

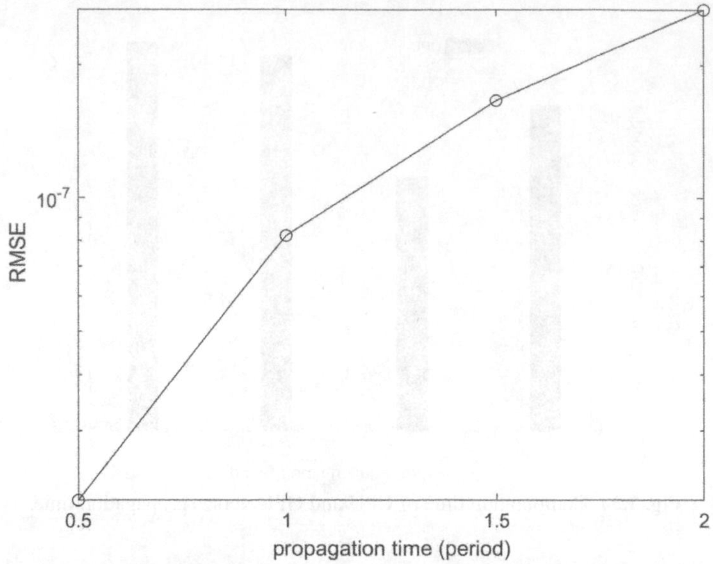

Fig. 7.35 Difference of propagation results obtained by CPU and GPU versus propagation time.

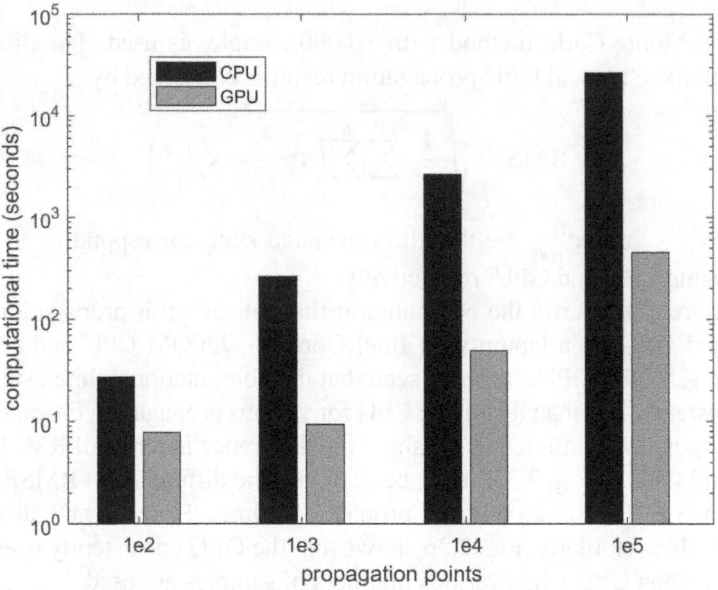

Fig. 7.36 Computation time of CPU and GPU versus propagation points.

is a network-based computing that provides a way to use resources of multiple computers. The scalability of the cloud computing is remarkable and can be used to propagate the uncertainty of many space objects.

MapReduce algorithm (Dean and Ghemawat 2008) is a powerful distributed data processing algorithm. It follows the concept "Divide and Conquer". Four steps are involved in the MapReduce algorithm, Split, Map, Shuffle, and Reduce, as shown in Fig. 7.37. The Map function takes datasets and splits them into smaller datasets. Then, these smaller datasets are processed in parallel. The output of the Map function is a set of Key-Value pairs. The shuffle function takes into the Key-Value pairs, merges and sorts all Key-Value pairs that have the same key value. The output of the shuffle function is Key-Value (list) pairs. The Reduce function takes the Key-Value (list) pairs and performs the reduction operation. The output of the Reduce function is also Key-Value pairs. Note that customized operations can be implemented in the Map function and Reduce function.

For example, the input was split into multiple parts, e.g., four parts. If 20 space objects are propagated and each object has N_p points to represent the uncertainty, each part contains $5N_p$ points. Each part of the data is consumed by a mapper. The mapper implements the orbit propagation algorithm. Specifically, each point is propagated via Eq. (7.1). The output is the key-value pairs. The Key Value is represented by the space object identifier. In the shuffling phase, the output of mappers is consolidated into relevant data according to

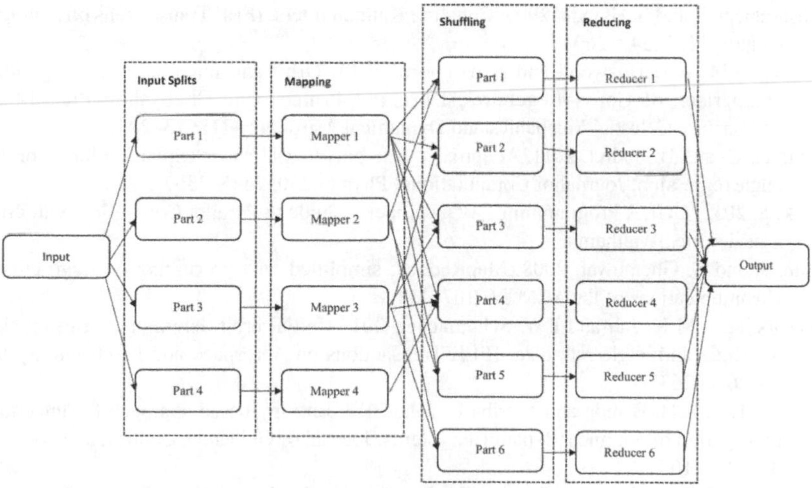

Fig. 7.37 Map-Reduce framework of uncertainty propagation via the point-based method.

the key value. Finally, in the reducing phase, the output of the shuffling phase is aggregated for each space object. The uncertainty for each space object is obtained as the output.

Note that the cloud computing and the GPU technology can be used together. The GPU technology can be embedded in the framework of cloud computing as an accelerator.

7.8 Summary

To build realistic uncertainty propagation systems, the computational complexity and the accuracy, as well as the cost of resources, need to be considered. It is more reliable to use the Monte Carlo based uncertainty propagation if sufficient resources are available. However, if there are limited resources, various efficient algorithms, such as the gPC, aPC and GMM based algorithms, can be applied. If only the mean and covariance are concerned, the direct grid-based orbit propagation method can be used. How to adaptively choose different algorithms needs further investigations.

References

Ahlfeld, R., B. Belkouchi and F. Montomoli. 2016. SAMBA: Sparse approximation of moment-based arbitrary polynomial chaos. Journal of Computational Physics 320: 1–16.

Alspach, D. and H. Sorenson. 1972. Nonlinear Bayesian estimation using Gaussian sum approximations. IEEE Transactions on Automatic Control 17: 439–448.

Arasaratnam, I. and S. Haykin. 2009. Cubature Kalman filters. IEEE Transactions on Automatic Control 54: 1254–1269.

Aristoff, J.M., J.T. Horwood and A.B. Poore. 2014. Orbit and uncertainty propagation: a comparison of Gauss–Legendre-, Dormand–Prince-, and Chebyshev–Picard-based approaches. Celestial Mechanics and Dynamical Astronomy 118: 13–28.

Blatman, G. and B. Sudret. 2011. Adaptive sparse polynomial chaos expansion based on least angle regression. Journal of Computational Physics 230: 2345–2367.

Cook, S. 2012. CUDA Programming: A Developer's Guide to Parallel Computing with GPUs. Elsevier Inc., Waltham.

Dean, J. and S. Ghemawat. 2008. MapReduce: simplified data processing on large clusters. Communications of the ACM 51: 107–113.

DeMars, K.J., M.K. Jah and P.W. Schumacher. 2012. Initial orbit determination using short-arc angle and angle rate data. IEEE Transactions on Aerospace and Electronic Systems 48: 2628–2637.

DeMars, K.J., R.H. Bishop and Moriba K. Jah. 2013. Entropy-based approach for uncertainty propagation of nonlinear dynamical systems. Journal of Guidance, Control, and Dynamics 36: 1047–1057.

Dormand, J.R. and P.J. Prince. 1980. A family of embedded Runge-Kutta formulae. Journal of Computational and Applied Mathematics 6: 19–26.

Dormand, J.R., M.E.A. El-Mikkawy and P.J. Prince. 1987. High-order embedded Runge-Kutta-Nystrom formulae. IMA Journal of Numerical Analysis 7: 423–430.

Dutta, P. and R. Bhattacharya. 2010. Nonlinear estimation of hypersonic state trajectories in Bayesian framework with polynomial chaos. Journal of Guidance, Control, and Dynamics 33: 1765–1778.

Elgohary, T.A., J.L. Junkins and S.N. Atluri. 2015. An RBF-collocation algorithm for orbit propagation. AAS/AIAA Space Flight Mechanics Meeting. USA 155: AAS 15–359.

Farnocchia, D., G. Tommei, A. Milani and A. Rossi. 2010. Innovative methods of correlation and orbit determination for space debris. Celestial Mechanics and Dynamical Astronomy 107: 169–185.

Fujimoto, K. and D.J. Scheeres. 2012. Correlation of optical observations of earth-orbiting objects and initial orbit determination. Journal of Guidance, Control and Dynamics 35: 208–221.

Horwood, J.T. and A.B. Poore. 2011. Adaptive Gaussian sum filters for space surveillance. IEEE Transactions on Automatic Control 56: 1777–1790.

Horwood, J.T., N.D. Aragon and A.B. Poore. 2011. Gaussian sum filters for space surveillance: theory and simulations. Journal of Guidance, Control, and Dynamics 34: 1839–1851.

Hou, T.Y., W. Luo, B. Rozovskii and H. Zhou. 2006. Wiener Chaos expansions and numerical solutions of randomly forced equations of fluid mechanics. Journal of Computational Physics 216: 687–706.

Hussein, I.I., W.T.R. Christopher, M. Mercurio and M.P. Wilkins. 2018. Probabilistic admissible region for multihypothesis filter initialization. Journal of Guidance, Control, and Dynamics 41: 710–724.

Ito, K. and K. Xiong. 2000. Gaussian filters for nonlinear filtering problems. IEEE Transactions on Automatic Control 45: 910–927.

Jazwinski, A. 1970. Stochastic Processing and Filtering Theory. Academic Press, New York.

Jia, B., M. Xin and Y. Cheng. 2011. Sparse Gauss-Hermite quadrature filter with application to spacecraft attitude estimation. Journal of Guidance, Control, and Dynamics 34: 367–379.

Jia, B., M. Xin and Y. Cheng. 2012. Sparse-grid quadrature nonlinear filtering. Automatica 48: 327–341.

Jia, B., M. Xin and Y. Cheng. 2013. High-degree cubature Kalman filter. Automatica 49: 510–518.

Jia, B. and M. Xin. 2017. Orbital uncertainty propagation using positive weighted compact quadrature rule. Journal of Spacecraft and Rockets 54. 683–697.

Jia, B. and M. Xin. 2018. Arbitrary polynomial chaos for short-arc orbital uncertainty propagation. 2018 IEEE Conference on Decision and Control. USA (in press).

Julier, S.J. and J.K. Uhlmann. 2004. Unscented filtering and nonlinear estimation. Proceedings of IEEE 92: 401–422.

Kullback, S. 1997. Information theory and statistics. Dover Publications, Inc., Mineola, New York.

Li, J. and D. Xiu. 2009. A generalized polynomial chaos based ensemble Kalman filter with high accuracy. Journal of Computational Physics 228: 5454–5469.

Luo, W. 2006. Wiener chaos expansion and numerical solutions of stochastic partial differential equations. Ph.D. Thesis, California Institute of Technology, Pasadena, California.

Marelli, S. and B. Sudret. 2015. UQLab user Manual-Polynomial Chaos Expansions, Report UQLab-V0.9-104, Chair of Risk, Safety & Uncertainty Quantification, Zurich, Switzerland.

Murphy, T.S., B. Flewelling and M.J. Holzinger. 2015. Particle and matched filtering using admissible regions. AAS/AIAA Space Flight Mechanics Meeting. USA 155: AAS 15–253.

Nickolls, J., I. Buck, M. Garland and K. Skadron. 2008. Scalable parallel programming with CUDA. Queue 6: 40–53.

Oladyshkin, S. and W. Nowak. 2012. Data-driven uncertainty quantification using the arbitrary polynomial chaos expansion. Reliability Engineering & System Safety 106: 179–190.

Park, R.S. and D.J. Scheeres. 2006. Nonlinear mapping of Gaussian statistics: theory and applications to spacecraft trajectory design. Journal of Guidance, Control, and Dynamics 29: 1367–1375.

Prabhakar, A., J. Fisher and R. Bhattacharya. 2010. Polynomial chaos-based analysis of probabilistic uncertainty in hypersonic flight dynamics. Journal of Guidance, Control, and Dynamics 33: 222–234.

Prince, P.J. and J.R. Dormand. 1981. High order embedded Runge-Kutta formulae. Journal of Computational and Applied Mathematics 7: 67–75.

Runnalls, A.R. 2007. Kullback-Leibler approach to Gaussian mixture reduction. IEEE Transactions on Aerospace and Electronic Systems 43: 989–999.

Sankaran, S. and A.L. Marsden. 2011. A stochastic collocation method for uncertainty quantification and propagation in cardiovascular simulations. Journal of Biomechanical Engineering 133: 031001-031001-12.

Shampine, L.F. and M.K. Gordon. 1975. Computer solution of ordinary differential equations: the initial value problem. W.H. Freeman & Co. Ltd., London.

Sorenson, H.W. and D.L. Alspach. 1971. Recursive Bayesian estimation using Gaussian sums. Automatica 7: 465–479.

Sun, Y. and M. Kumar. 2016. Uncertainty propagation in orbital mechanics via tensor decomposition. Celestial Mechanics and Dynamical Astronomy 124: 269–294.

Terejanu, G., P. Singla, T. Singh. and P.D. Scott. 2008. Uncertainty propagation for nonlinear dynamic systems using Gaussian mixture models. Journal of Guidance, Control, and Dynamics 31: 1623–1633.

Vallado, D.A. 2013. Fundamentals of Astrodynamics and Applications. 4th Edition, Microcosm Press, EI Segundo, California.

Wan, X. and G.E. Karniadakis. 2006. Multi-element generalized polynomial chaos for arbitrary probability measures. SIAM Journal on Scientific Computing 28: 901–928.

Williams, J.L. 2003. Gaussian mixture reduction for tracking multiple maneuvering targets in clutter. M.S. Thesis, Air Force Institute of Technology, Wright-Patterson Air Force Base, Ohio.

Witteveen, J. and G. Iaccarino. 2010. Simplex elements stochastic collocation for uncertainty propagation in robust design optimization. 48th AIAA Aerospace Sciences Meeting. USA, AIAA 2010–1313.

Wojtkiewicz, S., M. Eldred, R. Field, A. Urbina and J. Red-Horse. 2001. Uncertainty quantification in large computational engineering models. 19th AIAA Applied Aerodynamics Conference. USA, AIAA 2001–1455.

Xiu, D. and G.E. Karniadakis. 2002. The Wiener–Askey polynomial chaos for stochastic differential equations. SIAM Journal on Scientific Computing 24: 619–644.

Xiu, D. and G.E. Karniadakis. 2003. Modeling uncertainty in flow simulations via generalized polynomial chaos. Journal of Computational Physics 187: 137–167.

Xiu, D. 2007. Efficient collocational approach for parametric uncertainty analysis. Communications in Computational Physics 2: 293–309.

Yu, Z., P. Cui and M. Ni. 2016. A polynomial chaos based square-root Kalman filter for Mars entry navigation. Aerospace Science and Technology 51: 192–202.

Application Tracking and Navigation 8

Nonlinear estimation is widely used in many engineering fields, such as robotics, guidance, navigation, and control. In this chapter, we present two applications of the grid-based nonlinear filters in target tracking and relative navigation.

8.1 Single Target Tracking

Target or object tracking is essential to many tasks, such as sensor resource allocation, navigation, and control. In general, the target/object should be detected or represented before tracking. In computer vision, the object/target, is usually represented by a feature vector. In aerospace engineering, however, the target/object is often represented as a point mass and the physical descriptor of the target/object is often ignored in tracking tasks. This assumption is valid since the size of the target/object is far smaller than the moving range of the target/object. In computer vision, this assumption is no longer valid, though. In this chapter, we will treat the target/object as a point mass. But, it can be extended, as the Kalman filter is used in the computer vision as well.

Target tracking has been intensively studied for many years. It includes many research aspects, such as single target tracking, multiple target tracking, data association and track fusion, sensor modeling, and performance evaluation. In this chapter, we mainly focus on the space object tracking using the grid-based methods.

Example 8.1: Tracking a maneuvering target

In this example, the goal is to track a target executing a maneuvering turn in a two-dimensional space with unknown and time-varying turn rate. The target dynamics is highly nonlinear due to the unknown turn rate. It has been used

as a benchmark problem to test the performance of different nonlinear filters (Arasaratnam and Haykin 2009, Jia et al. 2013).
The discrete-time dynamic equation of the target motion is given by:

$$\mathbf{x}_k = \begin{bmatrix} 1 & \dfrac{\sin(\omega_{k-1}\Delta t)}{\omega_{k-1}} & 0 & \dfrac{\cos(\omega_{k-1}\Delta t)-1}{\omega_{k-1}} & 0 \\ 0 & \cos(\omega_{k-1}\Delta t) & 0 & -\sin(\omega_{k-1}\Delta t) & 0 \\ 0 & \dfrac{1-\cos(\omega_{k-1}\Delta t)}{\omega_{k-1}} & 1 & \dfrac{\sin(\omega_{k-1}\Delta t)}{\omega_{k-1}} & 0 \\ 0 & \sin(\omega_{k-1}\Delta t) & 0 & \cos(\omega_{k-1}\Delta t) & 0 \\ 0 & 0 & 0 & 0 & 1 \end{bmatrix} \mathbf{x}_{k-1} + \mathbf{v}_{k-1} \quad (8.1)$$

where $\mathbf{x}_k = [x_k, \dot{x}_k, y_k, \dot{y}_k, \omega_k]^T$; $[x_k, y_k]$ and $[\dot{x}_k, \dot{y}_k]$ are the position and velocity at the time k, respectively; Δt is the time interval between two consecutive measurements; ω_{k-1} is the unknown turn rate at the time k–1; $\mathbf{v}_{k-1}$ is the white Gaussian process noise with zero mean and covariance $\mathbf{Q}_{k-1}$,

$$\mathbf{Q}_{k-1} = \begin{bmatrix} \Delta t^3/3 & \Delta t^2/2 & 0 & 0 & 0 \\ \Delta t^2/2 & \Delta t & 0 & 0 & 0 \\ 0 & 0 & \Delta t^3/3 & \Delta t^2/2 & 0 \\ 0 & 0 & \Delta t^2/2 & \Delta t & 0 \\ 0 & 0 & 0 & 0 & 1.75\times10^{-4}\Delta t \end{bmatrix} \quad (8.2)$$

The measurements are the range and angle given by

$$\mathbf{y}_k = \begin{pmatrix} \sqrt{x_k^2 + y_k^2} \\ \operatorname{atan2}(y_k, x_k) \end{pmatrix} + \mathbf{n}_k \quad (8.3)$$

where atan2 is the four-quadrant inverse tangent function; $\mathbf{n}_k$ is the white Gaussian measurement noise with zero mean and covariance $\mathbf{R}_k = \mathrm{diag}([1000\ \mathrm{m}^2, 100\ \mathrm{mrad}^2])$. The sampling interval is $\Delta t = 1s$.

The simulation results are based on 100 Monte Carlo runs. The initial estimate $\hat{\mathbf{x}}_0$ is generated randomly from the normal distribution $N(\hat{\mathbf{x}}_0; \mathbf{x}_0, \mathbf{P}_0)$ with $\mathbf{x}_0$ being the true initial state

$$\mathbf{x}_0 = [1000\ \mathrm{m}, 300\ \mathrm{m/s}, 1000\ \mathrm{m}, 0, -3^0/\mathrm{s}]^T$$

and $\mathbf{P}_0$ being the initial covariance

$$\mathbf{P}_0 = \mathrm{diag}([100\ \mathrm{m}^2, 10\ \mathrm{m}^2/\mathrm{s}^2, 100\ \mathrm{m}^2, 10\ \mathrm{m}^2/\mathrm{s}^2, 100\ \mathrm{mrad}^2/\mathrm{s}^2])$$

The simulation runs for 100 seconds. The root mean square error (RMSE) is used to evaluate the performance of different filters. The RMSE for the position estimate is given by

$$\mathrm{RMSE}_{\mathrm{pos}}(k) = \sqrt{\frac{1}{N_{mc}} \sum_{i=1}^{N_{mc}} \left((x_k - \hat{x}_k)^2 + (y_k - \hat{y}_k)^2 + (z_k - \hat{z}_k)^2 \right)} \quad (8.4)$$

where N_{mc} is the number of Monte Carlo runs. The RMSE for the velocity and turn rate estimates can be defined similarly.

The unscented Kalman filter (UKF), (3rd-degree and 5th-degree) cubature Kalman filter, Gauss-Hermite quadrature filter (GHQF), and the particle filter (PF) with 1500 particles are compared. The UKF parameter κ is chosen to be $\kappa = 1$, $\kappa = 2$, and the suggested value of $\kappa = 3 - n$. The GHQF is based on the 5th-degree Gauss Hermite quadrature (GHQ) rule constructed from the three-point univariate GHQ (Jia et al. 2011). Note that in the simulation setup, all the filters use the same dataset to test the performance. The computational complexity is proportional to the number of points used in different filters, which is summarized in Table 8.1.

From Fig. 8.1 and Fig. 8.2, it can be seen that the 5th-degree CKF achieves lower RMSE for the position estimate than the 3rd-degree CKF, and UKF with the tested parameters. In addition, the 5th-degree CKF has a very close performance to the GHQF. However, the 5th-degree CKF uses much less number of points than GHQF, as shown in Table 8.1. The RMSE of the velocity, as shown in Fig. 8.3 and Fig. 8.4, follows the same trend. The RMSEs of the turn rate, as shown in Fig. 8.5 are very close for different filters. The PF with 1500 particles does not give a good performance for this application. The results contain large errors compared with GHQF or the 5th-degree CKF. Although the PF can be improved by increasing the number of particles or using some advanced techniques, the computational complexity will increase. Hence, the 5th-degree CKF is the best choice in terms of the computational complexity and the estimation accuracy.

The next example is the space object tracking problem. Space object tracking plays an important role in routinely maintaining the space object catalog for space situational awareness. It is also an important component for other tasks, such as conjunction analysis and maneuvering detection. The space object tracking problem has been researched for years and many tools are available (Vetter 2007).

For surveillance of the earth orbit, all space objects involved are concerned. It is different from the traditional tracking scenario, which tracks targets of interested without considering other targets. To track space objects routinely, multiple sensors are available. There are different sensor networks for space

Table 8.1 Number of points for different filters.

Filters	Number of Points
UKF	11
3rd-degree CKF	10
5th-degree CKF	51
GHQF	243
PF	1500

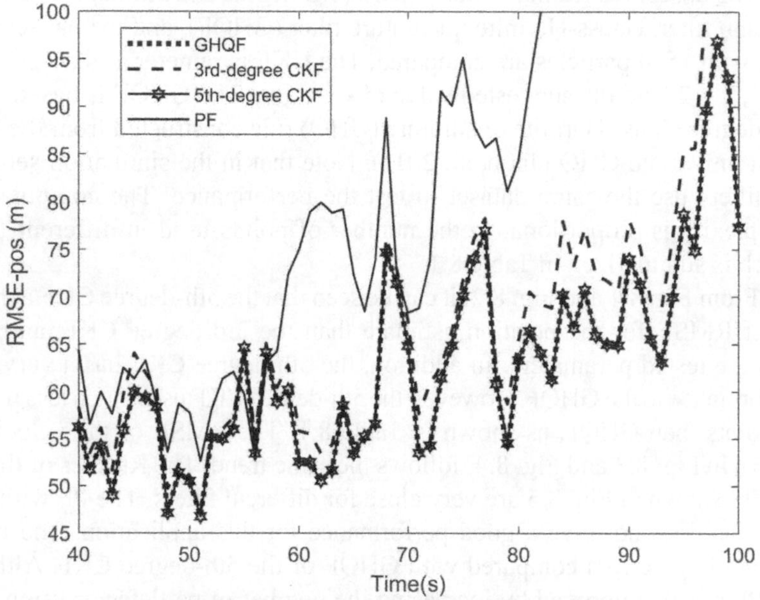

Fig. 8.1 RMSE of position for CKFs, GHQF, and PF.

surveillance, such as the space surveillance network (SSN), which uses 20 to 30 primary sensors globally (Weeden et al. 2010, Vallado and Griesbach 2011). Two main types of sensors are RADAR and electro-optical (EO) sensors. Roughly speaking, the EO sensor can observe objects far from the Earth while the RADAR sensors are mainly used for objects near the Earth. In addition, the EO sensor, in general, is less costly than the RADAR sensor.

To digest observations and maintain the accuracy of the catalog, the recursive nonlinear filtering is the primary way to update the trajectory information of space objects. To use the recursive nonlinear filtering, the model should be built first. Next, we review the space object equation of motion and measurement equation using an EO sensor and its constraints.

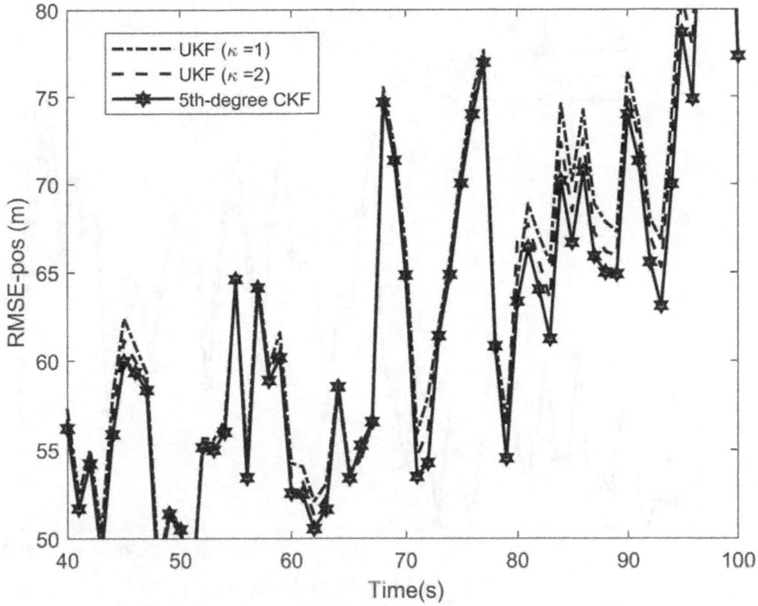

Fig. 8.2 RMSE of position for UKFs and the 5th-degree CKF.

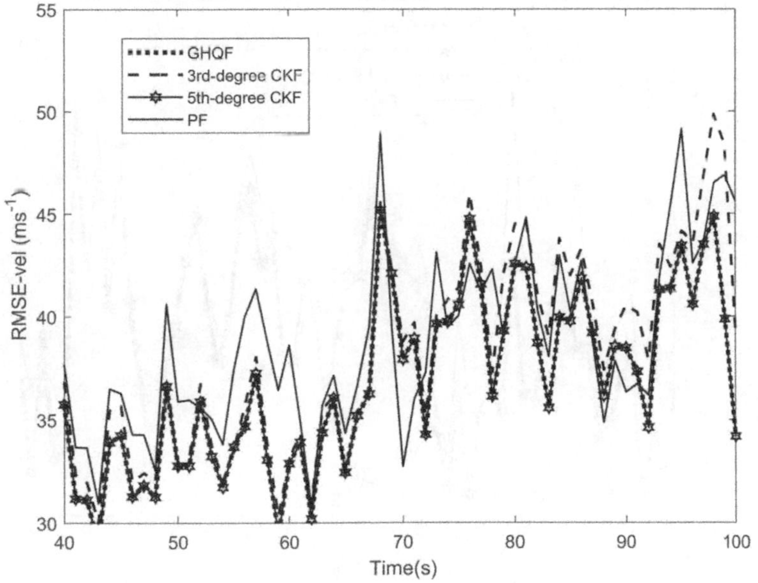

Fig. 8.3 RMSE of velocity for CKFs, GHQF, and PF.

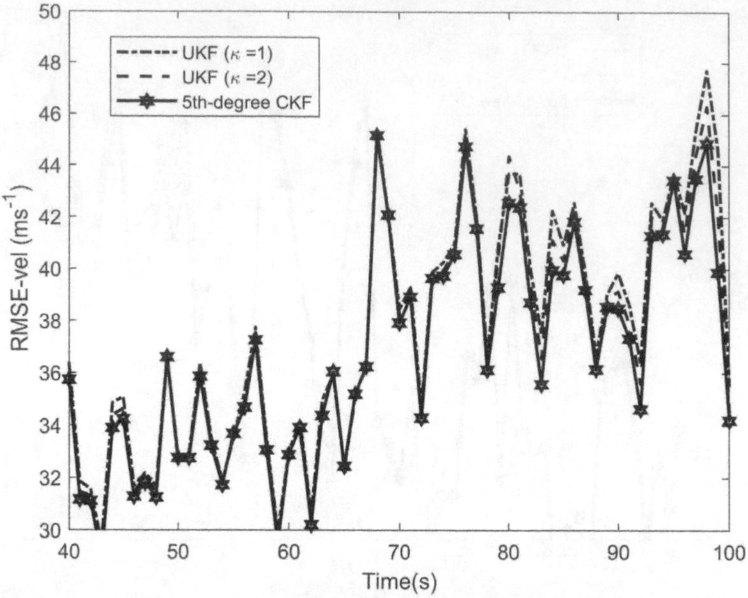

Fig. 8.4 RMSE of velocity for UKFs and the 5th-degree CKF.

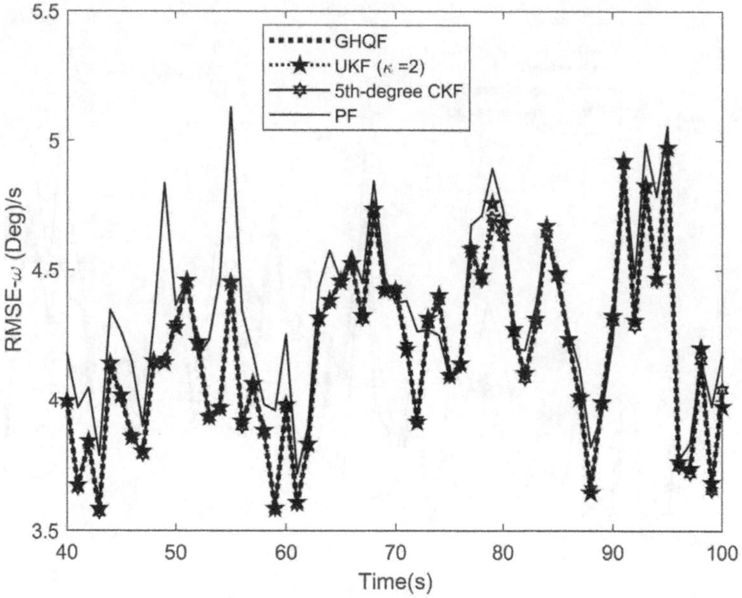

Fig. 8.5 RMSE of turn rate for GHQF, UKF, the 5th-degree CKF, and PF.

Dynamic equation

The dynamic equation of the near-earth space object is given by (Vallado 2013)

$$\ddot{\mathbf{x}} = -\frac{\mu}{r^3}\mathbf{x} + \mathbf{a}_{J_2} + \mathbf{v} \tag{8.5}$$

where $\mathbf{x} = [x, y, z]^T$ is the position of the space object in the inertial coordinate frame (I-J-K), μ is the standard gravitational constant, $r = \sqrt{x^2 + y^2 + z^2}$. $\mathbf{v}$ is the white Gaussian process noise, and $\mathbf{a}_{J_2}$ is the J_2 perturbation,

$$\mathbf{a}_{J_2} = -\frac{3}{2}J_2\left(\frac{R_E}{r}\right)^2 \cdot \frac{\mu}{r^3} \cdot \left[x\left(1-5\frac{z^2}{r^2}\right), y\left(1-5\frac{z^2}{r^2}\right), z\left(3-5\frac{z^2}{r^2}\right)\right]^T \tag{8.6}$$

where R_E is the radius of the earth and J_2 is a constant.

Note that Eq. (8.5) does not consider drag and solar radiation pressure for simplicity. But, these parameters can be incorporated when the mass and surface area of the space object are given.

Electro-optical sensor measurement equation

The ground EO sensor measurement can be described by

$$\begin{cases} az = \tan^{-1}(\rho_e/\rho_n) + n_{az} \\ el = \tan^{-1}\left(\rho_u\Big/\sqrt{\rho_e^2 + \rho_n^2}\right) + n_{el} \end{cases} \tag{8.7}$$

where the azimuth (*az*), the elevation (*el*), and the range $\boldsymbol{\rho} = [\rho_u, \rho_e, \rho_n]^T$ can be measured by the optical sensor on the ground with respect to the local observer coordinate system, ($\hat{\mathbf{u}} - \hat{\mathbf{e}} - \hat{\mathbf{n}}$; "up, east, and north"). n_{az} and n_{el} are white Gaussian measurement noise. The covariance of the measurement noise is assumed to be diag ([10arc sec, 10arc sec]^2).

The geometry of the observation model is shown in Fig. 8.6. The range can be related to the position vector in the inertial frame (**I-J-K**) by the coordinate transformation given by

$$\begin{bmatrix} \rho_u \\ \rho_e \\ \rho_n \end{bmatrix} = \begin{bmatrix} \cos\lambda & 0 & \sin\lambda \\ 0 & 1 & 0 \\ -\sin\lambda & 0 & \cos\lambda \end{bmatrix} \begin{bmatrix} \cos\theta & \sin\theta & 0 \\ -\sin\theta & \cos\theta & 0 \\ 0 & 0 & 1 \end{bmatrix} \begin{bmatrix} x - \|\mathbf{R}\|\cos\lambda\cos\theta \\ y - \|\mathbf{R}\|\cos\lambda\sin\theta \\ z - \|\mathbf{R}\|\sin\lambda \end{bmatrix} \tag{8.8}$$

where $\|\mathbf{R}\| = 6378.1363\,\text{km}$ is the Earth radius; λ and θ are the latitude and local sidereal time of the observer respectively.

Measurements from an EO sensor will be unavailable when the line-of-sight path between the sensor and the space object is blocked by the Earth. The

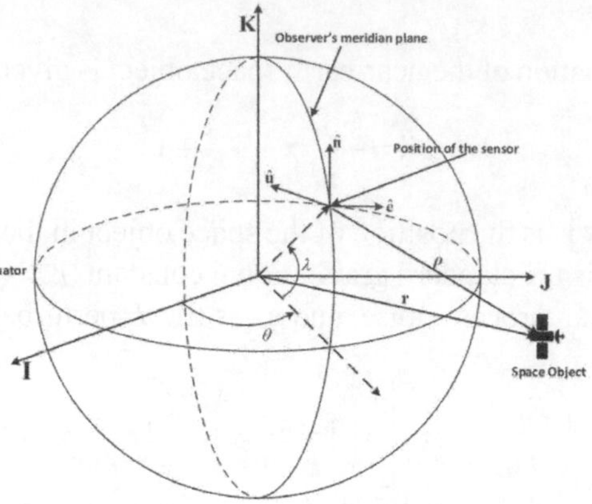

Fig. 8.6 Illustration of the observing geometry.

condition of the Earth blockage is examined between the distance function D_α and the radius of the Earth $R_E = \|\mathbf{R}\|$ (Teixeira et al. 2008). If $D_\alpha < R_E$, where

$$D_\alpha = \sqrt{[(1-\alpha)x_s + \alpha x]^2 + [(1-\alpha)y_s + \alpha y]^2 + [(1-\alpha)z_s + \alpha z]^2} \quad (8.9)$$

then the measurement from the sensor to the space object will be blocked by the Earth. The location of the sensor and space object is denoted as $[x_s, y_s, z_s]^T$ and $[x, y, z]^T$, respectively. $\alpha \in [0,1]$ is a parameter. The minimum of D_α is achieved at $\alpha = \alpha^*$, where α^* is given by

$$\alpha^* = -\frac{x_s(x - x_s) + y_s(y - y_s) + z_s(z - z_s)}{(x - x_s)^2 + (y - y_s)^2 + (z - z_s)^2} \quad (8.10)$$

Thus, the system first examines whether $\alpha^* \in [0,1]$ and then checks the Earth blockage condition $D_\alpha < R_E$.

Another geometric constraint for the EO sensor is that the *dark background* (night) is required. To satisfy a dark background, the angle (as illustrated in Fig. 8.7) between the vector from the Earth's center to the sensor and the vector from the Earth's center to the Sun (θ_n) should be greater than 102 degrees (Hobson 2015). The angle θ_n can be calculated by the law of cosines with the coordinates of the Sun, the sensor, and the space object.

The terrestrial visibility of space objects also needs to be considered. The geometrical constraint for terrestrial visibility is shown in Fig. 8.8.

The angle subtended by the region of the Earth's surface is defined as θ_v. The angle corresponding to a visible space object is calculated by

$$\theta_v = 2\left(\arccos\left(\frac{r_\oplus \cos(\varphi_{\min})}{r_{so}}\right) - \varphi_{\min}\right) \quad (8.11)$$

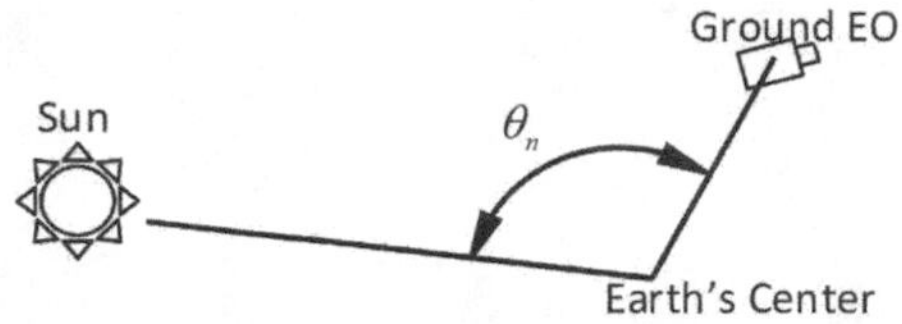

Fig. 8.7 Illustration of the dark background check.

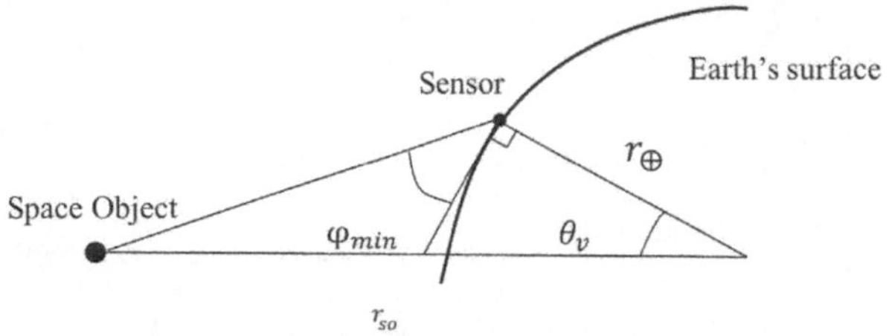

Fig. 8.8 Illustration of geometrical constraint.

where $r_\oplus$ is the Earth radius. $\varphi_{\min}$ is the minimum elevation in order to observe the space object by the ground sensor. r_{so} is the distance from the Earth center to the space object (Hobson 2015).

The Earth's surface from which the space object can be observed due to the terrestrial visibility is represented by the solid angle Ω and it is given by

$$\Omega = 4\pi \sin^2\left(\frac{\theta_v}{4}\right) \tag{8.12}$$

We assume that the Earth is a perfect sphere to build these constraints.

Besides the ground sensors, there are space sensors. Compared to the ground sensors, the visibility constraints of the space sensors are fewer. It is highly possible in the future to deploy cubic satellites to observe the space objects.

Example 8.2: Single space object tracking

Assume that the coordinates of the latitude, longitude, and altitude of the EO sensor are 20°, –10°, 1000 m, respectively. It can be represented in the Earth-centered Inertial (ECI) coordinate frame as $\mathbf{x}_s = [-3064.1, -5154.0, 2170.2]$ km and the space object is represented as $\mathbf{x}$ = [–27828.9136 km, –31685.0205 km, 3.51107 km, 2.30981 km/s, –2.02866 km/s, 0.00192 km/s]^T.

Using Eqs. (8.9), and (8.10), we have $\alpha^* = -0.1573 \notin [0,1]$. Hence, the Earth does NOT block the space object. Assuming the epoch is June 1, 2006, 00:00:00, we have $\theta_n > 102°$, which means nightfall is good for this space object. Using Eq. (8.11), the angle $\theta_v = 61.8385$ with $\varphi_{\min} = 20°$. By checking the latitude and longitude corresponding to the EO sensor, we can be sure that the sensor can observe the space object.

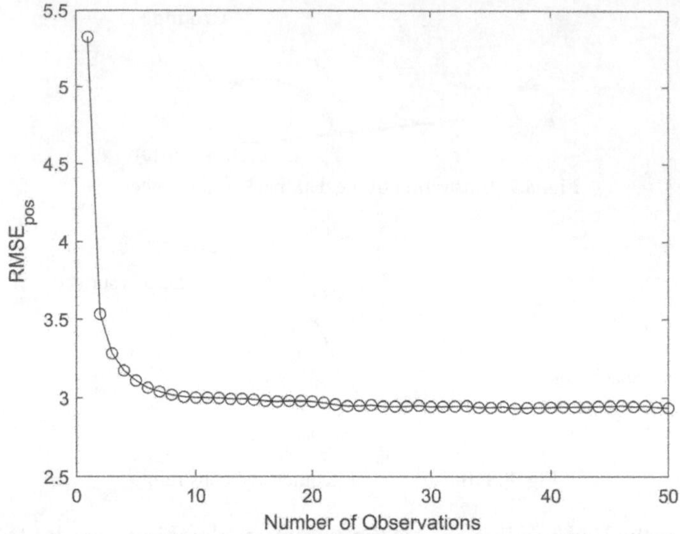

Fig. 8.9 RMSE of position.

For convenience, we assume the variance of the process noise in Eq. (8.5) is given by

$$Q_c = \text{diag}\left(\left[0,0,0,1\times\exp(-12),1\times\exp(-12),1\times\exp(-12)\right]\right)$$

The true state is obtained by using Eq. (8.5). The 3rd-degree spherical-radial cubature Kalman filter (CKF) is used to provide estimated results. The process noise is discretized and the sampling time is 10 seconds. The discretization has been discussed in many papers (Axelsson and Gustafsson 2015, Frogerais et al. 2012). The measurement interval is also 10 seconds.

The RMSE is used to evaluate the performance of CKF. 100 Monte Carlo simulations are conducted. The RMSE corresponding to the position is shown in Fig. 8.9. It can be seen that the RMSE becomes smaller with the increase of the number of observations, which demonstrate the effectiveness of the tracking algorithm.

Note that, if the physical parameters of a space object, such as the mass and area, are available, accurate perturbations can be calculated, such as the drag and solar radiation pressure.

The EKF has been compared to the grid-based Gaussian approximation filters, for example, UKF, in some references (Gaebler et al. 2012). Roughly speaking, the grid-based Gaussian approximation filter has advantages in some challenging scenarios. It is an alternative method compared to the classical EKF in the space object tracking.

8.2 Multiple Target Tracking

One main difference between the multiple target tracking and the single target tracking is how to associate measurements to existing targets. In this chapter, we introduce some typical multiple target tracking algorithms, such as the nearest neighbor filter and probabilistic data association filter.

For the multiple target or space object tracking problem, the measurement-to-track association has to be resolved, as shown in Fig. 8.10. In this section, the global nearest neighbor (GNN) and joint probabilistic data association (JPDA) algorithms are introduced to solve the measurement-to-track association problem.

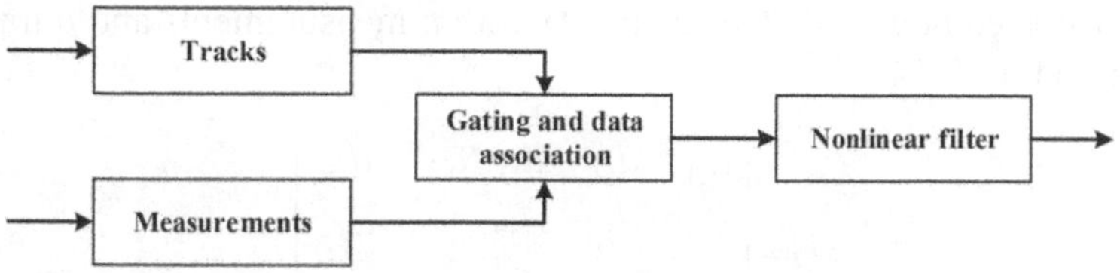

Fig. 8.10 Flowchart of the multiple target tracking.

8.2.1 Nearest Neighbor Filter

The nearest neighbor filter selects the closest neighbor as the consecutive measurement in the track. Instead of using the Euclidean distance, the Mahalanobis distance is used to evaluate the distance from a measurement to the distribution of the predicted measurement. It is assumed that the true measurement is close to the set of observations described by the distribution of the predicted measurement. Mahalanobis distance is given by

$$\sqrt{(\mathbf{y}_k - \hat{\mathbf{y}}_k)^T (\mathbf{P}_{yy,k})^{-1} (\mathbf{y}_k - \hat{\mathbf{y}}_k)} \tag{8.13}$$

In addition, it is assumed that each target generates at maximum one measurement. Hence, it is possible that the same measurement is used into more than one track if the nearest neighbor filter is used in the multiple target tracking scenario. To address this issue, the nearest neighbor filter can be extended to multiple target tracking by using a global least squares integer optimization. This variant is often named as global nearest neighbor (GNN) filter.

Before applying the GNN and JPDA algorithms, the gating procedure can be performed over the measurements in order to reduce the computational time and complexity. The typical ellipsoidal gate region is given by

$$\left(\mathbf{y}_k^m - \hat{\mathbf{y}}_{k|k-1}^l\right)^T \mathbf{M}^{-1} \left(\mathbf{y}_k^m - \hat{\mathbf{y}}_{k|k-1}^l\right) \le \eta^2$$

where $\mathbf{M}$ is the covariance matrix corresponding to $\mathbf{y}_k^m - \hat{\mathbf{y}}_{k|k-1}^l$ and η is the threshold. $\mathbf{y}_k^m$ is the mth measurement at time k and $\hat{\mathbf{y}}_{k|k-1}^l$ is the predicted measurement by the lth track.

To use the GNN algorithm, the distance is defined as $_{ml} = \left(\mathbf{y}_k^m - \hat{\mathbf{y}}_{k|k-1}^l\right)$ $\mathbf{M}^{-1}\left(\mathbf{y}_k^m - \hat{\mathbf{y}}_{k|k-1}^l\right)$. For each measurement from the sensor observation $\mathbf{y}_k^m$, the goal of the GNN algorithm is to choose assignments to minimize the distance. When a single measurement is gated to a single track, it is easy to associate the measurement with the track. For closely spaced targets, it is very likely that multiple measurements fall in a single gate. Simply assigning observations to tracks using minimization of distance could give a wrong assignment solution (Konstantinova et al. 2003). Hence, the validation matrix $\mathbf{D}$ should be constructed and used. This matrix $\mathbf{D}$ with n measurements and p tracks is shown in Eq. (8.14).

$$\mathbf{D} = \begin{bmatrix} d_{11} & d_{12} & d_{13} & \cdots & d_{1p} \\ d_{21} & d_{22} & d_{23} & \cdots & d_{2p} \\ \vdots & \vdots & \vdots & \ddots & \vdots \\ d_{n1} & d_{n2} & d_{n3} & \cdots & d_{np} \end{bmatrix} \tag{8.14}$$

Recall that, for d_{ml}, ‘m’ is used to denote the mth measurement and ‘l’ denotes the track. The goal of the GNN is to find the assignment solution that minimizes the summed total distance using the validation matrix. The Munkres algorithm (Bourgeois and Lassalle 1971) can be used to find the measurement to track association pairs based on the validation matrix.

8.2.2 Probabilistic Data Association Filter

Probabilistic Data Association Filter (PDAF) analyzes association probabilities of measurements and targets for each scan (measurement period). Specifically, PDAF calculates association probabilities for all measurements inside the validation gate. The procedure is given in Fig. 8.11 (Jwo et al. 2013).

Note that the PDAF does not include the initialization algorithms. The clutter is assumed uniformly distributed in the measurement space and the uncertainty of the state is represented by the Gaussian distribution. The PDAF treats all but one measure in the validation region as clutter. It suffers the track coalescence problem. Different from the PDAF, the joint probabilistic data association filter (JPDAF) treats the interacting targets as a cluster. The JPDAF calculates the association probability for all measurements and the active targets in the cluster. Hence, one measurement is used to update more tracks if the measurement is within more than one validation gates.

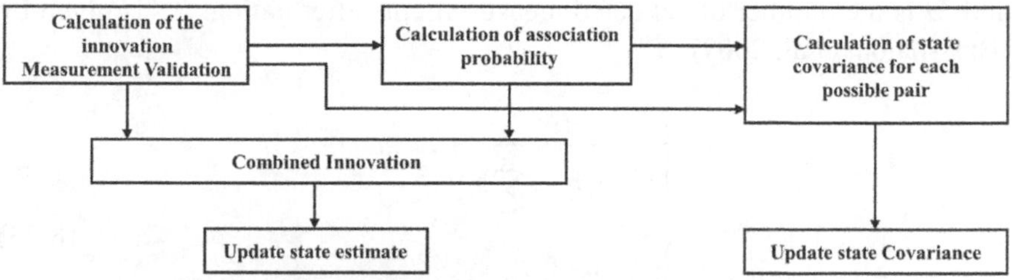

Fig. 8.11 Procedure of PDAF.

Similarly, the probabilistic data association algorithm can be used to solve the measurement-to-track association problem. For convenience, $DA^{m,l}$ is used to denote the event that the lth space object track is associated with the mth measurement. Under this specific association, the innovation $\tilde{\mathbf{y}}_k^{m,l}$ and innovation covariance $\mathbf{S}_k^l$ are given by

$$\tilde{\mathbf{y}}_k^{m,l} = \mathbf{y}_k^m - \boldsymbol{h}^l\left(\hat{\mathbf{x}}_{k|k-1}^l\right) \tag{8.15}$$

$$\mathbf{S}_k^l = \mathbf{H}_k^l \mathbf{P}_{k|k-1}^l \left(\mathbf{H}_k^l\right)^T + \mathbf{R}^l \tag{8.16}$$

where $\mathbf{y}_k^m$ is the mth measurement, the superscript 'l' denotes the lth track, and $\boldsymbol{H}^l$ is the Jacobian matrix of $\boldsymbol{h}^l(\cdot)$. Equation (8.16) can be rewritten by using Eq. (5.75). Assume the probability that the data association $DA^{m,l}$ is correct is $\beta^{m,l}$ and the probability that there is no measurement corresponding to the lth space object is $\beta^{0,l}$. The update procedure using the JPDA filter (JPDAF) for the lth space object can be written as (Kamal et al. 2013)

$$\hat{\mathbf{x}}_{k|k}^l = \hat{\mathbf{x}}_{k|k-1}^l + \mathbf{K}_k^l \tilde{\mathbf{v}}_k^l \tag{8.17}$$

$$\mathbf{P}_{k|k}^l = \mathbf{P}_{k|k-1}^l - \left(1-\beta^{0,l}\right)\mathbf{K}_k^l \mathbf{S}_k^l \mathbf{K}_k^{l\mathrm{T}} + \mathbf{K}_k^l \tilde{\mathbf{P}}_k^l \mathbf{K}_k^{l\mathrm{T}} \tag{8.18}$$

where

$$\mathbf{K}_k^l = \mathbf{P}_{k|k-1,xy}^l \left(\mathbf{S}_k^l\right)^{-1} \tag{8.19}$$

$$\tilde{\mathbf{v}}_k^l = \sum_{m=1}^{M} \beta^{m,l} \tilde{\mathbf{y}}_k^{m,l} = \mathbf{v}_k^l - \left(1-\beta^{0,l}\right)\boldsymbol{h}^l\left(\hat{\mathbf{x}}_{k|k-1}^l\right) \tag{8.20}$$

$$\mathbf{v}_k^l = \sum_{m=1}^{M} \beta^{m,l} \mathbf{y}_k^m \tag{8.21}$$

$$\tilde{\mathbf{P}}_k^l = \left(\sum_{m=1}^{M} \beta^{m,l} \tilde{\mathbf{y}}_k^{m,l} \left(\tilde{\mathbf{y}}_k^{m,l}\right)^{\mathrm{T}}\right) - \tilde{\mathbf{v}}_k^l \left(\tilde{\mathbf{v}}_k^l\right)^{\mathrm{T}} \tag{8.22}$$

and M is the number of validated measurements after gating. $\beta^{m,l}$ is given by (Bar-shalom et al. 2009)

$$\beta^{m,l} = \begin{cases} \dfrac{L^{m,l}}{1 - P_D P_G + \sum_{j=1}^{M} L^{j,l}}, & m = 1, \cdots, M \\ \dfrac{1 - P_D P_G}{1 - P_D P_G + \sum_{j=1}^{M} L^{j,l}}, & m = 0 \end{cases} \tag{8.23}$$

where P_D and P_G are the space object detection probability and gate probability, respectively. In addition,

$$L^{m,l} = \frac{N\left(\mathbf{y}^m; \hat{\mathbf{y}}^l_{k|k-1}, \mathbf{S}^l_k\right) P_D}{\tilde{\lambda}} \tag{8.24}$$

Note that $\hat{\mathbf{y}}^l_{k|k-1}$ is the predicted measurement. $\tilde{\lambda}$ is the density of the spatial Poisson process that models the clutter.

Besides the GNN and JPDA algorithms, the probability hypothesis density filter is another popular method (Vo and Ma 2006). There are some hybrid filters, such as the set joint probabilistic data association filter (Svensson et al. 2011, Crouse 2013). Since the JPDAF may have track coalescence problem, the improved version of the JPDAF was proposed in (Blom and Bloem 2000). Various JPDA algorithms have been implemented (Crouse 2017). The grid-based filters can collaborate with various JPDA algorithms to solve the multiple space object tracking problems.

The multiple hypothesis filter and the random finite set based filter have attracted great attention in multiple target tracking applications. The random finite set method uses a finite set of single target states to represent the multi-target state. Many such multi-target tracking algorithms are proposed in this framework, such as the probability hypothesis density filter (PHDF) (Vo and Ma 2006) and cardinalized probability hypothesis density filter (CPHDF) (Vo et al. 2007). More details can be found in (Vo et al. 2015). Note that in many multiple target tracking algorithms, the grid-based Gaussian approximation filter plays a fundamental role in estimation when the measurement and track are associated.

Many multiple target-tracking algorithms have been applied in space object tracking. In (Jia et al. 2016a), the JPDAF and the cubature Kalman filter were used. The cubature Kalman filter and the multiple hypothesis tracking algorithm were also attempted in (Jia et al. 2016b). To select the best multiple space object tracking algorithm is problem dependent. Benefiting from the development of data processing techniques, it is helpful to integrate the multiple target tracking techniques with the parallel computing techniques, such as GPU and cloud computing.

8.3 Spacecraft Relative Navigation

Relative navigation of a spacecraft is of utmost importance to various space missions, such as rendezvous, docking, and formation flying (Kim et al. 2007). Many navigation technologies have been utilized to determine the relative attitude and orbit between the spacecraft.

Although global positioning systems (GPSs) can be used in many relative navigation problems (Psiaki and Mohiuddin 2007), it is hard to use them for missions in which the GPS signal is difficult to obtain, such as some deep space missions. In addition, the errors in GPS signals due to multipath and atmospheric ionization may deteriorate the estimation accuracy. Hence, many self-dependent measurement systems have been proposed. The inner-formation gravity measurement satellite system was used in (Dang and Zhang 2011), and laser-radar-based relative navigation was proposed in (Wang et al. 2011). Vision-based navigation (VISNAV) and tracking system, which is a self-dependent system, has recently drawn much attention (Chen and Xu 2009, Gunnam et al. 2002, Alonso et al. 2000, Kim et al. 2007, Johnson et al. 2007). In (Chen and Xu 2009), double line-of-sight (LoS) measurement and the extended Kalman filter (EKF) were investigated for far- or medium-range autonomous rendezvous. The observability of the double LoS measuring system is better than that of the single LoS measuring system. In addition, the relative position and velocity estimation accuracy can be improved using the double LoS measurements. A sensor implementation of the VISNAV system was discussed in (Gunnam et al. 2002). A robust and efficient VISNAV algorithm was developed (Alonso et al. 2000) through an optimal observer design and was analyzed using the Lyapunov and contraction mapping approach. There are many advantages of the VISNAV system, such as a small sensor size and a wide sensor field of view. More features can be found in the references (Gunnam et al. 2002, Alonso et al. 2000). The vision-based airborne relative navigation and tracking problem (Johnson et al. 2007, Valasek et al. 2005, Dobrokhodov et al. 2008, Zhang et al. 2010) was investigated as well. A VISNAV system was developed in (Valasek et al. 2005) to address the precise and reliable probe-and-drogue autonomous aerial refueling problem for unmanned aerial vehicles (UAVs). A new autonomous moving target tracking algorithm was proposed in (Dobrokhodov et al. 2008) to estimate the geographic coordinates, speed, and heading of the target using the vision system outputs of the UAV. In (Zhang et al. 2010), the digital elevation map of the area of flight was combined with the analysis of the imaging geometry of the UAV's camera to control the accumulation of the estimation error. The VISNAV system has also been used in relative motion estimation (Oh and Johnson 2007) and mobile robot relative navigation problems (Ohya et al. 1998).

The vision-based relative navigation of two spacecraft (chief spacecraft and deputy spacecraft) is considered in this section. In the VISNAV system, beacons (specific light sources) are utilized to achieve a selective vision and sensors based on position-sensing diodes placed in the focal plane are used to locate the beacons (Kim et al. 2007, Gunnam et al. 2002, Alonso et al. 2000). From the LoS measurement, the relative navigation problem includes estimation of the relative attitude and the relative orbit.

Additionally, gyro biases of the chief and the deputy spacecraft can be determined or corrected as well. The total dimension of the estimation system is 15 and thus it is a high-dimensional estimation problem. For the attitude estimation, there are generally two different categories of parameters to represent the relative attitude: constrained parameters such as the unit quaternion, and unconstrained parameters such as the Euler angles, the Rodrigues parameters, the modified Rodrigues parameters (MRP) (Crassidis et al. 2007, Crassidis and Markley 2003), and the generalized Rodrigues parameters (GRPs) (Crassidis and Markley 2003). The quaternion is widely used to represent the attitude because it is free of singularities and the quaternion kinematic equation is bilinear. However, the unit-norm constraint of the quaternion is often violated in the standard filtering process (Crassidis and Markley 2003). A common approach to overcome this problem is to use the quaternion for global nonsingular attitude representation and a set of unconstrained parameters for local attitude representation and filtering (Crassidis et al. 2007, Crassidis and Markley 2003).

In this section, the relative attitude kinematics and the relative orbit dynamics of two spacecraft, as well as the vision-based measurement model, are briefly reviewed.

Relative attitude kinematics

The relative attitude described by the relative quaternion and its kinematics is given by (Kim et al. 2007)

$$\mathbf{q}_k = \Omega\left(\boldsymbol{\omega}_{d,k-1}\right)\Gamma\left(\boldsymbol{\omega}_{c,k-1}\right)\mathbf{q}_{k-1} \tag{8.25}$$

with

$$\Omega\left(\omega_{d,k-1}\right)=\begin{bmatrix}\cos\left(\frac{\left\|\omega_{d,k-1}\right\|\Delta t}{2}\right)I_{3\times3}-\left[\psi_{k-1}\times\right] & \psi_{k-1}\\ -\left(\psi_{k-1}\right)^{T} & \cos\left(\frac{\left\|\omega_{d,k-1}\right\|\Delta t}{2}\right)\end{bmatrix} \tag{8.26}$$

$$\Gamma\left(\omega_{c,k-1}\right)=\begin{bmatrix}\cos\left(\frac{\left\|\omega_{c,k-1}\right\|\Delta t}{2}\right)I_{3\times3}-\left[\zeta_{k-1}\times\right] & \zeta_{k-1}\\ -\left(\zeta_{k-1}\right)^{T} & \cos\left(\frac{\left\|\omega_{c,k-1}\right\|\Delta t}{2}\right)\end{bmatrix} \tag{8.27}$$

and

$$\psi_{k-1}=\frac{\sin\left(\frac{\left\|\boldsymbol{\omega}_{d,k-1}\right\|\Delta t}{2}\right)\boldsymbol{\omega}_{d,k-1}}{\left\|\boldsymbol{\omega}_{d,k-1}\right\|} \tag{8.28}$$

$$\zeta_{k-1}=\frac{\sin\left(\frac{\left\|\boldsymbol{\omega}_{c,k-1}\right\|\Delta t}{2}\right)\boldsymbol{\omega}_{c,k-1}}{\left\|\boldsymbol{\omega}_{c,k-1}\right\|} \tag{8.29}$$

where $\boldsymbol{\omega}_{c,k-1}$ and $\boldsymbol{\omega}_{d,k-1}$ are angular velocities of the chief and the deputy at the time $k-1$, respectively. Δt is the sample interval and $[\psi_{k-1}\times]$ is a cross product matrix. Assume that $\psi_{k-1}=\left[\psi_{1,k-1},\psi_{2,k-1},\psi_{3,k-1}\right]^{T}$ and $[\psi_{k-1}\times]$ is defined by

$$\left[\psi_{k-1}\times\right]=\begin{bmatrix}0 & -\psi_{3,k-1} & \psi_{2,k-1}\\ \psi_{3,k-1} & 0 & -\psi_{1,k-1}\\ -\psi_{2,k-1} & \psi_{1,k-1} & 0\end{bmatrix}. \tag{8.30}$$

The model used to measure the angular velocity is given by (Crassidis and Markley 2003)

$$\tilde{\boldsymbol{\omega}}=\boldsymbol{\omega}+\boldsymbol{\beta}+\boldsymbol{\eta}_{v} \tag{8.31}$$

$$\dot{\boldsymbol{\beta}}=\boldsymbol{\eta}_{u} \tag{8.32}$$

where $\tilde{\boldsymbol{\omega}}$ and $\boldsymbol{\omega}$ are the continuous-time measured and true angular velocity, respectively. $\boldsymbol{\beta}$ is the gyro bias, and $\boldsymbol{\eta}_v$ and $\boldsymbol{\eta}_u$ are independent white Gaussian noise with zero mean and covariance $\sigma_v^2 I_{3\times3}$ and $\sigma_u^2 I_{3\times3}$, respectively.

In the standard filtering formulation, given the post-update gyro drift rate $\hat{\boldsymbol{\beta}}_{k-1}^{+}$, the estimated angular velocity $\hat{\boldsymbol{\omega}}_{k-1}$ is given by (Crassidis and Markley 2003, Kim et al. 2007)

$$\hat{\boldsymbol{\omega}}_{k-1}^{+} = \tilde{\boldsymbol{\omega}}_{k-1} - \hat{\boldsymbol{\beta}}_{k-1}^{+} \tag{8.33}$$

where $\hat{\boldsymbol{\omega}}_{k-1}$ is the measured angular velocity at time $k-1$.
The prediction of the gyro drift is given by

$$\hat{\boldsymbol{\beta}}_{k}^{-} = \hat{\boldsymbol{\beta}}_{k-1}^{+} \tag{8.34}$$

Relative orbital dynamic equations

In this section, we consider a circular or near circular orbit. In this case, Clohessy-Wiltshire equations can be used to describe the relative orbital dynamics (Vallado 2001)

$$\ddot{x} - 2\omega_r \dot{y} - 3\omega_r^2 x = 0 \tag{8.35}$$

$$\ddot{y} + 2\omega_r \dot{x} = 0 \tag{8.36}$$

$$\ddot{z} + \omega_r^2 z = 0 \tag{8.37}$$

where $[x, y, z]^T$ and $[\dot{x}, \dot{y}, \dot{z}]^T$ are the relative position and velocity vector in the Hill coordinate frame, respectively and ω_r is the mean orbital rate, given by

$$\omega_r = \sqrt{\frac{\mu}{a^3}} \tag{8.38}$$

with μ and a being the standard gravitational parameter and the radius of the chief, respectively. The radius of the chief is assumed known.

The hill coordinate system is illustrated in Fig. 8.12. The vectors $\mathbf{r}_c$ and $\mathbf{r}_d$ denote the position of the chief and the deputy in the inertial coordinate frame (I-J-K), respectively. The x_{ob} axis of the Hill coordinate frame is along the radial direction. The z_{ob} is parallel to the orbit momentum vector. The y_{ob} axis completes the triad. The vector $\Delta\mathbf{r} = (x, y, z)$ denotes the relative position in the Hill coordinate frame.

Define the state variables of the relative orbit as $\mathbf{x}_r = [x, y, z, \dot{x}, \dot{y}, \dot{z}]^T$. Considering the perturbation accelerations, the state equation becomes

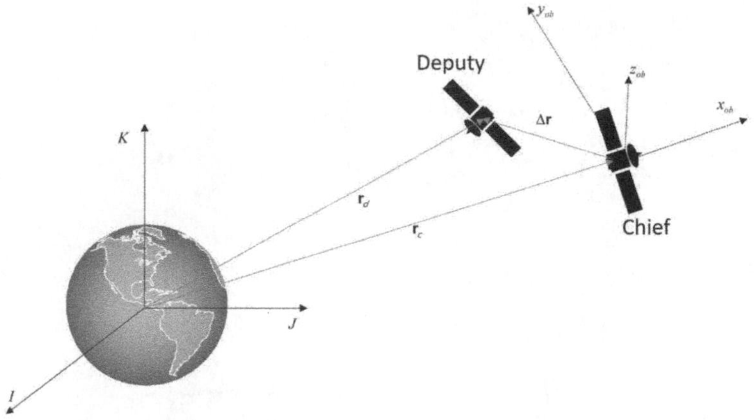

Fig. 8.12 Illustration of the Hill coordinate frame.

$$\dot{\mathbf{x}}_r = f(\mathbf{x}_r) = \begin{bmatrix} \dot{x} \\ \dot{y} \\ \dot{z} \\ 2\omega_r \dot{y} + 3\omega_r^2 x + v_x \\ -2\omega_r \dot{x} + v_y \\ -\omega_r^2 z + v_z \end{bmatrix} \tag{8.39}$$

where v_x, v_y, and v_z are the Gaussian noise with variances of σ_x^2, σ_y^2, and σ_z^2, respectively.

Vision-based measurement model

The schematic of the vision-based relative navigation system is illustrated by Fig. 8.13. The location of the sensor in the deputy coordinate system is assumed known. The locations of the beacons are assumed known in the Hill coordinate frame ($x_{ob} - y_{ob} - z_{ob}$), which is fixed on the chief. The location of the ith beacon (big star) is denoted as (X_i, Y_i, Z_i), and the relative position between the chief and the deputy is denoted as (x, y, z), both of which are defined in the hill coordinate frame.

It is assumed that the chief spacecraft frame coincides with the Hill frame. Also, the chief attitude is assumed known or can be estimated separately with sufficient accuracy. The image space is defined by $(x_{im} - y_{im} - z_{im})$, which is fixed on the deputy with the z_{im} axis pointing toward the boresight. The location of the image point $(\bar{x}_i, \bar{y}_i)$ (small star) in the image space corresponding to the ith beacon can be directly determined by the collinearity equation as follows.

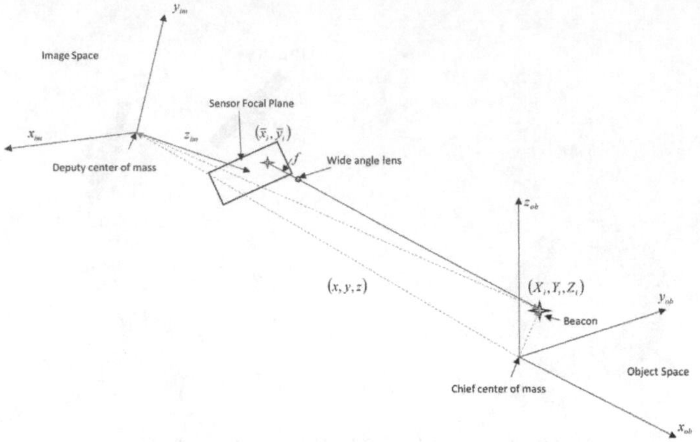

Fig. 8.13 Schematic of the vision-based relative navigation system.

$$\bar{x}_i = -f\frac{A_{11}(X_i - x) + A_{12}(Y_i - y) + A_{13}(Z_i - z)}{A_{31}(X_i - x) + A_{32}(Y_i - y) + A_{33}(Z_i - z)} \tag{8.40}$$

$$\bar{y}_i = -f\frac{A_{21}(X_i - x) + A_{22}(Y_i - y) + A_{23}(Z_i - z)}{A_{31}(X_i - x) + A_{32}(Y_i - y) + A_{33}(Z_i - z)} \tag{8.41}$$

where *f* is the given focal length of the wide-angle lens. $A_{jk}\ (j,k = 1,2,3)$ are the unknown elements of the relative attitude matrix A from the object space to the image space. (X_i, Y_i, Z_i) is the known object space location of the *i*th beacon. The vision-based measurement can be modeled as (Kim et al. 2007)

$$\tilde{\mathbf{b}}_i = A(\mathbf{q})\mathbf{r}_i + \mathbf{n}_i \qquad i = 1,2,\cdots,N_s \tag{8.42}$$

with $A(\mathbf{q}) = \Theta^T(\mathbf{q})\psi(\mathbf{q})$,

$$\Theta(\mathbf{q}) = \begin{bmatrix} q_4 I_{3\times3} + [\boldsymbol{\rho}\times] \\ -\boldsymbol{\rho}^T \end{bmatrix} \tag{8.43}$$

$$\psi(\mathbf{q}) = \begin{bmatrix} q_4 I_{3\times3} - [\boldsymbol{\rho}\times] \\ -\boldsymbol{\rho}^T \end{bmatrix} \tag{8.44}$$

where $\mathbf{q} = \left[\boldsymbol{\rho}^T, q_4\right]^T$ is the relative quaternion in which $\boldsymbol{\rho}$ is the vector part and q_4 is the scalar part; N_s is the number of sensors; $\mathbf{n}_i$ is white Gaussian noise. $\tilde{\mathbf{b}}_i$ denotes the measured value by the *i*th sensor and $\mathbf{r}_i$ is given by (Kim et al. 2007)

$$\mathbf{r}_i = \frac{1}{\sqrt{(X_i - x)^2 + (Y_i - y)^2 + (Z_i - z)^2}} \begin{bmatrix} X_i - x \\ Y_i - y \\ Z_i - z \end{bmatrix} \tag{8.45}$$

Relative attitude and orbit estimation

The state vector of the relative navigation, including the relative attitude and the relative orbit is given by $\mathbf{x} = \left[\mathbf{x}_a^T, \mathbf{x}_r^T\right]^T$ where $\mathbf{x}_a \in \mathbb{R}^9$ and $\mathbf{x}_r \in \mathbb{R}^6$ denote the relative attitude vector and the relative orbit vector, respectively. Many parameters can be used to represent the three-axis attitude (Crassidis and Markley 2003, Cheng and Crassidis 2010, Crassidis and Junkins 2004, Jia et al. 2012, Jia et al. 2011). In this chapter, unconstrained MRPs are used to represent attitude errors in the filtering algorithm. Given the attitude error represented by MPRs $\boldsymbol{\delta\rho}$, the error quaternion $\boldsymbol{\delta q} = [\boldsymbol{\delta\rho}^T, \delta p_4]$ is given by

$$\delta q_4 = \frac{-a_m \|\boldsymbol{\delta p}\|^2 + f_c\sqrt{f_c^2 + (1 - a_m)\|\boldsymbol{\delta p}\|^2}}{f_c^2 + \|\boldsymbol{\delta p}\|^2} \tag{8.46}$$

$$\boldsymbol{\delta\rho} = f_c^{-1}(a_m + \delta q_4)\boldsymbol{\delta p} \tag{8.47}$$

where a_m and f_c are two parameters. In this section, we use $a_m = 1$ and $f_c = 4$. Given the error quaternion $\boldsymbol{\delta q}$, the MRPs $\boldsymbol{\delta p}$ is given by

$$\boldsymbol{\delta p} = f_c \frac{\boldsymbol{\delta\rho}}{a_m + \delta q_4} \tag{8.48}$$

Define the state of the relative attitude at time k as $\hat{\mathbf{x}}_{a,k} = \left[\boldsymbol{\delta p}_k^T, \boldsymbol{\beta}_{c,k}^T, \boldsymbol{\beta}_{d,k}^T\right]^T$, where $\boldsymbol{\delta p}_k$, $\boldsymbol{\beta}_{c,k}$, and $\boldsymbol{\beta}_{d,k}$ are the attitude error, the gyro bias of the chief, and the gyro bias of the deputy, respectively.

Relative navigation algorithm

Given the initial estimate $\hat{\mathbf{x}}_0 = \left[\hat{\mathbf{x}}_{a,0}^T, \hat{\mathbf{x}}_{r,0}^T\right]^T$, $\hat{\mathbf{q}}_0$, and the initial covariance $\mathbf{P}_0$, the grid-based Gaussian approximation relative navigation filtering can be summarized as follows.

Prediction

(1) The transformed points $\xi_{k-1|k-1}(i) = \left[\left(\xi_{a,k-1|k-1}(i)\right)^T, \left(\xi_{r,k-1|k-1}(i)\right)^T\right]^T$ are calculated by Eq. (4.18). Note that $\xi_{a,k-1|k-1}(i) = \left[\left(\xi_{a,k-1|k-1}^{\boldsymbol{\delta p}}(i)\right)^T,\right.$

$\left(\xi_{a,k-1|k-1}^{\hat{\beta}_c}(i)\right)^T, \left(\xi_{a,k-1|k-1}^{\hat{\beta}_d}(i)\right)^T$ with i being the point index and $\xi_{a,k-1|k-1}^{\delta\mathbf{p}}(1) = \mathbf{0}_{3\times 1}$ (Crassidis and Markley 2003); The subscripts 'a' and 'r' denote the points corresponding to the attitude state and relative orbital state, respectively. The superscripts '$\delta\mathbf{p}$', '$\hat{\boldsymbol{\beta}}_c$', and '$\hat{\boldsymbol{\beta}}_d$' denote the points corresponding to the estimated MRPs, chief gyro bias, and the deputy bias, respectively. $\left(\xi_{r,k-1|k-1}(i)\right)^T$ are the transformed points corresponding to the six-dimensional state $\mathbf{x}_r$. $\xi_{a,k-1|k-1}^{\delta\mathbf{p}}(i)$ are then transformed into error quaternion $\delta\mathbf{q}_{k-1|k-1}^{\text{pre}}(i)$ by Eqs. (8.46) and (8.47). The superscript 'pre' denotes the prediction step.

(2) The transformed quaternions $\hat{\mathbf{q}}_{k-1|k-1}^{\text{pre}}(i)$ can be obtained by $\hat{\mathbf{q}}_{k-1|k-1}^{\text{pre}}(i) = \delta\hat{\mathbf{q}}_{k-1|k-1}^{\text{pre}}(i) \odot \hat{\mathbf{q}}_{k-1|k-1}$, where $\odot$ denotes the quaternion product.

(3) The predicted quaternions $\hat{\mathbf{q}}_{k|k-1}^{\text{pre}}(i)$ are predicted by Eq. (8.25). Note, $\hat{\boldsymbol{\omega}}_{c,k-1}$ and $\hat{\boldsymbol{\omega}}_{d,k-1}$ are obtained by Eq. (8.33).

(4) Error quaternions $\delta\mathbf{q}_{k|k-1}^{\text{pre}}(i)$ are calculated by $\delta\mathbf{q}_{k|k-1}^{\text{pre}}(i) = \hat{\mathbf{q}}_{k|k-1}^{\text{pre}}(i) \odot \left(\hat{\mathbf{q}}_{k|k-1}^{\text{pre}}(1)\right)^{-1}$. Then, the predicted points $\xi_{a,k|k-1}^{\delta\mathbf{p}}(i)$ can be calculated by Eq. (8.48). The predicted points $\xi_{a,k|k-1}^{\hat{\beta}_c}(i)$ and $\xi_{a,k|k-1}^{\hat{\beta}_d}(i)$ are given by Eq. (8.34) and $\xi_{r,k|k-1}(i)$ is obtained by the relative orbit dynamics in Eqs. (8.35)–(8.37). Hence, $\xi_{k|k-1}(i) = \left[\left(\xi_{a,k|k-1}(i)\right)^T, \left(\xi_{r,k|k-1}(i)\right)^T\right]^T$ can be obtained.

(5) The mean and covariance are calculated by Eqs. (4.16)–(4.17). Then the first three mean values are transformed into the error quaternion $\delta\mathbf{q}_{k|k-1}$. The predicted quaternion $\hat{\mathbf{q}}_{k|k-1}$ is obtained by $\hat{\mathbf{q}}_{k|k-1} = \delta\mathbf{q}_{k|k-1} \odot \hat{\mathbf{q}}_{k|k-1}^{\text{pre}}(1)$.

Update

(1) Similarly, the transformed points $\tilde{\xi}_{k|k-1}(i) = \left[\left(\tilde{\xi}_{a,k|k-1}(i)\right)^T, \left(\tilde{\xi}_{r,k|k-1}(i)\right)^T\right]^T$ are generated by Eq. (4.24). Then $\tilde{\xi}_{a,k|k-1}^{\delta\mathbf{p}}(i)$ can be transformed into the error quaternion $\delta\mathbf{q}_{k|k-1}^{\text{upd}}(i)$ by Eqs. (8.46) and (8.47). The superscript '*upd*' denotes the update step; $\tilde{\xi}_{a,k|k-1}^{\delta\mathbf{p}}(1) = \mathbf{0}_{3\times 1}$ (Crassidis and Markley 2003).

(2) Calculate quaternions $\hat{\mathbf{q}}_{k|k-1}^{\text{upd}}(i) = \delta\hat{\mathbf{q}}_{k|k-1}^{\text{upd}}(i) \odot \hat{\mathbf{q}}_{k|k-1}$.

(3) Given the measurement values, the state $\hat{\mathbf{x}}_{k|k}$ and covariance $\mathbf{P}_{k|k}$ can be updated by Eqs. (4.19)–(4.20) with $\hat{\mathbf{q}}_{k|k-1}^{\text{upd}}(i)$, $\tilde{\xi}_{k|k-1}(i)$, $\tilde{\mathbf{b}}_i$ and $\mathbf{R}_k$.

(4) Calculate the error quaternion $\delta\mathbf{q}_{k|k}$ by Eqs. (8.46) and (8.47), and the updated quaternion $\hat{\mathbf{q}}_{k|k} = \delta\mathbf{q}_{k|k} \odot \hat{\mathbf{q}}_{k|k-1}$.

Approximation of the process noise

The covariance of the discrete-time noise process, $\mathbf{Q}$, used in filtering is intractable because it depends on the attitude matrix (Kim et al. 2007). A numerically approximated discrete-time process noise can be obtained as follows.

The error-state dynamics of the relative navigation problem is given by (Kim et al. 2007)

$$\Delta\dot{\mathbf{x}} = \mathbf{F}\Delta\mathbf{x} + \mathbf{G}\mathbf{v}_e \tag{8.49}$$

where $\Delta\mathbf{x}$ is the error state vector with $\Delta\mathbf{x} = \left[\boldsymbol{\delta}\mathbf{p}, \boldsymbol{\delta\beta}_c, \boldsymbol{\delta\beta}_d, \boldsymbol{\delta}\mathbf{x}_r\right]^T$. $\boldsymbol{\delta}\mathbf{p}$, $\boldsymbol{\delta\beta}_c$, $\boldsymbol{\delta\beta}_d$ and $\boldsymbol{\delta}\mathbf{x}_r$ are the attitude error, the gyro bias error of the chief, the gyro bias error of the deputy, and the error of relative position and velocity, respectively.

$$\mathbf{F} = \begin{bmatrix} -[\hat{\boldsymbol{\omega}}_d \times] & A(\hat{\mathbf{q}}) & -I_{3x3} & \mathbf{0}_{3\times6} \\ \mathbf{0}_{3\times3} & \mathbf{0}_{3\times3} & \mathbf{0}_{3\times3} & \mathbf{0}_{3\times6} \\ \mathbf{0}_{3\times3} & \mathbf{0}_{3\times3} & \mathbf{0}_{3\times3} & \mathbf{0}_{3\times6} \\ \mathbf{0}_{6\times3} & \mathbf{0}_{6\times3} & \mathbf{0}_{6\times3} & \left.\dfrac{\partial f(\mathbf{x}_r)}{\partial \mathbf{x}_r}\right|_{\hat{\mathbf{x}}_r} \end{bmatrix} \tag{8.50}$$

$$\mathbf{G} = \begin{bmatrix} A(\hat{\mathbf{q}}) & -I_{3\times3} & \mathbf{0}_{3\times3} & \mathbf{0}_{3\times3} & \mathbf{0}_{3\times3} \\ \mathbf{0}_{3\times3} & \mathbf{0}_{3\times3} & I_{3\times3} & \mathbf{0}_{3\times3} & \mathbf{0}_{3\times3} \\ \mathbf{0}_{3\times3} & \mathbf{0}_{3\times3} & \mathbf{0}_{3\times3} & I_{3\times3} & \mathbf{0}_{3\times3} \\ \mathbf{0}_{3\times3} & \mathbf{0}_{3\times3} & \mathbf{0}_{3\times3} & \mathbf{0}_{3\times3} & \mathbf{0}_{3\times3} \\ \mathbf{0}_{3\times3} & \mathbf{0}_{3\times3} & \mathbf{0}_{3\times3} & \mathbf{0}_{3\times3} & I_{3\times3} \end{bmatrix} \tag{8.51}$$

The covariance of the process noise $\mathbf{v}_e$ is given by

$$\mathbf{Q}_c = \mathrm{diag}\left(\left[\left(\sigma_{cv}^2 I_{3\times3}\right), \left(\sigma_{dv}^2 I_{3\times3}\right), \left(\sigma_{cu}^2 I_{3\times3}\right), \left(\sigma_{du}^2 I_{3\times3}\right), \sigma_x^2, \sigma_y^2, \sigma_z^2\right]\right) \tag{8.52}$$

where $\sigma_{cv}^2 I_{3\times3}$ and $\sigma_{cu}^2 I_{3\times3}$ are covariance matrices corresponding to gyro noises in Eqs. (8.31) and (8.32) of the chief, respectively. Similarly, $\sigma_{dv}^2 I_{3\times3}$ and $\sigma_{du}^2 I_{3\times3}$ are covariance matrices corresponding to gyro noises of the deputy. σ_x^2, σ_y^2, and σ_z^2 are covariances of Gaussian process noises in the relative orbit dynamics (Eq. 8.39).
Then, the approximated discrete-time $\mathbf{Q}$ is given by (Kim et al. 2007)

$$\mathbf{Q} = \mathbf{B}_{22}^T \mathbf{B}_{12} \tag{8.53}$$

where $\mathbf{B}_{12}$ and $\mathbf{B}_{22}$ can be calculated from the following equations.

$$e^{\mathbf{A}} = \begin{bmatrix} \mathbf{B}_{11} & \mathbf{B}_{12} \\ \mathbf{B}_{21} & \mathbf{B}_{22} \end{bmatrix} \tag{8.54}$$

with

$$\mathbf{A} = \begin{bmatrix} -\mathbf{F} & \mathbf{G}\mathbf{Q}_c\mathbf{G}^T \\ \mathbf{0} & \mathbf{F}^T \end{bmatrix} \Delta t \tag{8.55}$$

$e^{(\cdot)}$ is the matrix exponential operator and the discrete-time process noise at the time $k-1$ can be obtained using Eq. (8.53).

Simulation results are presented to compare the SGQF with the EKF, the UKF, and the 3rd-degree CKF. In the simulation scenario, the chief orbit radius is assumed to be 7,278,136 m. The initial states are generated randomly by adding random errors with the normal distribution $N(\mathbf{0},\mathbf{P}_0)$ to the initial true value $\mathbf{x}_0 = \left[\mathbf{q}_0^T, \boldsymbol{\beta}_{c,0}^T, \boldsymbol{\beta}_{d,0}^T, \mathbf{x}_{r,0}^T\right]^T$ where $\mathbf{q}_0 = [-0.011108, 0.707019, 0.58552, 0.396443029419108]^T$, $\mathbf{x}_{r,0}$ = $[200\text{m}, 98.3471\text{m}, 200\text{m}, 0.05\,\text{m/s}, -0.4067\,\text{m/s}, 0.05\,\text{m/s}]^T$, $\boldsymbol{\beta}_{c,0}$ = $[1,1,1]^T$ deg/hour, and $\boldsymbol{\beta}_{d,0}$ = $[1,1,1]^T$ deg/hour. $\mathbf{P}_0$ is the initial covariance. The initial covariance of the attitude is $\text{diag}\left(\left[15^\circ, 15^\circ, 15^\circ\right]^2\right)$. The initial covariance matrices of the gyro biases for the chief and the deputy are assumed to be the same and are diag([10°/h, 10°/h, 10°/h]^2). The initial covariances of the position and velocity are diag([10 m , 10 m, 10 m]^2) and diag([0.01 m/s, 0.01 m/s, 0.01 m/s]^2), respectively.

To achieve bounded relative orbit, the following constraint for the initial values should be satisfied (Vallado 2001)

$$\dot{y}_0 = -2\omega_r x_0 \tag{8.56}$$

and

$$y_0 = 2\dot{x}_0 / \omega_r \tag{8.57}$$

where ω_r is given by Eq. (8.38).

The true angular velocities of the chief and deputy are given by $\boldsymbol{\omega}_c = [0, 0.001, -0.001]^T$ rad/s and $\boldsymbol{\omega}_d = [-0.001, 0, 0.001]^T$ rad/s, respectively. The standard deviations of the gyro biases for the chief and the deputy are $\sigma_{cu} = \sigma_{du} = \sqrt{10} \times 10^{-10}$ rad/s$^{3/2}$ and $\sigma_{cv} = \sigma_{dv} = \sqrt{10} \times 10^{-5}$ rad/s$^{3/2}$, respectively. Four beacons are used in the simulation. The locations of the beacons are listed in Table 8.2.

In the filters, the fourth-order Runge-Kutta method is used to propagate the state $\mathbf{x}_r$ of the relative orbit. The covariance in the measurement equation is $R_i = (0.0005°)^2$.

Table 8.2 Locations of beacons (meter).

	X_i	Y_i	Z_i
Beacon 1	0.5	0.5	0
Beacon 2	–0.5	0.5	0
Beacon 3	0.5	–0.5	0
Beacon 4	–0.5	–0.5	0

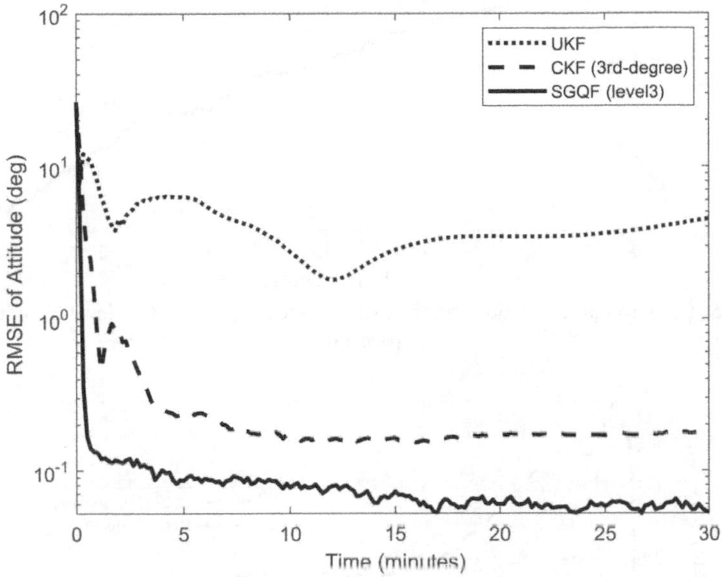

Fig. 8.14 The attitude estimation error of UKF, CKF (3rd-degree), and SGQF (level-3).

The simulation time is 30 minutes and the sampling time is 1 second. The measurement period is 10 seconds. The following simulation results are based on 50 Monte Carlo runs. The absolute estimation errors of each state are shown in Fig. 8.14–Fig. 8.18. The norm of attitude estimation error (represented by Euler angles and calculated from the estimated quaternion and true quaternion) is shown in Fig. 8.14. The result of EKF is not shown because it cannot provide acceptable results. Both the 3rd-degree CKF and level-3 SGQF give better performance than the UKF. The performance of the 3rd-degree CKF is worse than that of the level-3 SGQF. Also, the level-3 SGQF converges much faster than the 3rd-degree CKF, which is a typical level-2 filter. In the following figures, the performance of UKF is not shown due to its inferior performance. From Fig. 8.15 and Fig. 8.16, the level-3 SGQF can obtain acceptable results for estimating the gyro biases of the chief and the deputy while 3rd-degree CKF cannot.

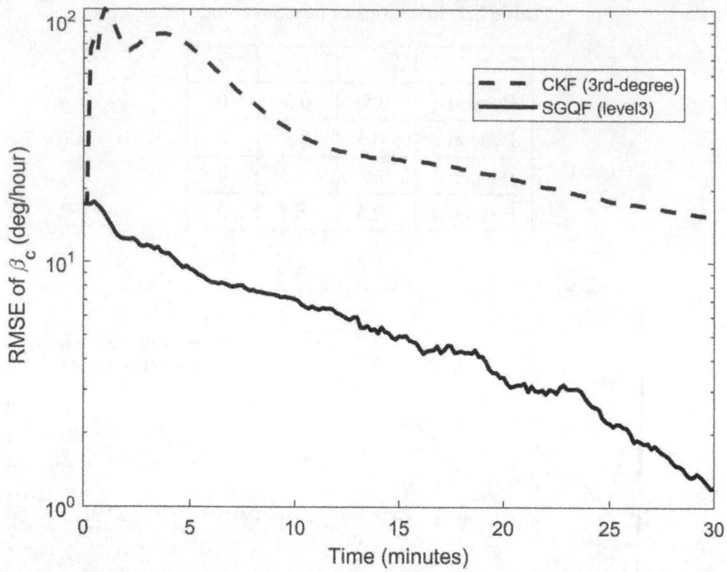

Fig. 8.15 The gyro bias estimation error of CKF (3rd-degree) and SGQF (level-3) for the chief spacecraft.

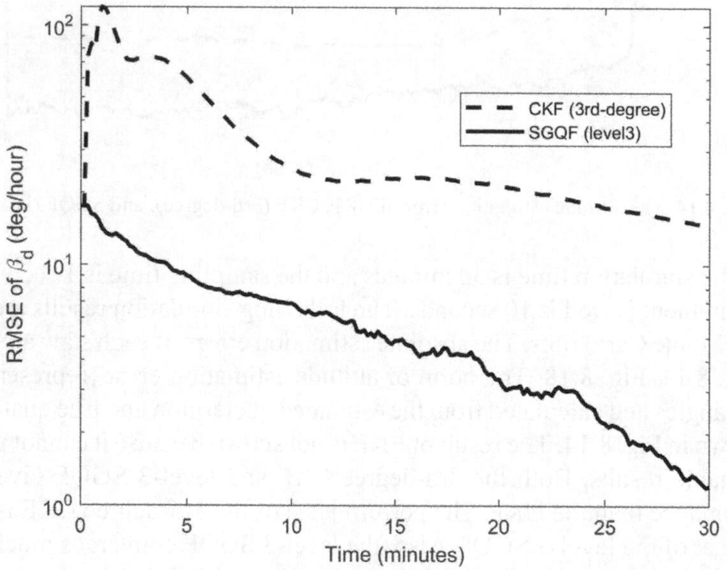

Fig. 8.16 The gyro bias estimation error of CKF (3rd-degree) and SGQF (level-3) for the deputy spacecraft.

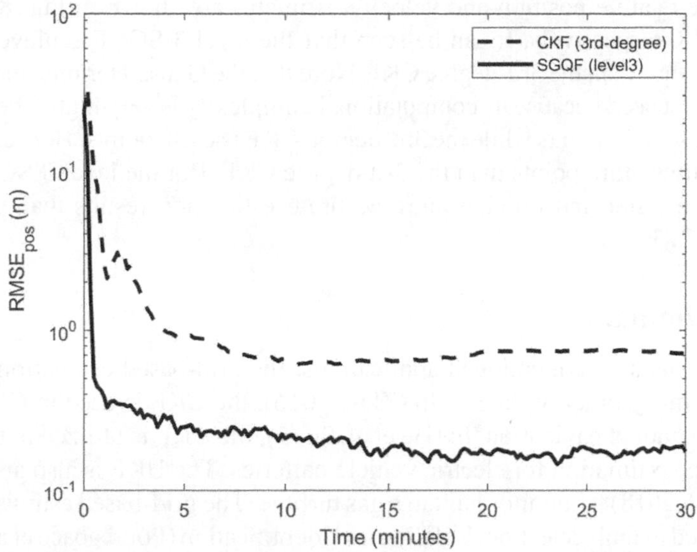

Fig. 8.17 The relative position estimation error of CKF (3rd-degree) and SGQF (level-3).

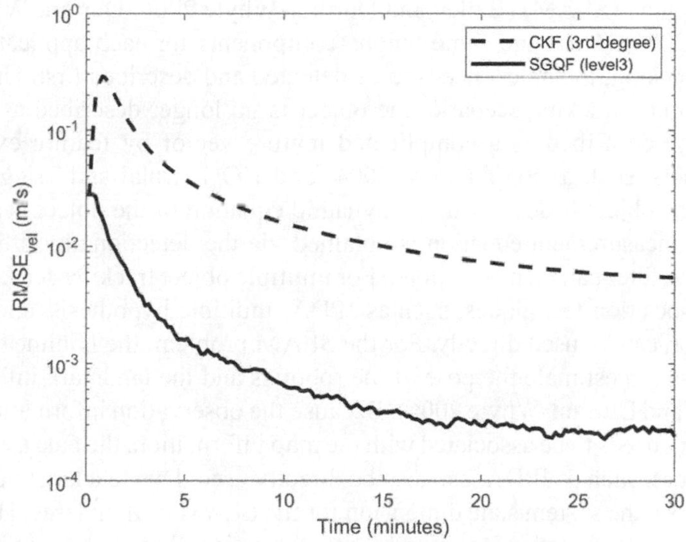

Fig. 8.18 The relative velocity estimation error of CKF (3rd-degree), and SGQF (level-3).

The relative position and velocity estimates are shown in Fig. 8.17 and Fig. 8.18, respectively. It can be seen that the level-3 SGQF achieves better performance than the 3rd-degree CKF. Note that the Gauss-Hermite quadrature filter is not used because its computational complexity is very high. The level-3 SGQF uses 451 points while the 3rd-degree CKF uses 30 points. Hence, level-3 SGQF uses more points than the 3rd-degree CKF. But the level-3 SGQF can converge faster and provide more accurate estimation results than the 3rd-degree CKF.

8.4 Summary

Besides the above-mentioned applications, the grid-based estimation can be used in many other problems. In (Zhao 2016), the CKF is used in GPS/IMU tightly-coupled navigation. In (He et al. 2013), the UKF is utilized in the state of charge estimation for electric vehicle batteries. The UKF is also applied in (Lu et al. 2018) to monitor fault in a gas turbine. The grid-based estimation has been used in fault detection, isolation and identification (Pourbabaee et al. 2016, Chatterjee et al. 2015, Xiong et al. 2007). Nonlinear filters have tremendous applications in video tracking (Bewley et al. 2016) and simultaneous localization and mapping (SLAM) (Bailey and Durrant-Whyte 2006, Durrant-Whyte and Bailey 2006). There are some unique components for each application. For video tracking, the object needs to be detected and described first. Unlike the typical radar tracking scenario, the object is no longer described as a point. It can be described as a complicated feature vector by feature extraction algorithms, such as SIFT (Lowe 2004) and HOG (Dalal and Triggs 2005). When the object is described, the dynamic equation of the object is assumed and the measurement equation is obtained via the detection algorithms. The nonlinear filter can then be applied. For multiple object tracking scenarios, the data association techniques, such as JPDA, multiple hypothesis, and nearest neighbor, can be used directly. For the SLAM problem, the nonlinear filter is often used to estimate the pose of the robotics and the landmark information (Bailey and Durrant-Whyte 2006). Because the observation information of the landmark needs to be associated with the map information, the data association techniques, such as JPDA, can also be directly used. Due to a large number of landmarks, the system state dimension for the SLAM is often large. Hence, an efficient implementation is required. The information filter can be used in SLAM to improve the computational efficiency (Bailey and Durrant-Whyte 2006). There are numerous other promising applications using grid-based estimation techniques. Due to the inherent connections among various grid-based nonlinear estimation techniques, it offers a great flexibility for the designer to select these methods for different applications.

Estimation is also an indispensable component in the feedback control system. Although the estimation and control problems are often solved separately, the performance of the control system greatly depends on the accuracy of the state estimation. The object tracking and sensor resource allocation problems can be viewed as an estimation and control problem, respectively. The grid-based Gaussian approximation filters can be integrated with many other techniques in different areas, such as machine learning for modeling of dynamic systems, system identification, and fault detection. We can foresee the broad impact of the grid-based estimation techniques as a fundamental component in many different fields.

References

Alonso, R., J.L. Crassidis and J. Junkins. 2000. Vision-based relative navigation for formation flying of spacecraft. AIAA Guidance, Navigation, and Control Conference and Exhibit. USA, AIAA-2000–4439.

Arasaratnam, I. and S. Haykin. 2009. Cubature Kalman filter. IEEE Transactions on Automatic Control 54: 1254–1269.

Axelsson, P. and F. Gustafsson. 2015. Discrete-time solutions to the continuous-time differential Lyapunov equation with applications to Kalman filtering. IEEE Transactions on Automatic Control 60: 632–643.

Bailey, T. and H. Durrant-Whyte. 2006. Simultaneous localization and mapping (SLAM): Part II. IEEE Robotics & Automation Magazine 13: 108–117.

Bar-Shalom, Y., F. Daum and J. Huang. 2009. The probabilistic data association filter. IEEE Control Systems Magazine 29: 82–100.

Bewley, A., Z. Ge, L. Ott, F. Ramos and B. Upcroft. 2016. Simple online and realtime tracking. IEEE International Conference on Image Processing. USA, 3464–3468.

Blom, H.A.P. and E.A. Bloem. 2000. Probabilistic data association avoiding track coalescence. IEEE Transactions on Automatic Control 45: 247–259.

Bourgeois, F. and J. Lassalle. 1971. An extension of the Munkres algorithm for the assignment problem to rectangular matrices. Communications of the ACM 14: 802–804.

Chatterjee, S., S. Sadhu and T.K. Ghoshal. 2015. Fault detection and identification of non-linear hybrid system using self-switched sigma point filter bank. IET Control Theory & Applications 9: 1093–1102.

Chen, T. and S. Xu. 2009. Double line-of-sight measuring relative navigation for spacecraft autonomous rendezvous. Acta Astronautica 67: 122–134.

Cheng, Y. and J.L. Crassidis. 2010. Particle filtering for attitude estimation using a minimal local-error representation. Journal of Guidance, Control, and Dynamics 33: 1305–1310.

Crassidis, J.L. and F.L. Markley. 2003. Unscented filtering for spacecraft attitude estimation. Journal of Guidance, Control, and Dynamics 26: 536–542.

Crassidis, J.L. and J.L. Junkins. 2004. Optimal Estimation of Dynamic Systems. CRC Press, Boca Raton, Florida.

Crassidis, J.L., F.L. Markley and Y. Cheng. 2007. Survey of nonlinear attitude estimation methods. Journal of Guidance, Control, and Dynamics 30: 12–28.

Crouse, D.F. 2013. Advances in displaying uncertain estimates of multiple targets. Proceedings of SPIE, Signal Processing, Sensor Fusion, and Target Recognition XXII. USA 8745: 874504-1.

Crouse, D.F. 2017. The tracker component library: free routines for rapid prototyping. IEEE Aerospace and Electronic Systems Magazine 32: 18–27.

Dalal, N. and B. Triggs. 2005. Histograms of oriented gradients for human detection. IEEE Computer Society Conference on Computer Vision and Pattern Recognition. USA.

Dang, Z. and Y. Zhang. 2011. Relative position and attitude estimation for inner-formation gravity measurement satellite system. Acta Astronautica 69: 514–525.

Dobrokhodov, V.N., I.I. Kaminer, K.D. Jones and R. Ghabcheloo. 2008. Vision-based tracking and motion estimation for moving targets using unmanned air vehicles. Journal of Guidance, Control, and Dynamics 31: 907–917.

Durrant-Whyte, H. and T. Bailey. 2006. Simultaneous localized and mapping: Part I. IEEE Robotics & Automation Magazine 13: 99–110.

Frogerais, P., J. Bellanger and L. Senhadji. 2012. Various ways to compute the continuous-discrete extended Kalman filter. IEEE Transactions on Automatic Control 57: 1000–1004.

Gaebler, J., S. Hur-Diaz and R. Carpenter. 2012. Comparison of sigma-point and extended Kalman filters on a realistic orbit determination scenario. The Journal of the Astronautical Sciences 59: 301–307.

Gunnam, K.K., D.C. Hughes, J.L. Junkins and N. Kehtarnavaz. 2002. A vision-based DSP embedded navigation sensor. IEEE Sensors Journal 2: 428–442.

He, W., N. Williard, C. Chen and M. Pecht. 2013. State of charge estimation for electric vehicle batteries using unscented Kalman filtering. Microelectronics Reliability 53: 840–847.

Hobson, T.A. 2015. Sensor management for enhanced catalogue maintenance of resident space objects. Ph.D. Thesis, The University of Queensland, Brisbane, Australia.

Jia, B., M. Xin and Y. Cheng. 2011. Sparse Gauss-Hermite quadrature filter with application to spacecraft attitude estimation. Journal of Guidance, Control, and Dynamics 34: 367–379.

Jia, B., M. Xin and Y. Cheng. 2012. Anisotropic sparse Gauss-Hermite quadrature filter. Journal of Guidance, Control, and Dynamics 35: 1014–1022.

Jia, B., M. Xin and Y. Cheng. 2013. High-degree cubature Kalman filter. Automatica 49: 510–518.

Jia, B., K.D. Pham, E. Blasch, G. Chen, D. Shen and Z. Wang. 2016a. Multiple space object tracking via a space-based optical sensor. IEEE Aerospace Conference. USA, 1–10.

Jia, B., K.D. Pham, E. Blasch, D. Shen, Z. Wang and G. Chen. 2016b. Cooperative space object tracking using space-based optical sensors via consensus-based filters. IEEE Transactions on Aerospace and Electronic Systems 52: 1908–1936.

Johnson, E.N., A.J. Calise, Y. Watanabe, J. Ha and J.C. Neidhoefer. 2007. Real-time vision-based relative aircraft navigation. Journal of Aerospace Computing, Information, and Communication 4: 707–738.

Jwo, D.J., F.C. Chung and K.L. Yu. 2013. GPS/INS integration accuracy enhancement using the interacting multiple model nonlinear filters. Journal of Applied Research and Technology 11: 496–509.

Kamal, A.T., J.A. Farrell and A.K. Roy-Chowdhury. 2013. Information consensus for distributed multiple-target tracking. IEEE Conference on Computer Vision and Pattern Recognition. USA, 2403–2410.

Kim, S., J.L. Crassidis, Y. Cheng and A.M. Fosbury. 2007. Kalman filtering for relative spacecraft attitude and position estimation. Journal of Guidance, Control, and Dynamics 30: 133–143.

Konstantinova, P., A. Udvarev and T. Semerdjiev. 2003. A study of a target tracking algorithm using global nearest neighbor approach. Proceedings of the International Conference on Computer Systems and Technologies. Bulgaria, 290–295.

Lowe, D.G. 2004. Distinctive image features from scale-invariant key points. International Journal of Computer Vision 60: 91–110.

Lu, F., Y. Wang, J. Huang, Y. Huang and X. Qiu. 2018. Fusing unscented Kalman filter for performance monitoring and fault accommodation in gas turbine. Proceedings of the Institution of Mechanical Engineers, Part G: Journal of Aerospace Engineering 232: 556–570.

Oh, S. and E.N. Johnson. 2007. Relative motion estimation for vision-based formation flight using unscented Kalman filter. AIAA Guidance, Navigation and Control Conference and Exhibit. USA, 2007–6866.

Ohya, I., A. Kosaka and A. Kak. 1998. Vision-based navigation by a mobile robot with obstacle avoidance using single-camera vision and ultrasonic sensing. IEEE Transactions on Robotics and Automation 14: 969–978.

Pourbabaee, B., N. Meskin and K. Khorasani. 2016. Sensor fault detection, isolation, and identification using multiple-model-based hybrid Kalman filter for gas turbine engines. IEEE Transactions on Control Systems Technology 24: 1184–1200.

Psiaki, M.L. and S. Mohiuddin. 2007. Modeling, analysis, and simulation of GPS carrier phase for spacecraft relative navigation. Journal of Guidance, Control, and Dynamics 30: 1628–1639.

Svensson, L., D. Svensson, M. Guerriero and P. Willett. 2011. Set JPDA filter for multitarget tracking. IEEE Transactions on Signal Processing 59: 4677–4691.

Teixeira, B.O.S., M.A. Santillo, R.S. Erwin and D.S. Bernstein. 2008. Spacecraft tracking using sampled-data Kalman filters. IEEE Control Systems 28: 78–94.

Valasek, J., K. Gunnam, J. Kimmett, J.L. Junkins, D. Hughes and M.D. Tandale. 2005. Vision-based sensor and navigation system for autonomous air refueling. Journal of Guidance, Control, and Dynamics 28: 979–989.

Vallado, D.A. and J.D. Griesbach. 2011. Simulating space surveillance networks. The AAS/AIAA Astrodynamics Specialist Conference. USA, AAS, 11–580.

Vallado, D.A. 2013. Fundamentals of Astrodynamics and Applications. 4th Edition. Microcosm Press, EI Segundo, California.

Vetter, J.R. 2007. Fifty years of orbit determination. Johns Hopkins APL Technical Digest 27: 239.

Vo, Ba-ngu and W. Ma. 2006. The Gaussian mixture probability hypothesis density filter. IEEE Transactions on Signal Processing 54: 4091–4104.

Vo, Ba-ngu, M. Mallick, Y. Bar-Shalom, S. Coraluppi, R. Osborne, R. Mahler, Ba-Tuong, Vo. 2015. Multitarget tracking. pp. 1–25. *In*: Webster, J.G. (ed.). Wiley Encyclopedia of Electrical and Electronics Engineering. John Wiley & Sons, Inc., Hoboken.

Vo, B.-T., B.-N. Vo and A. Cantoni. 2007. Analytic implementations of the cardinalized probability hypothesis density filter. IEEE Transactions on Signal Processing 55: 3553–3567.

Wang, X., D. Gong, L. Xu, X. Shao and D. Duan. 2011. Laser radar based relative navigation using improved adaptive Huber filter. Acta Astronautica 68: 1872–1880.

Weeden, B., P. Cefola and J. Sankaran. 2010. Global space situational awareness sensors. Advanced Maui Optical and Space Surveillance Conference. USA.

Xiong, K., C.W. Chan and H.Y. Zhang. 2007. Detection of satellite attitude sensor faults using the UKF. IEEE Transactions on Aerospace and Electronic Systems 43: 480–491.

Zhang, J., Y. Wu, W. Liu and X. Chen. 2010. Novel approach to position and orientation estimation in vision-based UAV navigation. IEEE Transactions on Aerospace and Electronic Systems 46: 687–700.

Zhao, Y. 2016. Performance evaluation of Cubature Kalman filter in a GPS/IMU tightly-coupled navigation system. Signal Processing 119: 67–79.

Index